An Environmental History of Russia

The former Soviet empire spanned eleven time zones and contained half the world's forests; vast deposits of oil, gas, and coal; various ores; major rivers such as the Volga, Don, and Angara; and extensive biodiversity. These resources and animals, as well as the people who lived in the former Soviet Union – Slavs, Armenians, Georgians, Azeris, Kazakhs, and Tajiks; indigenous Nenets and Chukchi – were threatened by environmental degradation and extensive pollution. This environmental history of the former Soviet Union explores the impact that state economic development programs had on the environment. The authors consider the impact of Bolshevik ideology on the establishment of an extensive system of nature preserves, the effect of Stalinist practices of industrialization and collectivization on nature, the rise of public involvement under Khrushchev and Brezhnev, and changes to policies and practices with the rise of Gorbachev and the breakup of the Soviet Union.

Paul Josephson is a professor of history at Colby College. A specialist in big science and technology in the twentieth century, he is the author of nine books on the history of science and technology and on human-nature interactions.

Nicolai Dronin is a senior researcher in the department of geography at Moscow State University.

Aleh Cherp is a professor of environmental sciences and policy at the Central European University in Budapest, Hungary, and associate professor at Lund University, Sweden.

Ruben Mnatsakanian is a professor of environmental sciences and policy at the Central European University and head of the Collaborating Centre of the Global Environmental Outlook Project launched by the United Nations Environmental Programme.

Dmitry Efremenko is the head of the sociology department at the Institute for Scientific Information on Social Sciences at the Russian Academy of Sciences.

Vladislav Larin is a senior analyst, writer, and researcher for the Russian Academy of Sciences' journal, *Energy: Economics, Technology, Ecology*.

Studies in Environment and History

Editors

Donald Worster, *Universiy of Kansas*
J. R. McNeill, *Georgetown University*

Editor Emeritus

Alfred W. Crosby, *University of Texas at Austin*

Other Books in the Series

Donald Worster *Nature's Economy: A History of Ecological Ideas, second edition*
Kenneth F. Kiple *The Caribbean Slave: A Biological History*
Alfred W. Crosby *Ecological Imperialism: The Biological Expansion of Europe, 900–1900, Second Edition*
Arthur F. McEvoy *The Fisherman's Problem: Ecology and Law in the California Fisheries, 1850–1980*
Robert Harms *Games Against Nature: An Eco-Cultural History of the Nunu of Equatorial Africa*
Warren Dean *Brazil and the Struggle for Rubber: A Study in Environmental History*
Samuel P. Hays *Beauty, Health, and Permanence: Environmental Politics in the United States, 1955–1985*
Donald Worster *The Ends of the Earth: Perspectives on Modern Environmental History*
Michael Williams *Americans and Their Forests: A Historical Geography*
Timothy Silver *A New Face on the Countryside: Indians, Colonists, and Slaves in the South Atlantic Forests, 1500–1800*
Theodore Steinberg *Nature Incorporated: Industrialization and the Waters of New England*
J. R. McNeill *The Mountains of the Mediterranean World: An Environmental History*
Elinor G. K. Melville *A Plague of Sheep: Environmental Consequences of he Conquest of Mexico*
Richard H. Grove *Green Imperialism: Colonial Expansion, Tropical Island Edens and the Origins of Environmentalism, 1600–1860*
Mark Elvin and Tsui'jung Liu *Sediments of Time: Environment and Society in Chinese History*
Robert B. Marks *Tigers, Rice, Silk, and Silt: Environment and Economy in Late Imperial South China*
Thomas Dunlap *Nature and the English Diaspora*
Andrew Isenberg *The Destruction of the Bison: An Environmental History*
Edmund Russell *War and Nature: Fighting Humans and Insects with Chemicals from World War I to Silent Spring*
Judith Shapiro *Mao's War Against Nature: Politics and the Environment in Revolutionary China*
Adam Rome *The Bulldozer in the Countryside: Suburban Sprawl and the Rise of American Environmentalism*
Nancy J. Jacobs *Environment, Power, and Injustice: A South African History*
Matthew D. Evenden *Fish versus Power: An Environmental History of the Fraser River*

(continued after Index)

An Environmental History of Russia

PAUL JOSEPHSON
Colby College

NICOLAI DRONIN
Moscow State University

ALEH CHERP
Central European University

RUBEN MNATSAKANIAN
Central European University

DMITRY EFREMENKO
Russian Academy of Sciences

VLADISLAV LARIN
Russian Academy of Sciences

CAMBRIDGE UNIVERSITY PRESS
Cambridge, New York, Melbourne, Madrid, Cape Town,
Singapore, São Paulo, Delhi, Mexico City

Cambridge University Press
32 Avenue of the Americas, New York, NY 10013-2473, USA

www.cambridge.org
Information on this title: www.cambridge.org/9780521689724

© Paul Josephson, Nicolai Dronin, Aleh Cherp, Ruben Mnatsakanian,
Dmitry Efremenko, and Vladislav Larin 2013

This publication is in copyright. Subject to statutory exception
and to the provisions of relevant collective licensing agreements,
no reproduction of any part may take place without the written
permission of Cambridge University Press.

First published 2013

Printed in the United States of America

A catalog record for this publication is available from the British Library.

Library of Congress Cataloging in Publication Data
Josephson, Paul R.
An environmental history of Russia / Paul Josephson . . . [et al.].
  p. cm. – (Studies in environment and history)
Includes bibliographical references and index.
ISBN 978-0-521-86958-4 (hardback) – ISBN 978-0-521-68972-4 (pbk.)
1. Human ecology – Russia (Federation) – History. 2. Indigenous peoples – Ecology – Russia (Federation) – History. 3. Environmental degradation – Russia (Federation) – History. 4. Environmental policy – Russia (Federation) – History. 5. Russia (Federation) – Environmental conditions. I. Title.
GF602.7.J67 2013
304.20947–dc23   2012042749

ISBN 978-0-521-86958-4 Hardback
ISBN 978-0-521-68972-4 Paperback

Cambridge University Press has no responsibility for the persistence or accuracy of URLs for external or third-party Internet Web sites referred to in this publication and does not guarantee that any content on such Web sites is, or will remain, accurate or appropriate.

Contents

	Introduction: From Geographic Determinism to Political and Economic Factors	*page* 1
1	From Imperial to Socialist Nature Preservation: Environmental Protection and Resource Development in the Russian Empire, 1861–1925	23
2	Stalinism: Creating the Socialist Industrial, Urban, and Agricultural Environment	71
3	The Khrushchev Reforms, Environmental Politics, and the Awakening of Environmentalism, 1953–1964	136
4	Developed Socialism, Environmental Degradation, and the Time of Economic "Stagnation," 1964–1985	184
5	Gorbachev's Reforms, Glasnost, and Econationalism	254
	Conclusion: After the Breakup of the Soviet Union: Inheriting the Environmental Legacy	287
Index		321

Under Stalin, the nation embraced a series of large-scale projects to transform nature to operate in a "machine-like" – and planned – way, including the creation of forest defense belts to prevent any future draught. "Land, you will never suffer from draught."

Introduction

From Geographic Determinism to Political and Economic Factors

The Soviet empire stretched 8,000 kilometers from Europe to the Pacific Ocean and 5,000 kilometers from the Arctic Ocean south to the Asian continent of Persia, Iraq, Iran, Afghanistan, India, and China. The vegetation, climate, and natural resources of this vast nation had remarkable diversity. In some respects one could claim that this was the wealthiest nation in the world, if only it could manage its resources rationally. A taiga consisting largely of boreal forest contained roughly one-half of the world's forests. Its major rivers – the Don, Dnieper, and Volga west of the Ural Mountains, the Ob, Irtysh, Lena, Angara, and Amur in Siberia – have total annual flow that rival those of the other great rivers of the world. Reserves of oil, gas, and coal; of iron, magnesium, manganese; bauxite (aluminum), gold, and platinum, often located in the frigid Arctic or Siberia; and other ores and minerals are among the richest in the world.

Yet, both the tsarist and Soviet governments largely mismanaged these resources, developed them in a haphazard fashion that contributed to their waste and profligate use, and took insufficient measures either to prevent extensive pollution or to engage in remediation once they discovered the severity of pollution problems, in spite of the fact of a long tradition of what we would call today ecological thought among scholars in the empire. Scientists under the Romanov dynasty failed to convince government officials, businesspeople, and even their own colleagues to adopt modern scientific management techniques to protect resources and ensure their availability for present and future generations ("conservation"); Nicholas II, the last Romanov, was more consumed by other issues, including pressure to reform the government in the direction of a constitutional monarchy, a war against Japan, and World War I.

The Bolshevik government, by contrast, embraced scientists' contributions to identify and develop those resources. A number of projects to develop water resources, construct hydroelectric power stations, expand and protect arable land, and so on that had languished in the Tsarist era found an enthusiastic audience. At the same time, leading ecologists, zoologists, and other specialists, joined by writers and compilers of local lore, successfully pressed the government to establish a national network of nature preserves – called *zapovedniks* – that reflected at times contradictory views about whether it was possible to "manage" nature, if wilderness had inherent value, and even if it was possible to establish in any society a system of inviolable reserves as if untouched by human hands. By the time of Joseph Stalin's rule (1929–1953), the nation had adopted breakneck policies for economic and military development that put zapovedniks at great risk and accelerated resource use and misuse to such an extent that the peoples of the former Soviet Union will face significant problems of pollution and degradation for years to come.

This book is a response to the need for a comprehensive environmental history of the Soviet Union. No such work exists that covers the institutions, actors, and ideology behind the great Soviet experiment in modernity – and in the great achievements and failures of Soviet modernization programs in agriculture, industry, and nature transformation. Soviet leaders, specialists, and workers accomplished a great deal over the course of Soviet history. They set out, from the point of view of economy, society, and nature, to create something different, something revolutionary, something entirely modern in response to Tsarist economic mismanagement and what they believed was capitalist exploitation of the worker. In their nature transformation, industrial and agricultural programs they shared accomplishments with the capitalist nations. Although the Soviet experience has much in common with the rest of the world, the Soviet exaggeration of modernity tells us something important not only about Russian history, but about the relationship between nature and the West. In achieving these goals, we draw on the strengths of the authors' expertise in history, policy, economics, agricultural economics, geography, and literature.

It may indeed be that the industrial revolutions in capitalist Europe and North America and the socialist revolution in the Soviet Union share the features of profligate resource use and extensive pollution. They also share the important consideration that even when they are based on well-intended efforts to protect national security, promote public health, and improve the daily lives of citizens, they often have unanticipated and extensive social and environmental costs. Paradoxically, the socialist nations promised to use and protect those resources in the name of the people to limit those costs of industrialization. Indeed, the environmental problems in socialist nations such as China, the former Soviet Union, and the allies of Soviet power in Eastern Europe were, overall, much more significant than in capitalist nations where the motivation to develop them came largely in the pursuit of profit motive. One reason may be that the people under socialism were largely silenced by their leaders from speaking openly and actively about environmentalism, whereas those in capitalist nations were able to engage in visible public campaigns to protect the environment owing to the expansion of civic culture throughout the twentieth century. However, over the course of Soviet history, the attitudes of leaders toward the environment and public involvement in environmental issues evolved significantly, actually permitting and encouraging public discussion in a number of spheres, and this involvement reveals a dynamism about environmentalism in Soviet society that is discussed in this book.

Still, the absence of a well-developed civic culture to promote full consideration of environmental issues in the former Soviet Union may be the major factor that led to such significant environmental degradation. Environmentalists (mostly biologists and writers) sought to protect a series of precious nature preserves from encroachment and quietly lobbied Communist Party and government officials to be aware of the economic potential of conservation and preservation. But they faced great odds in advancing any kind of "environmentalist" agenda, including in some periods the threat of arrest, loss of careers, and even execution.

The people of the Russian empire and the Soviet Union lived under absolute rulers for centuries, first the tsars and then the Soviet leaders from Lenin to Gorbachev. There were significant differences in the two regimes, of course, one being an autocracy, the other a communist authoritarian government. Yet, in more ways than one, continuity existed in conservation, preservation, resource management, and nature transformation movements of imperial Russia and the Soviet Union. First, the state owned the vast natural resources in both empires, although private ownership and individual initiative were crucially important to the economy during the tsarist era. Still, the state was a major engine of economic development and assiduously avoided measures to protect natural resources and limit air and water pollution with concern that these measures might interfere with development or were unnecessary. It poorly supported enforcement of laws, and sought insufficient penalties for lawbreakers.

Second, these were peasant societies, at least through the late 1950s, when Soviet leaders claimed that the nation had become urban with a working class whose size substantially exceeded that of the peasantry. At turns, leaders ignored the peasants, considered them incapable of a modern world view needed for industrialization, perhaps even viewed them as enemies of modernity – and opponents of efforts to improve on or tame nature; under Stalin they declared a war against them in the collectivization effort. Thus, a striking mismatch between the goals of the state and those of the citizen prevailed, no matter how government officials claimed best to understand those goals, express them for citizens, and define who the citizens were. This was the result, once again, of the weak development of civic culture, and few individuals had the mind set or training to question officials – had they been permitted to do so. Still, healthy scientific and technical debates often percolated on how to deal with various resource, pollution, and other problems.

Third, although having a strong scientific tradition, including in the life sciences and ecology, both the Tsarist and Soviet government adopted policies that left scientists often isolated from their colleagues in the West. This isolation grew pronounced in the Stalin era, even though this was a socialist regime whose leaders claimed to have embraced an internationalist scientific doctrine, Marxism, and whose policies contributed to the rapid expansion of the scientific enterprise. The political leadership in both systems mistrusted independent academic expertise and sought generally to control the intelligentsia. In spite of these controls and policies, the environmental sciences and environmental movements remained vital, and the activities of environmentalists, loosely defined, largely paralleled those of environmentalists in North America and Europe.

There is a danger of attributing to climate or geography overriding importance in explaining the environmental history of the former Soviet Union. Yet, the challenges of climate and geography (see section on "Physical Geography and Ecosystems of the Soviet Union") presented significant challenges to resource development, as did the great distances between those resources and population centers. The high costs of developing resources in Siberia, the Far North, and the Far East certainly – and unfortunately – encouraged practices with significant impacts on the environment. Both governments were plagued by the challenges of great spatial dispersion of people, resources, nature, and the need to develop infrastructure to master those distances.

Of course, there were significant differences between the two regimes. After the Russian Revolution, nascent ecological science expanded rapidly during the social upheaval and political experimentation of the 1920s. Officials, scientists, and engineers worked out an ambitious national electrification program. They charted the construction of modern hydroelectric power stations. They embarked on an ambitious program to build on the few existing nature preserves to establish scores of them, many of which still exist. Yet, during the Stalin era, state-mandated programs for collectivization of agriculture, rapid industrialization, and autarky ensured that economic development was the sine qua non of decision making. Those who stood in the way of the programs – wittingly or unwittingly – were often labeled "wreckers." The "wreckers" included some of the nation's most able biologists, forestry and fisheries specialists, agronomists, and ecologists. Officials and ideologues came to consider nature itself an "enemy of the people" for refusing to buckle under to plans for rapid economic growth, and many of them believed that nature preserves were a waste of energy and resources. The emphasis was on heavy industry at the expense of the consumer, health, and housing sectors. Subsequent leaders adopted more rational policies toward the utilization of natural resources and introduced a number of environmental constraints on development, but were unable to change considerably the environmentally destructive momentum of the planned economy. Yet, the Stalinist legacy of mismanagement of resources, haphazard disposal of hazardous waste, and inadequate regulations persists into the twenty-first century.

A problem for ideologues, planners, specialists, and party officials in the Soviet Union was that for such spheres of human activity as nature conservation Marx, Engels, and Lenin had not enunciated clear positions, although clearly they saw the future world as one in which an industrial ethos prevailed.[1] In the absence of a classic Marxist position, self-appointed defenders of the proletariat, many of them of working-class origin, many of them with only rudimentary education yet considered "red specialists," condemned as "bourgeois" any traditional field, especially if they did not understand it. Of course, many others, perhaps most others, had moderate views of the human's place as a part of nature that resembled that in other countries of the world.

Whatever the continuity and change, the result, on the eve of the twenty-first century, was a new nation, the Russian Federation, still

---

[1] For an effort to find ecological thought in Marx, see John Bellamy Foster, "Marx's Ecology in Historical Perspective," http://pubs.socialistreviewindex.org.uk/isj96/foster.htm.

almost as wealthy as its predecessors in terms of natural resources, but with extensive environmental problems, and the Newly Independent States of the former Soviet Union with equally pressing problems. In this book – intended both as a survey and an original study – we explore the environmental history of the former Soviet Union. In each chapter we include discussion of major trends, actors, ideas, and institutions. Each chapter describes and evaluates change in the environment itself as a result of human action, and how changes in the environment had an impact on human activities.

Each chapter has a major narrative story, yet covers similar issues. We evaluate the significance of environmental issues from resource management to pollution abatement in a society in which the state was a major actor, where the economy was centrally planned, and where, because of the overriding centralization of bureaucracies and organizations, virtually all projects became costly, large-scale, resource-intensive projects. We consider the challenges in managing resources scientifically and in getting users to pay attention to regulations in these circumstances. We explore the nature of ecology as much as a social movement as a scientific field.

What Is Environmental History in This Book?

In this environmental history we tend to focus on economic and political factors more than may currently be the fashion in the field of environmental history. But as Douglas Weiner pointed out, environmental history is a big tent, an interdisciplinary approach to understanding human–nature interactions. These interactions clearly are not one-directional. Humans do not stand outside of or above nature to make rational, value-free judgments about how "nature" functions. In spite of a number of attempts that scientists and others pursued to see nature as an empirical object, it is best to understand it as a site of human and other interactions in all ecosystems, in cities and in the countryside, in forests and meadows, and in plowed land and seeming wilderness, and also to realize that political, economic, and other factors shape our attitudes toward nature and what we strive to make of it. We cannot deny the role of history in understanding ecology. Rather, as Weiner and others have argued, nature is a social construct, not some "real world" that exists independent of us.[2]

---

[2] Douglas R. Weiner, "A Death-Defying Attempt to Articulate a Coherent Definition of Environmental History," *Environmental History*, vol. 10, no. 3 (July 2005), pp. 404–420.

The reason for the focus on economic and political factors in an environmental history of the Soviet Union is simple: in the Soviet Union, state actors were the crucial individuals in shaping policies and behaviors and in apprehending what nature was. Political and ideological desiderata about what role nature should play in the construction of socialism, how people ought to react to the challenges of developing resources, and whether there were limits to human power all played out against the backdrop of the effort to create self-consciously a society that differed in so many respects from capitalist societies. Attitudes toward industry, agriculture, ecosystems, biodiversity, urban planning, and so on were shaped largely by political and economic concerns. Scientists of all fields, writers – whether representatives of official genres and approaches or not – and citizens understood "nature" both in regard to their personal relationships with nature and in regard to official attitudes about the Soviet polity and economy. Scientists who sought to temper industrialization, forestry, fishery, and agricultural programs for their potential risk to people and the environment, or who wished to prod the state to expand the designation of nature preserves, had to address political and economic concerns directly, or indirectly using careful language, even if they gave the appearance of writing about ecosystems as existing somehow divorced from broader social concerns. This is not to say that we ignore cultural and scientific components of environmental history that other specialists have stressed. Instead, when speaking about pollution problems, wilderness, "Virgin Lands," and so on, we argue that all of these issues and concepts were shaped to a great degree by political–economic concerns as well as by epistemological, ontological, and other concerns. Finally, we accept Weiner's argument that environmental history is precisely about power, about who "will control access to resources and amenities," what role experts played in determining the expected risks and benefits of one approach or another, and who actually made choices. Once one has made a choice about what is a fact and what is not, what constitutes the truth about nature and what does not, that person has made a deeply political choice because he or she excludes others from "the truth" if others do not accept that view.[3]

Perhaps to a greater extent in the Soviet Union than elsewhere, ruling elites designed and attempted to design entire landscapes in the name of science and progress – in the case of the Soviet Union, this would

---

[3] One of the most powerful enunciations of the view that epistemological choices are political ones is Donna Haraway, *Primate Visions* (New York: Routledge, 1989).

be an industrial landscape through and through – or, so we argue in this book. The prevailing view among officials in the Soviet Union about what constituted nature reflected a broad transformationist agenda to change nature for the better than it was under capitalism.

Large-Scale Projects and Large-Scale Bureaucracies

Another distinguishing feature of environmental history in the Soviet Union was the evolution of large-scale organizations concerned with studying nature, nature management, and nature transformation. The USSR relied heavily on centralized, large-scale projects and bureaucracies to force the pace of economic production, perhaps more so than in the United States with its Army Corps of Engineers, Bureau of Reclamation, Tennessee Valley Authority, or Bonneville Power Administration. These include military organizations and such massive economic ministries as the Ministry of Water Resources (Minvodkhoz), the Ministry of Electrification, the Ministry of Agriculture, and the Ministry of Middle Machine Building (the nuclear energy and weapons ministry) that commanded significant resources of manpower and capital and were allied unquestioningly with meeting state-mandated economic production targets to build dams, reclaim land, manage forests, produce food, and so on. The government also employed cheap, forced prison labor through the gulag system in such major geological engineering projects as the Baltic-White Sea Canal; the hydroelectric power stations on the Volga, Ob, and Angara Rivers; mining, road, and railroad; and forestry enterprises in the Far North and Siberia. We track the impact of these organizations and approaches across taiga and tundra, steppe and floodplains, forest and desert.

No ideas can rationally utilize – or "master," in the Soviet case – nature on their own. They require organizational and institutional actors. Thus, we consider the institutions – formal and informal – that molded and reflected environmental concerns: government, scientific, nongovernmental, regulatory, and so on. In the Soviet Union, institutions, especially those ministries and trusts connected with directing economic activities, were the key players. We explore how ecologists, planners, policy makers, and citizens worked through, with, and against these institutions.

Each chapter explores central ideas about the environment and ecology, how scientists, government officials, and citizens viewed those ideas, and the institutions and bureaucracies they created that had an impact on the environmental history of the Soviet Union. The starting point for explaining environmental change is to understand the concepts, paradigms, and attitudes that guided Soviet society in its relationship to the environment; how they reflected cultural, political, and economic values; life; and how they are reflected in the practices of resource management, pollution, pollution abatement, and regulation; in scientific studies; and even in literary and artistic activity. As elsewhere, in the former Soviet Union, they were shaped by Enlightenment thought – that is, that humans can understand, control, and even improve on nature. But, as many historians have noted, a kind of geological or climatic determinism shaped the way Russians – from scientists and bureaucrats to peasants – viewed "nature." For them, it was vast, rich, yet unforgiving, something to be mastered, perhaps with science, and without a doubt by Bolshevik certainty. Under Soviet power, with fulfillment of the plan target the only judge of success or failure, resource use accelerated, worker safety was ignored, and pollution regulation and abatement were largely ignored.

Concepts as "biodiversity" and "ecosystem" are relatively recent ideas, largely of the second half of the twentieth century, although they existed in some form or another from the mid-nineteenth century. Hence we must be careful not to use a term that has recently acquired significance in a nineteenth or an early twentieth century context. Still, representatives of the Russian intelligentsia long ago advanced ideas about the interaction between humans and nature – of humans as part of nature, and not as a species above it – and understood that human activities would have a direct impact on other species and their habitats. The biogeochemist Vladimir Vernadsky developed the notion of the noosphere on the eve of the Russian Revolution according to which humans are a large-scale geophysical force who must understand human–nature interactions on the basis of study of the past to ensure the future. Vernadsky explored the human development of ferrous and nonferrous resources and countless artificial chemical combinations, geoengineering of rivers, seas, and oceans.

Another important theme in this book concerns state–society relations, and these comments refer not only to the case at hand, the Soviet Union, but to Germany, France, the United Kingdom – England, the United States, and other modern states. The modern state – government structures and bureaucracies and the officials representing the government and its official policies and ideology – has had an increasingly central role in environmental history in a variety of ways since the early eighteenth century, for example, when Peter the Great ordered surveys of Russian forests and issued proclamations to assign them solely to the tsar's use for the navy and other military purposes. As a polity, a nation, and a sovereign political entity with a monopoly on power, the state can influence and direct activities that will shape the environment. Although generally seeing resources of fish, forest, furs, ore, and so on for their utility and direct benefit to state power and hesitating to establish rules to regulate their use or set aside lands for nature preserves or parks, officials have not been enemies of nature. They have understandably put military or economic programs ahead of nature protection in the name of national security and employment. But they have also worked with scientists, writers, and journalists and other individuals establish limits and prevent profligate use. They set aside forests; established seasons on hunting; regulated the use of waterways among competing interests; and eventually set aside national parks, wilderness areas, and other "objects" of human culture. The Soviet Union was no different in all of these ways.

By the mid-nineteenth century, governments around the world had chartered scientific societies and set aside funds for research and development activities, including the establishment of institutions that had direct and indirect relationships with environmental concerns. Hence, when we speak about the Soviet regime and its policies, we should recall that other states have pursued similar paths, encountered similar obstacles to rational resource management, and sought similar paths, policies, and laws to protect biodiversity. The relationship between the state and society – the bureaucrats and officials, the workers and peasants, the intelligentsia, merchants, businesspeople, and so on – must be seen therefore along a continuum: in some cases and in some periods, the state more actively engages in what we consider today to be environmentally sound practices, and at other times it supports practices that encourage profligate use of resources, damage ecosystems, and destroy biodiversity. We highlight here and there the ways in which the Soviet Union differed in its experiences, with brief reference to experiences in other settings. But the focus is the Soviet Union.

Similarly, the multifold relationship of the state to the environment as protector and exploiter exists in "society" as well. Private interests have long sought to develop resources toward the ends of profit, and have seen those resources as inexhaustible, pursued their exploitation rapidly to limit others' access to them, or perhaps simply misunderstood how plentiful they were, whereas, at the other end of a spectrum, naturalists and others have worried about what they perceive as the destruction of "pristine" nature, and in some societies have gained a reputation for standing in the way of "progress." Other members of society – indigenous people and peasants, for example, have also often been accused unfairly of being obstacles to progress. Generally, however, members of society occupy some middle grounds where they debate the extent of resources and how best to manage them.

In the late nineteenth century, the notion of "conservation" developed. Conservationists believe that resources of water, forest, fish, and so on are scarce or limited, but that, on the basis of rational scientific decisions, they can be managed among competing interests for present and all future generations. Preservationists believe that nature has intrinsic value and that glorious wilderness must be protected from human encroachment. In Russia and the Soviet Union, just as elsewhere, strong traditions of conservationism and preservation existed and played out. For example, after a terrible famine in 1891, a number of scientists urged the tsarist government to manage forest, water, and land resources more rationally to prevent any future such famine.

The chapters in this book generally follow the period of rule of the leaders of the USSR and the Russian Federation: Lenin, Stalin, Khrushchev, Brezhnev, and Gorbachev, with a final chapter highlighting the events and changes since the breakup of the Soviet Union. This chronology makes sense in that each leader had his own style, goal, and approaches, believed them to be better than his predecessor's, and committed to various changes and shifts in economy and ideology that had direct and long-term impacts on the environmental history of the nation, as well as discernible impact on environmental science and policy.

These men were symbols of how the society operated and functioned as it moved from being a largely agrarian economy (under Lenin) through rapid industrialization (Stalin), became increasing urban (Khrushchev), and with the labor force changing from largely illiterate or barely literate to highly educated, and with the economy requiring substantial expert input (Brezhnev and beyond), became "modern." Their eras reflected broader political, cultural, and economic values that in turn were reflected in changing attitudes toward resource use and exploitation. This chronological approach enables readers to use this text easily along with Russian and Soviet history texts.

A danger of a book on the environmental history of the Soviet Union is to tar the Soviet Union with the Cold War criticism of communism failed, while holding up "capitalist" nations – generally those of Western Europe and North America – as superior in all respects concerning environmental protection, regulation of pollution, and so on. One commentator referred to the legendary and unfortunately largely true despoliation of the Soviet environment as "ecocide."[4] It is accurate to argue that for political, ideological, and economic reasons, the Soviet Union embraced development programs that directly and indirectly had significant costs for the environment across all ecosystems for all people, flora, and fauna. The Stalinist plans for rapid industrialization and collectivization of agriculture that persisted in many ways and forms to the end of the regime required party officials, planners, managers, scientists, and engineers to push forward with certainty that any environmental or human costs – if they considered them at all – were well within an acceptable range, and that the concerns of ecologists and other "lovers" of nature were exaggerated, misplaced, outdated, and perhaps even a form of "wrecking" of glorious Soviet intentions to improve on nature while fulfilling plans. After the death of Stalin, the government took a more active role in nature protection, in regulating waste disposal and pollution, and in reshaping managers' wasteful and dangerous practices in statues and fines. An environmental movement burst forth like that in Europe and North America. And, as Stephen Brain points out, under Stalin, a kind of environmentalism existed that was manifested in the "Stalinist Plan to Transform Nature" in its programs to reforest vast regions of the nation.[5]

In many ways, the Western democracies followed the same paths of breakneck development and profligate use of natural resources, of destroyed ecosystems, and tardily adopted laws and regulations to remediate and limit future problems. Almost universally, modern states have embraced enlightenment thinking about the ability to improve on nature with the resulting costs of ecosystems irrevocably changed. They have pursued large-scale nature management, melioration, and other projects with long-term and often irreversible impacts. They have put job formation and other economic pressures ahead of environmental concerns. When scientists and naturalists have pressed them to set aside lands for parks or preserves, they have responded hesitatingly. When these same scientists have pushed for scientific management of natural resources in the name of "conservation," they have, once again, often permitted concerns about economic growth to prevail over concerns about conserving and managing resources scientifically. All of this is to say that the

---

[4] Murray Feshbach and Alfred Friendly, Jr., *Ecocide in the USSR* (New York: Basic Books, 1992).
[5] Stephen Brain, "Stalin's Environmentalism," *Russian Review*, vol. 69, no. 1 (2010), pp. 93–118. See also Brain's recent *Song of the Forest* (Pittsburgh: University of Pittsburgh Press, 2011).

Soviet Union resembled the European and North American democracies in many ways concerning environmental history.

Another focus of each chapter is the dynamic relationship between scientists, economic planners, political leaders, writers and artists, and especially peasants and workers in order to avoid a statist, top-down approach to this environmental history. For example, we examine the cultural, political, economic, and scientific forces that shaped life in the city and in the countryside: peasant and nomad land use, migration and resettlement, city construction and natural urban expansion, and so on. Even when autarky prevailed in the sciences, this relationship operated, although at a much lower level of energy. In other words, there are millions of actors, and our approach – and the materials we have gathered – enables us to consider their actions, behaviors, and beliefs.

In each chapter, therefore we give attention to bottom-up perspectives on the environment in the Soviet Union, not only top-down policies that reflect leaders' and planners' preferences and perspectives. For some periods, because of the nature of sources available and state control of the media, this was difficult. But one of the strengths of this study is its use of published and archival materials that have not been used in other studies. Still, we recall that in the Soviet Union the state owned all property, and its employees in government and the party determined how to manage and develop resources. In a word, and perhaps tautologically, more so than in other countries the Soviet state had a significant role in the environmental history of its empire.

At the end of this book we briefly discuss the ramifications of the breakup of the Soviet Union and the formation of the Russian Federation on the environmental issues and practices. This includes discussion of the Putin and Yeltsin regimes and of environmental issues in several other nations formed after the breakup. For example, when Russia's transition from a planned economy and authoritarian political system commenced in the 1990s, many observers argued that dismantling the centrally planned economy would automatically deliver environmental improvement and that the environmental legacies of socialism would fade away.

The reality has proved dramatically different. There have been new threats to sustainability, including the fire sale of resources, the restructuring of the economy that drastically reduced resources available for environmental protection, and President Putin's decision ultimately to disband the Russian Federation's Environmental Protection Agency. As for the panacea of public participation in environmental movements leading to improvements in the situation, citizens have lost interest in the environment since they have been distracted by political and economic problems. Nationalist movements based on environmental leanings contributed to the breakup of the Soviet Union, but "econationalism" has faded.

Physical Geography and Ecosystems of the Soviet Union

Although recognizing the impossibility of describing fully the multitudinous ecosystems, biodiversity, and physical geography of an empire that stretches from the Arctic circle to the desert and mountains of Central Asia; from the European plain to the Pacific Ocean; with mountains that range to 7,500 meters (the Pamir, Urals, and Carpathians) and some of the largest rivers in the world by annual flow volume (the Ob, Irtysh, Enisei, Lena, Angara, Don, Volga, and others), it is useful to describe the richness of the landscapes in the former Soviet Union and biodiversity within them on some level. We present here several regions to give an indication of biodiversity in a variety of settings in anticipation of further discussions of human and nature interaction. Later in the book, the reader will encounter other such descriptions when they are crucial to understanding the environmental history of the Soviet Union.

Generally, the geography of natural landscapes of the former Soviet Union follows two climatic factors: temperature and summer precipitation. Elevated topography prevents moist and warm air from the Atlantic and Indian Oceans from entering the eastern part of the region. The average annual precipitation over the territory is 490 millimeters, compared with 782 millimeters over the United States.[6] Eighty-two percent of the region is located in the continental climate zone (according to Köppen-Geiger classification),[7] receiving a low to moderate amount of rain. In Central Asia and the Caucasus, 70 percent and 16 percent of land, respectively, is in the arid and semiarid climate zones. Only 1 percent of the region, predominantly along the Black Sea coast, is in the humid temperate climatic zone, with high precipitation, warm and long summers, and mild winters, compared with 34 percent in the United States.[8]

---

[6] N. C. Field, "Environmental Quality and Land productivity: A Comparison of the Agricultural Land Base of the USSR and North America," *Canadian Geographer*, vol. 12 (1968), pp. 1–14.

[7] M. Kottek et al., "World Map of the Köppen-Geiger Climate Classification Updated," *Meteorologische Zeitschrift*, vol. 15 (2006), pp. 259–263.

[8] W. H. Parker, *The Superpowers. The United States and the Soviet Union Compared* (New York: Halsted Press, 1972).

Droughts are among the most important natural disasters affecting the Soviet Union. Droughts occur generally when a mass of dry arctic air sweeps down into the European part of Russia and forms an anticyclone. The anticyclone, being quasi stationary, causes the air mass to become drier. Along the southern and southwestern periphery of the anticyclone, dry and hot air spreads. An especially strong drought takes place when an anticyclone is fed by an air mass from an Azores anticyclone moving in from the West. Moving across Europe, the air mass loses its humidity and reaches European Russia virtually dry.[9] The droughts resulting from these large-scale atmospheric processes usually cover vast territories of Moldova, Ukraine, and Russia, including the Northern Caucasus, the Middle and Low Volga basin, the Urals, and periodically spread over the central chernozem (black soil) region and even the northern regions of European Russia. For example, the drought of 1946 covered 50 percent of total agricultural land of the Soviet Union. The scale and consequences of droughts can be catastrophic for the country. In the case of the most serious crop failures in agricultural production in the Soviet Union (1921, 1924, 1946, 1948, 1957, 1959, 1963, 1972, 1975, 1979, and 1981), droughts were responsible.[10] On several occasions, famine resulted.

Tundra[11]

Tundra is a kind of arctic desert, with winter temperatures at less than forty centigrade not out of the ordinary, with summers reaching only ten or fifteen degrees centigrade, and with frequent strong winds that lead to very low wind chill temperatures. In the tundra, vegetation consists of funguses, grasses, mosses, lichens, and low shrubs; the summers are too

---

[9] A. V. Protserov, "Zasukhi na Evropeiskoi Territorii SSSR," in *Agroklimaticheskie Usloviia Stepei Ukrainskoi SSR i Puti Ikh Ulucheniia* (Kiev: Academia Nauk Ukrainskoi SSR, 1950), pp. 17–22.

[10] Nikolai M. Dronin and Edward G. Bellinger, *Climate Dependence and Food Problems in Russia, 1900–1990: The Interaction of Climate and Agricultural Policy and Their Effect on Food Problems* (Budapest: Central European University Press, 2005).

[11] Sten Larson, "On the Influence of the Arctic Fox... on the Distribution of Arctic Birds," *Oikos*, vol. 11, fasc. 2 (1960), pp. 276–305; George Llano, "Utilization of Lichens in the Arctic and Subarctic," *Economic Botany*, vol. 10, no. 4 (October–December 1956), pp. 367–392; John Marr, "Ecology of the Forest-Tundra Ecotone on the East Coast of Hudson Bay," *Ecological Monographs*, vol. 18, no. 1 (January 1948), pp. 117–144; Albert Seeman, "Regions and Resources of Alaska," *Economic Geography*, vol. 13, no. 4 (October 1937), pp. 334–346; J. C. F. Tedrow and H. Harries, "Tundra Soil in Relation to Vegetation, Permafrost and Glaciation," *Oikos*, vol. 11, fasc. 2 (1960), pp. 237–249.

short and cold for trees to grow, although one occasionally encounters a few trees. Plant roots spread out on the surface to take in water and do not grow deep because of the constantly freezing and thawing soil that destroys the roots. Permafrost – permanently frozen subsoil – lies under the surface and is frozen to a great depth, to 600 or 700 meters. A brief summer thaw allows vegetation to grow in thawed soil that is 25 centimeters to 1 meter deep.

Tundra is located at the top of the Northern Hemisphere in Europe, Asia, and North America. Tundra covers 20 percent of the earth's surface just below the polar cap. The Siberian tundra is located in the northeastern part of Russia between 60 degrees to 80 degrees north latitude, and 70 degrees to 180 degrees east longitude. Not surprisingly, in the tundra, winters are long and summers are short. Summers are marshy from puddles called thermokarsts formed from melted snow and ice. Another tundra phenomenon is the pingo that forms when pools of water freeze underground and push the ground up into a hill; pingoes are 3 to 300 feet tall and a half-mile wide.

Two such regions of tundra are in the Nenets Autonomous Okrug and Sakha (Yakutia). For example, in the former, the Bol'shezemel'skaia tundra is a hilly moraine (sandy and pebbly soil left) between the Pechora, Usa, Ural, and Khei-Yaga Rivers. Although the rivers move rapidly in their upper reaches through valleys, as they move downstream, they slow and become still, except during the spring thaw when they move at very high speed and volume and spread several miles wide by their deltas.[12]

The animals of the tundra include fish; birds, such as ducks and geese that migrate thousands of kilometers to nest in the summer; such mammals as rabbits and other small rodents, as well as polar bears and reindeer, and insects. Animals have extra fat and thick fur to keep them warm, and most of them are low to the ground, and their arms, legs, tails and ears are small to keep from losing heat. Global warming has begun to have a very powerful impact on tundra, with melting and flooding already noted in some regions.

Yakut, Nenets, Saami, and Komi people lived in these regions with local economies and in small communities with relatively infrequent and short-term contact with Russian and other explorers until efforts to colonize the north commenced with vigor in the 1930s. Mining and smelting plants were built at this time and have had a significant impact on the

---

[12] The Pechora coal, oil, and gas basins are quite extensive and have been increasingly exploited with significant environmental impact.

environment. The increasingly urban environment of the tundra under Soviet power also contributed to pollution problems because of poor – or in many cases nonexistent – municipal garbage management practices. In addition, Soviet colonization meant increasing use of guns to hunt animals, leading to overhunting and some extinctions. The Russian Federation in the twenty-first century has begun to establish hunting and fishing limits and seasonal controls, to require licenses, and to put more effort into protecting national parks that were established in the previous century.[13]

Taiga[14]

The taiga, a northern coniferous boreal forest, lies south of the tundra; it occupies the north of European Russia and extends to cover much of Siberia and Far Eastern Russia, or nearly half of the Russian federation. Much of this region also has permafrost. Although the vast taiga zone is made up predominantly of coniferous trees, in some places, small-leaved trees, such as birch, poplar, aspen, and willow, grow. The forest zone of Russia also consists of a much smaller southern area of mixed coniferous–deciduous forest. In the northwest – Karelia, parts of Leningrad, and Arkhangelsk regions – pines cover the land, with significant fir, birch, and other trees, too. Pulp, paper, lumber, and wood products for domestic and export markets have long been important to the nations of the former Soviet Union, to Russia more than any of the others.

As a forest zone, taiga is also a high-risk agriculture zone. In the northern and northeastern parts of Russia, agriculture (primarily growing grasses and cereals for fodder) is limited by permafrost and insufficient growing degree days. The best lands in terms of productivity are located in the forest-steppe and steppe zone, a wide 1,500-kilometer zone south of fifty-five degree north latitude, which is flanked by the broad-leaved forests in the north and semi-deserts in the south. In the forest-steppe, tree stands alternate with open grasslands. In the European part of the region, these tree stands consist of oak and other broad-leaved species and nowadays are largely cleared for agriculture. In western Siberia, forest steppe, birch, and aspen are common, with the steppe zone much narrower and in strips. Further east, in Central and eastern Siberia, the

---

[13] http://www.blueplanetbiomes.org/siberian_tundra.htm.
[14] Robert Hoffmann, "The Meaning of the Word 'Taiga,'" *Ecology*, vol. 39, no. 3 (July 1958), pp. 540–541.

steppes are mostly localized in depressions and on the low parts of the slopes of mountain lands.

Steppe

Further south, the forest steppe yields to steppe grasslands on fertile chernozem soils (black soil rich in humus and covering much of Moldova, Ukraine, south of European Russia). Steppe is grassy plains land that stretches from Hungary to Mongolia, usually found between forest and desert, with short hot summers and long cold winters. With more rain, steppe would become forested, whereas trees in steppe grow mostly near rivers or other bodies of water. With less rain, steppe becomes desert. Thousands of seed plants fill steppe, but especially many different types of grasses. Although much of the steppe in western Russia has been plowed and forests cleared for agriculture, island-like remnants of forest-steppe habitat have been preserved in the Tsentral'no-Chernozemnyi Zapovednik (nature preserve, established in 1935), located in the Kursk region.[15] The nature reserve protects the world's last intact chernozem. According to one observer, each spring the steppe bursts into bloom of scores of flowers and colors – cowslip, pheasant's-eye, forget-me-nots, speedwells, feather grass, clary, and so on.[16]

The annual precipitation in the steppe is 250 to 400 millimeters versus 400 to 500 millimeters in the forest steppe. Smaller precipitation limits the tree species to forested ravines and valleys. Originally, the steppe subzone was used almost exclusively for grazing; however, later the tillable lands came almost entirely under cultivation. The nontillable lands, especially those in the Caspian Sea area, have annual amount of precipitation less than 250 millimeters and are similar to semi-deserts. These lands have limited use for grazing: even though the soils are relatively fertile, their total area under irrigation is less than 5 percent of the total area under agriculture in Russia.

Water shortage was the dominant factor limiting irrigation, as the rivers of the southern slope had been tapped out – exhausted – for municipal, industrial, and other agricultural needs. Stalin planned in the steppe to create a vital grain, fruit, and other agricultural products region on the foundation of an extensive irrigation and melioration program. As in the

---

[15] A. A. Gusev et al., "Tsentral'no-Chernozemnyi zapovednik," in *Zapovedniki Evropeiskoi Chasti RSFSR* (Moscow: Mysl', 1989), pp. 109–137.
[16] http://www.wild-russia.org/bioregion3/tsent-chern/3_tsen-chern.htm.

plains states in the United States, for the steppe, the principal threat to natural vegetation was extensive agricultural development in the effort to turn it into arable land.

Arid Regions

Water shortage is the principal characteristic for Central Asia (Kazakhstan, Uzbekistan, Kyrgyzstan, Tajikistan, and Turkmenistan), in which arid land, mountainous terrain, desert steppe, and desert predominate. In Kazakhstan, rain-fed production is mainly limited by the extreme shortage of precipitation, the dry, short, and warm growing period, and extremely cold and windy winters. In other republics of Central Asia, agriculture is generally restricted to oases and irrigated lands along the major rivers (Amu Daria and Syr Daria) and canals. Hence, the availability of water for irrigation is yet again the major factor constraining agriculture. Major crops include wheat, rice, and cotton. The oases are surrounded by deserts.

Positioned at the heart of the Eurasian continent, the key geographical feature of Central Asia is that all of its rivers drain centrally into its interior rather than outward into the external oceans, making it hydrologically almost self-contained. As such, its numerous rivers all terminate in low-lying basins – some large, such as the Balkash, Sistan, and Tarim depressions; and some small. Huge permanent inland lakes accumulate in some of these, such as Lake Balkash. In others, lakes appear during the flood season but then soak away to leave behind a vast marshland, as in the case of Sistan. In certain cases, the rivers simply end in deltas that just fizzle out in the desert, like the Murgab delta close to Merv, the Zerafshan near Samarkand, or the Chou to the north of the lower Syr Daria. By contrast, the Tarim basin is occupied by one of the world's largest deserts, the Taklamakan.

The most westerly basin in Central Asia also happens to be the largest in terms of its overall surface area – draining a region of just over 1.5 million square kilometers. It stretches eastward to the borders of China, India, and Pakistan and includes the whole of Tajikistan and Uzbekistan, most of Turkmenistan, and parts of Kazakhstan and Kyrgyzstan. Topographically, it divides between the mountain zone in the extreme east and the remaining desert plains, which slope imperceptibly toward the west. The Aral Sea, with the Khorezm oasis laying immediately to its south, occupies its western and lowest part.

The Aral Sea

The Aral Sea is, of course, not a sea but a terminal lake with no outfall (it is *endorheic*). It not only is extremely shallow but, as a result of its central Eurasian desert location, also is subject to intense evaporation during the hot summers. Its existence depends on a constant inflow of water from two rivers, the Amu Daria and the Syr Daria. Both rivers receive the majority of their water supply from the mountainous regions of the watershed. Some 43 percent of the Aral basin water supply arises in Tajikistan, 25 percent in Kyrgyzstan, and 18 percent in Afghanistan. But because of Soviet agricultural development programs, the Aral Sea is under extreme threat.

The authors express a debt of gratitude toward the scholars whose work in environmental history have made this text possible. Many of them the reader will meet in the footnotes. Suffice it to note here that such scholars as Marshall Goldman, Thane Gustafson, Charles Ziegler, and notably Douglas Weiner, have written extensively on various aspects of environmental policy and history.[17] Professor Weiner's two books, *Models of Nature* (1988) and *A Little Corner of Freedom* (1999), pointed the way for every step. Murray Feshbach has done more than any other scholar to alert us to the true costs of environmental degradation in the former Soviet Union.

Paul Josephson wrote the Introduction, Chapters 1 and 2, and edited the entire volume. Ruben Mnatsakian was the inspiration for this project. Aleh Cherp, Nicolai Dronin, Dmitry Efremenko, and Vladislav Larin contributed extensively and wrote most of Chapters 3, 4, and 5. Nicolai Dronin was also centrally engaged in the final editing of the book. Sergei Mirnyi wrote a contribution to the section on Chernobyl. Kim Losev and Yuri Golubchikov added significantly to Chapter 2.

---

[17] Jane Dawson, *Econationalism* (Durham: Duke, 1996); Marshall Goldman, *The Spoils of Progress* (Cambridge: MIT, 1972); Thane Gustafson, *Reform in Soviet Politics* (Cambridge: Cambridge University Press, 1981); Philip Pryde, *Conservation in the Soviet Union* (Cambridge: Cambridge University Press, 1972), *Environmental Management in the Soviet Union* (Cambridge: Cambridge University Press, 1991), ed., *Environmental Resources and Constraints in the Former Soviet* (Boulder: Westview Press, 1995); Douglas Weiner, *Models of Nature* (Bloomington: Indiana University Press, 1988), *A Little Corner of Freedom* (Berkeley: University of California Press, 1999); and Charles Ziegler, *Environmental Policy in the USSR* (Amherst: University of Massachusetts Press, 1987).

At various times, Karl Hall and Edward Bellinger were involved in editing and discussions. Nora Mzavanadze, a doctoral student in environmental policy, assisted in the preparation of the footnotes. In these notes and elsewhere in the book we have used a simplified form of the Library of Congress system of transliteration, in a number of cases dropping "soft signs" from the transliterated words, and in other cases using generally accepted spellings of places and names (Trotsky, not Trotskii; Yablokov, not Iablokov; Yamal, not Iamal). The authors thank the Central European University for its support. Special thanks goes to the Open Society Archives of the Central European University (Budapest, Hungary) for providing opportunities for one of the authors (Nicolai Dronin) to use its collection of Radio Free Europe of the Soviet media concerning environmental problems for the 1950s–1980s, whose rich materials enabled a deeper understanding of Soviet environmental history. Several colleagues kindly provided critical comments: Stephen Brain on the introduction, Jim Webb on the introduction and Chapter 5, and David Moon and Andy Bruno on Chapter 1, Laura Henry on the conclusion. The authors thank the two referees whose critical insights and suggestions have made this a better book, one of whom, Nicholas Breyfogle, agreed to be identified in these acknowledgements. To both referees: we are deeply grateful for your thoughtful and detailed comments. Finally, we acknowledge the generosity of Allan Gamborg of the Gamborg Gallery in Moscow, Russia, who provided the rights to the illustrations used in this book.

Vladimir Lenin, who led the Soviet Union until his death in 1924, presided over the expansion of the scientific enterprise, and saw technology, here the tractor, 100,000 of them, as the key to the modernization of the backward peasant.

I

From Imperial to Socialist Nature Preservation

Environmental Protection and Resource Development in the Russian Empire, 1861–1925

A series of tensions played out in the environmental history of the Russian empire. These tensions had an impact on policies, practices, institutions, and human–nature interactions in the Soviet period. One was top-down pressure to modernize, by which leading officials including such tsars as Peter the Great meant to westernize, in part through adopting enlightenment attitudes toward nature and landscape. By force of will – or force of military and political occupation – Russia would create modern industry and agriculture across an ever-expanding empire. A second tension existed precisely between the power of the state and the private sector, between autocracy and public participation. In the Russian Empire, given the central role of the state, which rivaled or exceeded that in other nations, any environmental concern – alarm about the health of the forest, worries about agricultural performance and quality of soils, or concerns of nascent conservation movements – played out against concerns about the power of the government. The state could be a force of modernization and reform yet also a brake on development through its policies. Granted, the tsarist state had fewer bureaucrats per capita than the major European states and relied on devolving administration on the local population – nobles, peasant communes, and so on. Still, regarding the environment, it had a crucial role. The tsars determined to expand the empire and tame the periphery, push back the frontier, settle the steppe, and create agriculture that met the needs of growing domestic markets and export. No longer would agriculture be subsistence. This required that arable land succumb to agronomy, that polar and subpolar regions reveal their secrets, that Siberia become a part of the patrimony of the tsars and contribute to the economy, and that nomadic and indigenous people in lightly settled areas give way to settlers.

Indeed, a third tension was the apparent – and real – vastness of lands leading to the genesis of a kind of frontier mentality among bureaucrats, scientists, and activists about the need to bring "civilization" and its modernities of technology, industry, agronomy, and large-scale land and water management practices to the natural environment, even if this meant overwhelming traditional practices and local inhabitants. Many leaders in a variety of fields came to see the peasant and the indigenous minorities as "backward" and in need of reform. Finally, a fourth major tension was between well-intended specialists who came to believe in the gospel of science to improve agriculture and industry, especially peasant practices, which they viewed as outmoded, and the peasants, who often mistrusted them. But they often approached reforms based on science with limited data or incomplete understandings of ecosystems by contemporary standards, even if Russian scientists were in the forefront of developing such understandings in the late nineteenth and early twentieth centuries. The limited data are not surprising given the early stages of development of fields from forestry to fisheries, and from agronomy to hydrology, and so on. These tensions were not unique to the tsarist empire or the Soviet regime but have served as the backdrop for environmental history in such diverse settings as Germany, the United States, France, and Brazil.

The Soviet empire occupied 22,402,200 square kilometers or roughly one-sixth of the Earth's land surface. The vegetation, climate, and natural resources of this vast nation were marked by their remarkable diversity. The location of raw materials, far from population centers, made their development a difficult proposition. Among both imperial and Soviet power, government officials, economic planners, and scientists and engineers encountered great challenges in developing rational policies to manage those resources; often pursued programs that led to profligate use of those resources and to extensive pollution; and failed to involve special experts, let alone the public, sufficiently in deciding policies. Peasants engaged in subsistence agriculture, employed slash-and-burn methods to establish farming, and often resisted efforts to improve its performance, although they were not opposed to innovation. The tsars were inconsistent in their nature protection efforts, seeking more to shield their own property from unwelcome intrusions and ensure access to materials needed for the military rather than to promote any kind of national conservation ideology. There was no counterpart, for example, to the

Progressive Era movement in the United States that President Teddy Roosevelt supported. Only belatedly did the last tsar, Nicholas II, turn to scientists for assistance, but this was in a desperate effort to wage World War I, not generally to seek out rational policies.

Furthermore, scientists lacked the resources to promote conservation to the extent that they desired. Still, the Imperial Academy of Sciences supported a series of extensive expeditions in the eighteenth century, and by late in the century they had established a small number of societies to support the study of fisheries, forests, water resources, and the mineral wealth of the country. Although they had to get a charter from the tsarist bureaucracy to establish these societies, these organizations were often marked by autonomy, and several of them grew quite large. A number of them embraced nature transformation projects such as those in other countries in an effort to improve communications, transport, agriculture, forestry, and fisheries. These included the draining of wetlands, the construction of canals and irrigation networks, and the building of roads, bridges, and dams. But most of the major projects had to wait until after the Russian Revolution, because the tsarist government hesitatingly supported these endeavors as too expensive or perhaps believed them to be unnecessary. Such progressive individuals as Count Sergei Witte, a proponent of state support for rapid industrialization, struggled mightily to gain funding for technological development; he succeeded in appropriating funds for the trans-Siberian Railroad as minister of finance but then was dismissed from the government.[1]

Another challenge for scientists was that "conservation" was in its infancy. As noted in the Introduction, conservation was the belief that through careful, scientifically based evaluation of the extent of resources and their management, present and future generations would always have access to natural resources. These resources included fish, forest, and water important to a variety of industries, the nation, and the citizenry.[2] Russian scientists had begun to develop notions of conservation of forest, soil, and other resources in an effort to improve the quality of life for

---

[1] Stephen Marks, *The Road to Power: The Trans-Siberian Railroad and the Colonization of Asian Russia, 1850–1917* (Ithaca: Cornell University Press, 1991).

[2] Samuel Hays, *Conservation and the Gospel of Efficiency* (Cambridge: Harvard University Press, 1959). See, for example, "Current Progress in Conservation Work," *Science*, vol. 29 (March 26, 1909), pp. 490–491; Charles Adams, "The Conservation of Predatory Animals," *Journal of Mammalogy*, vol. 6, no. 2 (May 1925), pp. 83–96; and Henry Graves, "The Conservation Problem of the Paper and Pulp Industry," *Scientific Monthly*, vol. 20, no. 3 (March 1925), pp. 225–235.

citizens in the empire, for example, in response to the famine of 1891–1892. The famine resulted from crop failure, itself a result of drought, economic policy, and backwardness in agriculture. The inadequacy of the tsarist government's response led scientists to push for measures to protect against any future such disaster.[3] Even though the state under Peter the Great promoted the first sustained efforts to protect forest and other tsars to secure such resources as fur-bearing animals, scientists did not believe these measures went for enough and called on the state to act much more systematically and forcefully.

In this chapter, we offer an environmental history of the tsarist regime and the early period of the Bolshevik Revolution. We discuss the relationship between people and nature before the modernization of Russia that commenced with Peter the Great at the beginning of the eighteenth century. We discuss the extent of fish and forest resources and the halting steps toward their rational use, including tardy and incomplete surveys of their extent, and the impact of agricultural practices on the environment. We conclude with a discussion of the challenges facing the first Bolshevik efforts to identify and manage resources, the political turmoil of civil war after the Russian Revolution, and the uneasy accommodation reached between the Bolsheviks and scientists that created the foundation for the development of modern ecology.

Many scientists believed, even if they mistrusted Lenin and the Bolsheviks, that the Bolsheviks offered a positive alternative to tsarist inconsistency. The Bolsheviks embraced big science and technology as the centerpiece of their economic development programs and soon embarked on those projects with the full support of scientists. In this atmosphere, scientists more actively pursued projects for managing forests, advancing agronomy and soil science, protecting flora and fauna, and pursuing melioration and reclamation than they had under the tsars. This suggests that, in more ways than one, there was great continuity in conservation, preservation, and nature transformation movements of imperial Russia and the Soviet Union, at the very least in the individuals engaged in conservation and preservation and in the attitudes of scientists about the importance of state support for their efforts.

---

[3] On the famine, see J. Y. Simms, "Economic Impact of the Russian Famine of 1891–92," *The Slavonic and East European Review*, vol. 60, no. 1 (January 1982), pp. 63–74. See also Francis Reeves, *Russia Then and Now, 1892–1917: My Mission to Russia During the Famine of 1891–1892* (New York, London, G.P. Putnam's Sons, 1917).

Nature and Society in Pre-Petrine Russia, 900–1700

From the period of the establishment of Rus' in the ninth and tenth centuries – the forerunner of the Moscovy of Ivan the Terrible in the 1500s – the Slavic peoples lived in the forest and in small villages. They hunted, gathered honey, trapped animals for their skins and furs, and engaged in trade. Although population densities were low, the people began to tame the forest, in some places pushing it back for agriculture. Trade between the Slavs and people of Eurasia for products from the forest – honey, furs, and pelts – increased pressure on the forest. Eventually, some individuals established small lumbering operations and beekeeping. Where beekeepers established apiaries, they felled more forest, and agricultural fields appeared. They also hunted and trapped for furs for clothing and trade. The Little Ice Age, an era of lower average temperature circa 1350 to 1850, increased the demand for fur, especially Siberian sable. The history of the conquest of Siberia has been called a "unitary trade expedition for sable."[4]

Although Russia accelerated the rapacious exploitation of vast deposits of coal for its metallurgical industry in the twentieth century – those of the Don, Kuznetsk, and Kansk-Achinsk regions served Soviet power – during the Moscovite and Petrine periods, the empire's smelting relied on charcoal produced from lumber more than on coal. Demand for charcoal grew in the Middle Ages to produce high temperatures in kilns and smelters to make ceramics and metals, respectively. The metal products were used for weapons, tools, carriages, and such agricultural instruments as plows. Trees were also the source of tar, resin, and other products, and increased demand for these products for export also put pressure on the forest. Maple, oak, elm, and hazel were used in the production of potash, whose production demanded large amounts of lumber in a complex process. Peasants also used linden in significant quantities. They stripped off the bark to make boxes and tools. The small number of people who lived off the forest in these ways limited their impact to local impacts and near trade routes.

Demand for lumber expanded rapidly in the 1580s with an increase in the production of ferrous metals for cannons and guns. Depending

---

[4] S. V. Kirikov, *Chelovek i Priroda Vostochno-Evropeiskoi Lesostepi v X – nachale XIX v.* (Moscow: Nauka, 1979) and *Antropogennye Factory v Istorii Razvitiia Sovremenennykh Ekosistem* (Moscow: Nauka, 1981); and E. Rekliu, S. V. Bakhrushin, *Ocherki po Istorii Kolonizatsii Sibiri v XVI I XVII vv.* (Moscow: M. and S. Sabashnikovykh 1927).

on the type and location of ore, for example, from swampy regions, the amount of charcoal needed grew significantly; only with charcoal could the metallurgists achieve the temperatures needed to melt the metal.[5] Still, until the eighteenth century, import of armaments – for example, Scottish cannons – played a greater role than domestic production. The metallurgical industry ultimately was centered near ore deposits in the Ural Mountain region. The forests nearby – the main source of charcoal – were denuded by the end of the nineteenth century, and this led to a shift in metallurgical industries based on coal and an increase in the importance of the Donets River basin ("Donbass") coal deposits in Ukraine.

In the Middle Ages, early thinking about the environment was practiced in the sacralization of nature, shamanism, animism, and other religious activities.[6] These practices helped peasants to comprehend an otherwise capricious and powerful nature that seemed beyond their control. Yet, the peasants had a comprehensive if quite imperfect knowledge of nature connected with their farming and other activities; during the Middle Ages, a landowners' calendar was produced that included ecological contents and admonishments. The peasants developed communes to share the risk for bad harvests given the difficulties of farming in the Russian climate.

Some historians believe that lands around monasteries that the monks protected and developed were in some ways the prototypes of contemporary nature preserves (called *zapovednik*); others point to the forerunners of zapovedniks in the various preserves of nobility. The Izmailovo Forest Park of Alexei Mikhailovich, at twelve square kilometers, was one of the first, if not the first, large forest park-preserve in Europe with an experimental-selection aspect. During the second half of the seventeenth century, workers at Izmailovo planted wheat, rye, oats, barley, and other crops in it. In a garden along the banks of the Serebrianka River, they cultivated barberries and different kinds of apple, pear, plum, and cherry trees. They experimented with various kinds of grasses and herbs used for medicinal purposes and a vineyard, and made efforts to grow mulberry trees, gooseberries, and lilac. They brought in stock-breeding cattle from Belgorod and built a small menagerie for wild animals. Noblemen and -women contributed to the construction of ideas about nature through their activities on their estates. They established gardens and

---

[5] S. G. Strumilin, *Istoriia Chernoi Metallurgii v SSSR. Feodal'nyi Period (1500–1860 gg.)*, I, (Moscow: 1954).

[6] B. E. Boreiko, *Ekologicheskie Traditsii, Dover'ia, Religioznye Vozzrezniia Slaviansikh i Drugikh Narodov*, 2nd ed., I (Kiev: Kievskii Ekolog-Kul'turnyi Tsentr, 1997), p. 224.

arbors, beehives, and so on, and not only at such well-known estates as Pushkin Mountains, Iasnaia Poliana, Muranovo, Tenishchevo, Marfino, Kikol'skoe, and Spassko-Lutovinovo. Among some individuals, the belief that man ought to possess and command nature held prominence.[7]

Conservation from Peter the Great to the Russian Revolution

The state controlled the forest from the thirteenth century, when the first tsars strengthened their property rights to permit inheritance through appropriate deeds. Russian princes kept their own preserves and protected them from outside encroachment with severe penalties for trespassing. Subsequently, much of the land remained the patrimony of the tsars, whereas the nobility gained power over much of what was left, leaving the church with lands around monasteries and peasants with virtually nothing. Forestry served largely local and immediate purposes, such as the sale of wood abroad (by the sixteenth century), mining and metallurgy (in the seventeenth century), and shipbuilding (in the eighteenth century, after Peter the Great set out to establish a navy).

In the Middle Ages, the forests provided a buffer for Russia against nomadic Mongols and Turks, who fought Slavs for control of Eurasia. The forest insulated scores of rivers and lakes from overfishing and protected beekeeping, fur, and other nascent industries. The process of pushing back the forest to create agricultural land was much slower in Russia than in America because it was such hard work. The land yielded food grudgingly in a short growing season. After cutting down trees, the peasants burned the underbrush. They practiced subsistence farming, often in cooperation with other families in communes, to spread the heavy labor and risk for harvest failure over a larger number of persons. The poor soils might produce adequate harvest for three years; the peasants then moved on and repeated the process to clear other wooded lands. Their choice of land was severely restricted by state, noble, and church control of the forests.

Because of their reliance on climate and natural resources, Slavs venerated nature and the forest, and even after converting to Christianity in the tenth century they maintained pagan festivals of nature worship. When Peter the Great (1672–1725) westernized Russia at the beginning

---

[7] *Prirodnye Sviatyni Rossii (po Materialam Knig Al. A. Grigor'ev Sviatyni Rossii)* (St. Petersburg: St. Petersburg University Izdatel'stvo, 2004); and A. A. Grigor'ev, *Prirodnye Sviatyni* (St. Petersburg: Obrazovanie, 1997). See also Ol'ga Yur'evna Elina, *Ot Tsarskikh Sadov do Sovetskikh Polei: Istoriia Sel'sko-Khoziaistvennykh Opytnykh Uchrezhdenii XVIII-20-e gody XX v.* 2 vol. (Moscow: IIEiT RAN, 2008).

of the eighteenth century, he contributed to a widening schism between Old Believers and a more modern church that had joined forces with the autocracy. As a result, the Old Believers sought refuge in the northern forests of contemporary Arkhangelsk and Vologda provinces. They managed to survive by establishing some pastures and fields, building small mills for their grain, and acquiring more land to secure solitude from neighboring peasants and the state.[8]

Peter the Great promulgated conservation measures that went beyond his own estates. At the height of the Russian Empire (during the Soviet period), the forest, more than half of it boreal, comprised 910 million hectares, slightly more than one-fifth of that in the world, with total wood mass at 80 billion cubic meters or one-third of that in the entire world, including 65.4 billion cubic meters or 58.2 percent of the world's coniferous material. Ninety-six percent of Soviet forests fell within Russia, and 96 percent of that total was under the jurisdiction of massive forestry concerns whose harvesting practices have resulted in long-term and in some cases irreversible environmental impact. Peter issued forest protection decrees, insisted on sustained-yield forestry and set aside protected forests, especially along waterways to prevent erosion and to protect oaks necessary for naval construction. After Peter, few tsars were interested in nature protection or resource management, except as concerned their own access to good hunting and fishing.

Peter the Great organized a forest administration for shipbuilding. To ensure adequate resources for the burgeoning military, Peter imported the German model for imperial forestry – as he generally had imported European science for his nation – in shipbuilding, medicine, anatomy, mathematics, and its institutions (schools and the Imperial Academy of Sciences, founded in 1725). Peter ordered surveys of all forests along rivers; rangers counted all trees with a diameter of forty centimeters or more. The inventory led to decrees prohibiting cutting in swamps and very dry areas, identifying disease-free trees for felling, and setting the time of cutting for shipbuilding in November. Peter strengthened the position of Russia on the White Sea, issuing a directive for the preservation of forests along the Dvina River near Arkhangelsk.[9]

---

[8] Robert O. Crummey, *The Old Believers and the World of the Antichrist* (Madison: University of Wisconsin Press, 1970).
[9] V. K. Teplyakov, Ye. P. Kuzmichev, D. M. Baumgartner, and R. L. Everet, *A History of Russian Forestry and Its Leaders* (Pullman, WA: Washington State University, 1998), pp. 2–4, 13.

During the next 150 years, others tsars continued Peter's efforts to manage the forest, although not necessarily systematically or rationally. In the case of the tsarist empire, "rational" meant the strengthening of such industries of interest to the state as mining, metallurgy, and shipbuilding. In 1782, Catherine the Great ordered surveys of huge tracts of Russia's forest. In comparison with the private forests, whose owners destroyed them seeking only profit, the state forests did relatively well. In the first decades of the nineteenth century, the government created a forest department independent of the navy. The department established state forests to support salt production (1818), shipbuilding (1829), monasteries (1832), city councils (1832), and horse breeding (1833). State-run mills were established between 1811 and 1837, and in 1839 the government established a forest corps with more than 700 employees responsible for inspection, management, conservation, use, and accounting. Forest research accompanied but was secondary to geographic, soil, and other expeditions through the nineteenth century.

The origins of modern state-sponsored forestry date to the early nineteenth century. The forestry department was founded in 1798, followed by a detailed code of laws in 1801; by the 1840s, the government included a budget line item for forest projects in sixteen provinces. By 1849, 2,650,000 hectares had been surveyed; by 1859, 3,100,000 hectares; and by 1914, 535,000,000 hectares, although less than 32 percent had been thoroughly investigated. There was a rapid increase under Soviet power, with 949,000,000 hectares already surveyed in 1930, rising to 1.131 billion hectares in 1958.[10]

Peter the Great forced engagement of foreigners and the westernization of the Russian Empire. This engagement had an impact on how Russians viewed themselves and their relationship with nature. Peter fought excessive religious influence and hoped to bring science to Russia through the founding of the Imperial Academy of Sciences and a series of schools. The Academy was staffed initially with European scholars. These scholars and others connected with expeditions of the Academy contributed to growing knowledge of the empire's great wealth of natural and mineral resources. Peter himself traveled to Europe to see with his own eyes and learn with his hands how to use modern tools of navigation, build ships, and understand what a microscope does. He imported Western science, ideas, customs, and dress (for example, the requirement that nobility shave their beards or face a "beard tax"). The major symbol of this

---

[10] Teplyakov et al., *A History of Russian Forestry and Its Leaders*, pp. 5–7.

westernization was the construction of St. Petersburg, Peter's new capital city on the Neva River and Gulf of Finland. He ordered thousands of peasants to the construction site – a swamp and delta subject to spring floods, summer swarms of mosquitoes and deep, dark, frozen winters. Thousands of peasants died in the process of draining the swamps and building Peter's "Venice of the North." In any event, throughout the Petrine period and from that time forward, scientists engaged actively in expeditions, scientific studies, and work on animal husbandry and hybridization. These expeditions were crucial in generating nationalist sentiment, indicating that nature exploration and nationalism were strongly linked.[11]

With the rise of cities, patterns of impact shifted. Urban centers have their own ecology – and public health problems – associated with rapid growth; supply of food and clean water; heavily polluting industries centered on mining, metallurgy, paper, and textile mills; and migrations of people seeking jobs. In the early stages of urbanization, migrants were usually illiterate peasants or farmers with few skills. Early cities grew rapidly, although in fits and starts, without any regulation. As a result, the remaining green regions of cities were quickly threatened by industrial encroachment. Unless the authorities took active measures to establish parks and gardens, preserve trees, and protect waterways, the urban ecology rapidly deteriorated under these pressures.

Many towns and villages in the Slavic world date to the medieval period and Middle Ages: Kyiv, Novogorod, the towns of the Golden Ring – Sergiev-Posad, Tver, Iaroslavl, Vladimir, and Suzdal – and Moscow. The rise of cities demanded a huge amount of wood for fuel; hearths and stoves were inefficient. Wood was also was used for illumination and for activating yeast and baking bread. During the Little Ice Age, the amount of wood used for one person may have risen to four to five cubic meters per person per year.

By the mid-seventeenth century, Moscow had become a city on the scale of several European cities. It was a home to tradesmen, merchants, soldiers, and officers, nobility, peasants, and foreigners. Because of mistrust of the foreigners, they were essentially required to remain in the "German Quarter" (*Nemetskaia sloboda*) of the city. By the mid-eighteenth century, Moscow had become a vast, disorganized beehive of activity. The end point of major trade routes and the site of huge mills, it

---

[11] Mark Bassin, "The Russian Geographical Society, the 'Amur Epoch,' and the Great Siberian Expedition, 1855–1863," *Annals of the Association of American Geographers*, vol. 73, no. 2 (1983), pp. 240–256.

was devastated by the bubonic plague as late as 1772, when rats carrying infected fleas journeyed up the Volga and Moscow Rivers in bolts of fabric to the mills. In the entire tsarist empire, there were but a handful of doctors assigned to deal with public health. When industrialization took off in the Russian Empire in the nineteenth century, the cities were the focus of this rapid change. They also became the locus of poverty, epidemic diseases of typhus and typhoid, and public health crises all tied to problems of inadequate clean water supply and dangerous factories. At the end of World War I, during the civil war and subsequent famine, the cities emptied of residents who sought food, fuel, and shelter elsewhere. St. Petersburg lost one-third of its inhabitants.

Natural Resources: Early Management Practices of Forests

In spite of Russia's vast resources, under the tsars, the forests were poorly managed and the workers poorly paid and equipped, so that the industries performed at a very low level, so much so that Russia even imported forest products from the United States and Japan in the nineteenth century. In broader society, zoologists and agronomists associated with Moscow University were among the first to raise what today would be called ecological concerns, especially as they related to improvements in agriculture. They argued that science might show the way for humans to manage and improve on nature, discarding the standard passive attitudes toward nature embodied in religious beliefs and traditional thinking about the tsar's patrimony; their views reflected Enlightenment thought developing in late eighteenth century Europe and America. How human society might be an active force was a source of debate in various nascent organizations from the Moscow Agricultural Society to the Free Economic Society (located in St. Petersburg and founded under Catherine the Great, 1762–1795).

Specialists made presentations to members of the Free Economic Society over the years that addressed what came to be considered ecological issues. These papers suggested the need for greater involvement of educated society in the management of resources and the improvement of agriculture. Just after its founding, in 1767, one member called for measures to protect and increase the forest cover.[12] Even before pressures to

---

[12] P. I. Richkov, "O Sberezhenii i Razmnozhenii Lesov," in *Trudy Vol'nogo Ekonomicheskogo Obshchestva, k Pooshchreniiu v Rossii Zemledeliia i Domostroitel'stva*, chap. VI (St. Petersburg: Free Economic Society, 1767), pp. 84–112.

end serfdom built up, scholars discussed how best address the poverty and uncertainty of life in the countryside through the application of modern ideas. They called for replacing three-field rotation and endless division of parcels among many peasants with a more productive system of many-field rotation (for example, a 1771 article by A. Bolotov, "O Razdelenii Polei").

Forestry science developed in Russia on the foundation of German forestry practices. German foresters sought to turn the forest into a factory based on monocultures of various species to maximize production of timber on the basis of limited forestry reserves. They sought to produce only large trees and to thin out those they deemed unsuitable for such things as masts. They applied mathematics to their art in one of the first efforts to develop a scientifically founded notion of sustainable yield. They founded the world's first forestry schools and journals. There were three problems with the German foresters' approach. First, thinning and eliminating underbrush deprived the forest of organic material needed to fertilize the soil and trees. Second, their mathematical model became almost as important as the reality of the forest. And third, the creation of monocultures, as we now know, was both expensive and ultimately not healthy for the forest because it became much more prone to disease. Yet, because forestry came of age when the tsars were seeking to modernize Russian practices, the tsars imported precisely the German model – and foresters from Germany. They dominated forestry education and management for decades. German specialists frequently visited Russia to share their work, and the major German publications on the topic were translated into Russian soon after they appeared in German.[13]

By the late nineteenth century, Russian foresters established national and provincial forestry societies, where they and their counterparts in the government discussed national resource questions. Although they were poor and overworked, they were enthusiastic. Members of the Imperial Moscow Society of Agriculture and the Free Economic Society joined foresters in the examination of Russia's forests. Many of them worried that Russia's forests were under great threat. Most logging took place in concentrated areas; no such thing as selective cutting existed. Various species of animals also faced destruction of their habitat. These specialists were part of a growing international conservation movement seeking

---

[13] For a discussion of the influence of German practices on Russian forestry, see Stephen Brain, *Song of the Forest: Russian Forestry and Stalin's Environmentalism, 1905–1953* (Pittsburgh: University of Pittsburgh Press, 2011), chap. 1.

to bring scientific management to natural resources. The major research centers were the St. Petersburg Forestry Institute, which put out yearbooks, and the Imperial Forestry Institute, whose scientists published at least thirty volumes of studies between 1898 and 1914 alone.[14]

Foresters, likes specialists in other disciplines, accumulated a vast array of data on the makeup and what they considered to be signs of health of the forest. Paradoxically, these data and publications indicated how little they understood about forest ecosystems, not how much. Specialists within the government bureaucracy, academic institutions, and independent societies established a series of publications to explore the health of the forest and promote rational policies for its future development. The journals *Lesnoi Zhurnal* and *Russkoe Lesnoe Delo* had wide readerships. *Trudy Opytnogo Lesnichestva*, a publication of the Ministry of the Interior and Government Property's forestry department, presented vast quantities of data, papers from various All-Russian Congresses of Foresters throughout the century, and the descriptive results of scientific expeditions.

But what did the data mean? Unfortunately, the foresters believed that the public had no understanding of the impending disaster for Russia's forests. A series of laws, statutes, and instructions issued over the decades reveals that the foresters and state representatives still could not determine how best to manage the trees. The Imperial forestry ministry eventually published an eighteen-volume study in an attempt to provide a basis for rational management and for standardization of timber trade. It also enabled local authorities to regulate cutting of forests, land sales, and tax collection. Yet, the compilation led to the proliferation of regulations through which it was hard to see the trees. In 1905, the state introduced a new forestry code consisting of 815 articles; in 1913, it produced another code of 431 articles. The codes did not have the desired effect: private forests were exploited without regard to the future, whereas the public forest was unable to meet growing demand. Even with regulations, such areas of important forest industries as those in Arkhangelsk and Vologda provinces fell outside the attention of professional foresters. They considered the south and the steppe as easier for logging and transport, even as they surveyed the north and its rivers.[15] Ultimately, the rapid settlement

---

[14] See *Izvestiia Imperatorskago Lesnago Instituta*, various. See also *Ezhegodnik Lesnogo Departmenta*, a publication of the Ministry of the Interior and Government Property's forestry department.

[15] V. K. Teplyakov et al. *History of Russian Forestry*, pp. 6–8, 26–29.

of the steppe and destruction of its forests led such scientists as Vasily Dokuchaev to promote conservation measures (as noted in Box 2).

In addition to narrowly scientific issues, the publications reflected growing interest in economic, policy, and conservation issues. The early twentieth-century publication *Lesopromyshlennik* (Industrial Forester) was a weekly journal of the forestry trade and industry. The port of Riga, Latvia, published *Lesnoi Rynok* that concerned growing international forestry trade. Those interested in the relationship between the farming and forestry perused *Sel'skoe Khoziaistvo i Lesovodstvo*; those in the importance of the healthy forest for hunting and trapping, the mid-nineteenth century newspaper *Gazeta Lesovodstva i Okhoty*; and those in the national spirit of the trees, *Lesnoi Dukh* which continued simply as *Les* until its collapse in the economic and political turmoil of 1918.[16] Monograph publications of the nineteenth and early twentieth centuries revealed several concerns among foresters and trends, but little consensus. Several publications dealt with issues of conservation. Others touched on economic and legal issues, including taxation, trade, and the development of new technologies to improve harvests that would be of interest to large concerns and individual foresters, and included handbooks.[17] Still other publications reflected the fact that the empire's forest, paper, and pulp industries expanded rapidly in the second half of the nineteenth century as part of industrialization; domestic and foreign markets assumed great importance.[18] Interest in the forests of Siberia and other distant parts of the country also grew.[19]

Still, Russian forestry practices remained in their infancy. The forest industry was haphazard, poorly mechanized, short on qualified workers or laborers, and weak on leadership. The government's contribution

---

[16] In "Imaginations of Destruction: The 'Forest Question' in Nineteenth Century Russian Culture," *Russian Review*, vol. 62, no. 1 (January 2003), pp. 91–118, Jane Costlow traces debates among writers, painters, and others over the feared destruction of the Russian forest.

[17] See, for example, F. Baur, *Lesnaia Taksatsiia*, trans. A. S. Shafranov (St. Petersburg, 1878); V. Vrangel, *Istoriia Lesnogo Zakonodatel'stva Rossiiskoi Imperii* (St. Petersburg, 1841); A. Bode, *Ruchnaia Kniga dlia Khoziaistvennago Obrashcheniia s Lesami*, trans. Ia. Grakhoi (St. Petersburg, 1843); and V. I. Denisov, *Lesa Rossii, ikh Eksploatatskii i Lesnaia Torgovlia* (St. Petersburg, 1911).

[18] N. Shelgunov and V. Greve, *Lesnaia Tekhnologiia* (St. Petersburg, 1858); A. N. Popov, *Lesnaia Tekhnologiia* (St. Petersburg, 1871).

[19] For example, V. E. Bokov, *Drevoobrabotyvaiushaia Promyshlennost' v Permskoi Gubernoii* (Perm, 1899); S. Iu. Rauner, *Gornye Lesa Turkestana i Znachenie ikh dlia Vodnago Khoziaistva Kraia* (St. Petersburg, 1901); Ivan Ozerov, *K Voprosu o Nashikh Severnykh Lesakh* (Moscow, 1911); and A. A. Strogii, *O Lesakh Sibiri* (St. Petersburg, 1911).

was limited to support of extensive surveys and did not include efforts to transform empirical knowledge into silvicultural practices. Nor were merchants interested in lumber activities. Theirs remained a cottage industry; mills were small and underpowered; and the peasant who toiled in the forest was unmotivated to do more than eke out an existence. Forestry schools were lucky to turn out 200 people annually with higher degrees related to forestry. On the eve of the revolution, there was just one Forest Institute (in St. Petersburg) and two other departments at agricultural institutes to train forestry specialists for all of those millions of hectares. In all, there were about 7,000 employees in the forestry ministry, fewer than one-third of whom had higher education.[20]

If state forests were in disarray, the forests of the peasants were in terrible condition. The peasants mistrusted the state either because they were deprived of ownership of forests or because they expected the state to arbitrarily confiscate what forests they owned. Under the circumstances, they understandably adopted short-term strategies to exploit the resources rather than longer term – and foreign to them – policies recommended by foresters and state representatives. Peasant forests were overharvested, leaving brush and erosion in their wake, especially close to the villages. The growth of population, the free grazing of farm animals, fires, inefficient construction, and the absence or flouting of laws contributed to this state of affairs. By the 1880s, a number of scientists and social commentators raised the specter of the destruction of the forests and the decimation of game within those forests and urged the government to act. Peter's forestry codes were updated several times late in the century, and hunting laws followed suit, although enforcement powers lagged, there were few rangers, and flora and fauna continued their rapid decline in central Russia. Several schools of conservation thought developed at this time that were based on international and national intellectual trends and biological thinking, including a strongly scientific basis.[21] In spite of a law in 1888 attempting to regulate forest use, between that year and 1908, 3 million hectares were very nearly clear cut, and the total area of forests in European Russia declined by 9 million hectares. According to M. A. Tsvetkov (1957), from 1696 to 1914, the forested areas of Kursk, Poltova, Voronezh, and Kharkiv provinces dropped from 18.4 percent

---

[20] A. I. Bovin, B. P. Tsepliaev, and D. T. Kovalin, *Lesnoe Khoziaistvo SSSR, 1917–1918* (Moscow-Leningrad: Goslesbumizdat, 1958), pp. 38–39.
[21] Douglas Weiner, *Models of Nature* (Bloomington: Indiana University Press, 1988), pp. 9–12.

to 6.8 percent and in Tombov and Penza from 46.1 percent to 16.2 percent. (See **Box 1**.) Only the northern provinces remained relatively untouched.[22]

During the last decade of the tsarist regime, a movement in favor of protection of forests emerged among professional foresters, who saw the solution to the steady decline in Russia's great forests in "the establishment of a single, rationally planned national forest economy, utilizing scientific methodologies and technical knowledge, managed in consensual association between center and localities and based on state ownership."[23] Because forest resources seemed so vast, there had been to this time little effort to promote conservation such as that under Gifford Pinchot in the United States in the 1890s and 1900s. Pinchot, the first chief of the U.S. Forest Service, strongly advocated conservation, by which he meant planned management and renewal of resources, which he believed would contribute to the general prosperity of the nation.[24] But because of the international trend toward conservation, and because of the 1891–1892 famine, conservation gained a number of Russian adherents. These individuals called for reforestation in the European parts of the empire, where farming and poaching had pushed back the forest and left remaining forest in poor condition and had also degraded soils.[25]

Agriculture and Environment in Tsarist Russia

Agriculture is crucial to environmental history because it centers on issues of land use, including the clearing of forests; the use of slash-and-burn methods; the plowing of meadows; the use of manure and chemical pesticides, herbicides, and fertilizers; hybridization of crops and animals; and the development of monocultures. Traditional ways of farming often fail

---

[22] Bovin, *Lesnoe Khoziaistvo SSSR*, pp. 6–8, 16–17.

[23] Brian Bonhomme, *Forests, Peasants and Revolutionaries* (Boulder, CO: East European Monographs, 2005), p. 58.

[24] For example, see Gifford Pinchot, "The Relation of Forests to Stream Control," *Annals of the American Academy of Political and Social Science*, vol. 31 (January 1908), pp. 219–227, and "The Foundations of Prosperity," *The North American Review*, vol. 188, no. 636 (November 1908), pp. 740–752.

[25] See P. V. Baranetskii, *Lesokhranenie* (St. Petersburg, 1880); Ia. Veinberg, *Les, Znachenie ego v Prirode i Mery ego Sokhranenie* (Moscow, 1884); *Doklad Vtoroi Kommissii po Voprosu o Merakh Bor'by Protiv Pochvivshagos v Lesakh* (Moscow, 1895); N. Iu. Shimanovskii, *Okhrana Lesa* (St. Petersburg: S. N. 1910); D. M. Zaitsev, *Nuzhny Li v Rossii Lesookhranitel'nye raboty* (St. Petersburg: S. N. 1911); N. I. Kuznetsov, *O Russkom Lese. Pochemu Dolzhno Berech Les* (St. Petersburg, 1913); and S. N. Arkhupov, various, on reforestation measures.

> **Box 1. Public Concern about Deforestation in Russia in the Late Nineteenth Century**
>
> David Moon shows that intense debates took place in Russian society about the potential impact of deforestation on climate. Supporters advanced two major hypotheses: whether climate change is progressive, caused by anthropogenic factors, mainly deforestation; or cyclical, with autogenic and human activity having only local impact. These debates spread far beyond the work of scientists. Anton Chekhov, in his unsuccessful vaudeville *The Wood Demon* (1889) and then in a very successful play, *Uncle Vanya* (1896), presented the issue through a personage devoted to forest conservation in the Russian Black Earth Region. In the first act of *Uncle Vanya*, Dr. Astrov makes an impassioned speech about the impact of deforestation on the environment:
>
> "Russian forests are crashing down under the axe, billions of trees are perishing, the habitats of wild animals and birds are being laid to waste, rivers are getting shallower and drying up, wonderful landscapes are disappearing never to return, and all because man is too lazy to have the sense to extract fuel from the earth.
>
> "(To Elena Andreevna) Isn't it true madam? You would need to be an unreasoning barbarian to burn such beauty in your stove, to ruin that which we are not able to create. Man has been endowed with reason and the power to create so that he can increase what has been given to him, but until now he has not created anything, only destroyed. There are fewer and fewer forests, rivers are drying up, wildlife is being exterminated, the climate is ruined, and with every passing day the land is becoming poorer and more disfigured.
>
> "(To Voinitskii) You are looking at me ironically and are thinking that everything I am saying is not serious, and... and, perhaps it is a bit eccentric, but when I walk past peasants' forests that I saved from felling or when I hear the sound of my own young forest, which I planted with my own hands, I am conscious that I have a little power over the climate, and that if in a thousand years man will be happy, then to a small extent I will have been responsible."[26]
>
> Dr. Astrov expressed the prevailing view that had emerged over the middle decades of the nineteenth century, that the "climate is ruined" as a consequence of human destruction of forests. However,
>
> *(continued)*

---

[26] Anton Chekhov, *Polnoe Sobranie Sochinenii i Pisem* (Moscow, 1972–1984), vol. 13, pp. 72–73.

> Box 1 (*continued*)
>
> other individuals soon advanced the argument that the steppe climate was no different from the past, and that human activity had not, or indeed could not, affect the climate. The latter view relied on existing observations, whereas the earlier view referred to memories of old-timers about the slightly more distant past and other such largely anecdotal evidence.[27]
>
> In a way, current debates on global warming indicate similar controversy, although among reputable scientists there is no doubt about the human sources of climate change.

to produce sufficient crops for market, let alone subsistence for a family, and often rely on methods that exhaust the soil within a few years because peasants may have accumulated sufficient experience over generations to meet local needs but not understood how best to rotate crops or use natural fertilizers. There is evidence, however, that peasants accumulated considerable knowledge over time, and that their techniques became damaging only under pressure of population increase or when they tried to introduce them in different environments.

As in the United States, where settlers had the impression of vast, nearly unlimited land virtually devoid of native inhabitants, so in Russia, the steppe was a frontier to be conquered. And as in the United States, where settlement led to wholesale changes in ecosystems and to the human populations in them, so in Russia, conquering of the steppe had long-term impacts on the environment and its inhabitants. The Russian state promoted taming of vast expanses of open grasslands that lay south and east of the steppe. This was, according to Willard Sunderland, a kind of imperialist empire building. Beginning with Ivan the Terrible and especially with Catherine the Great, large areas of steppe from the Danube to the North Caucasus were annexed, with the state consciously pursuing a civilizing mission. It sponsored the settlement of steppe by mostly Russian and Ukrainian agriculturalists to populate and cultivate lands that they saw as wild and empty. Of course, the settlers

---

[27] David Moon, "The Debate over Climate Change in the Steppe Region in Nineteenth-Century Russia," *Russian Review*, vol. 69 (April 2010), pp. 251–275, and "Agriculture and the Environment on the Steppes in the Nineteenth Century," in Nicholas Breyfogle, Abby Schrader, and Willard Sunderland, eds., *Peopling the Russian Periphery: Borderland Colonization in Eurasian History* (London: Routledge, 2007), pp. 81–105. See also Moon's *The Russian Peasantry, 1600–1930: The World the Peasants Made*, (Boston: Addison Wesley Longman, 1999).

encountered hardships of unfamiliar environments and interference by native, largely nomadic peoples who resisted efforts to colonize them and their land. Expeditions of scientists contributed to the taming of the steppe by locating and charting resources.[28]

Over the course of centuries, migration and settlement contributed to significant changes in wildlife, soils, forests, mixed forests, and steppe. David Moon notes that rich and varied wildlife filled the coniferous and mixed-forest belts before the decimation of their forest habitats with clearing for agriculture, and also because of hunting and trapping. Such larger mammals as elk, deer, bears, wolves, and wild boar, and such smaller fur-bearing animals as foxes, hare, beavers, mink, and sable lost their habitat. Deforestation accelerated with growth of populations. "In most provinces of the central non-black-earth, central black-earth, and mid-Volga regions, the area of land covered by forest was reduced by between a half and two-thirds over the eighteenth and nineteenth centuries," whereas steppe regions lost much of what woodland they had. Nobility, industrialists, and shipbuilders, of course, also contributed to deforestation. Deforestation occurred across southern Siberia, too, with a loss of animal life. In the grasslands, antelopes and wild horses were relatively plentiful until their range was ploughed up by peasant-settlers.[29]

Over the centuries, when most peasants lived in the forest-heartland, they grew cereals and kept livestock and supplemented their incomes with nonagricultural activities. They cleared forests to make arable land, meadows, and pastures. When they moved, they sought familiar environments. Moon writes, "Since most peasant-migrants aimed to continue farming, they settled in areas where the conditions were suitable for agriculture. Peasants did not, therefore, settle in large numbers in northern and eastern Siberia, where the permafrost made crop cultivation extremely difficult, or in the deserts of central Asia." The peasants adapted to new circumstances with technological change. Moon observes, "The traditional Russian horse-drawn wooden plough (sokha), which was satisfactory for turning over the light soils of the forested regions, was useless on the open steppes, where the heavy black earth was matted with the roots of the steppe grasses." The burning of grasses and scrub in the steppe, followed by plowing, overcropping, overgrazing, and other practices, both destroyed habitat for wild animals and contributed to erosion.

---

[28] Willard Sunderland, *Taming the Wild Field: Colonization and Empire on the Russian Steppe* (Ithaca: Cornell University Press, 2004).

[29] David Moon, "Peasant Migration and the Settlement of Russia's Frontiers, 1550–1897," *Historical Journal*, vol. 40, no. 4 (December 1997), pp. 859–893.

Together, deforestation and ill-advised agricultural practices likely contributed to drought.[30]

Peasants pushed agriculture as far as possible with existing technology, plowing, and fertilizing practices by the end of the nineteenth century. Policy makers and scientists worried about what they perceived as the backwardness of the peasant because agriculture was under great pressure to produce more. Rapid population growth, urbanization, and industrialization, plus growing export markets needed to provide investment for other sectors of the economy, all demanded change in practices. According to David Kerans, most peasants were unable intellectually to move beyond the stage of subsistence farming because of a world view – and climatic conditions – that did not include risk taking, technological exploration, and experimentation;[31] Moon and others dispute this negative view.

In any event, the seemingly inexhaustible amounts of land offered them no reason to improve cultivation practices since they felled forest, slashed and burned, planted, and moved on. Kerans also argues that peasants had limited competence as farmers, selected seeds poorly, and got the timing of sowing and reaping wrong. Peasants refused to risk anything more complex than their three-field system. The three-field system enabled peasants to operate with a limited view of the external world with a limited good – good land. Each peasant household endeavored to use land – and the soil in it – to its maximum because their use of it was temporary according to communal dictates.

The communes to which most peasants in the central Black Earth regions belonged were crucial in passing on traditional agricultural practices. The practices may have been inefficient by modern standards, but the commune enabled peasants to share the altogether too frequent risk for agriculture failure. Russian peasant agriculture was at constant risk for failure given the short growing season and the relatively limited regions where rainfall was sufficient. The commune members enforced a coordinated choice of crops, fertilizers, and the degree of exploitation of the soils. It also supervised the division of lands in families from generation to generation; this led unfortunately to the creation of small, narrow plots spread inefficiently around a village.[32]

---

[30] Moon, "Peasant Migration."
[31] David Kerans, *Mind and Labor on the Far in Black-Earth Russia, 1861–1914* (Budapest: Central European University Press, 2001).
[32] Among many other articles on the commune, see Boris Mironov, "The Peasant Commune After the Reforms of the 1860s," *Slavic Review*, vol. 44, no. 3 (Autumn 1985), pp. 438–467.

Because communes constantly redistributed the land and no household owned it, households had no long-term interest in fertilizing or upgrading soil, only in exhausting it.[33] Communes therefore acted in a very limited way to prevent soil exhaustion by imposing the three-field system, where one-third of land always lay fallow against exhaustion. As long as communal land use was the norm, then strictly coordinating choice and timing of crops served somewhat to protect the soil. Yet, it also prevented the development of agriculture that produced surplus for markets. Thus, the peasant commune had perhaps the greatest human influence on the environment before the industrial revolution and the rise of the city in Russia.

The commune was not an important institution in Ukraine and other regions outside the Central Black Earth region. Deforestation in these regions was not due to communal land practices. Rather, as the Russian empire colonized its territory, the peasants constantly pushed into the forest and the steppe by using slash-and-burn techniques. Trees were the most readily available fertilizer. In the late eighteenth century steppe region around Kharkiv, agriculture expanded significantly because of deforestation. Forest fires became an increasing problem in the late nineteenth century because of slash and burn practices. The situation became so bad in northern regions that wooden roofs disappeared to be replaced by straw roofs. And these were followed by more fires.[34]

Beginning in the late eighteenth century, reformers, most of whom were urban-based intelligentsia, turned their attention to the fragility of agriculture – and increasingly to the evils of serfdom. They hoped to bring what they believed were modern agricultural practices based on scientific understandings of soil, crop rotation, and so on to the countryside. Commune members continually frustrated these reformers, who defined success by surplus and profit and creating a market. The outsiders wanted to increase the productivity of the soil to support the projects of the state. But the peasants were not in the least concerned with these issues, but with family, subsistence, and communal support. In any event, the peasants were exhausting the land in central Russia, and remained at risk for crop failure.

When it acted, the Tsarist regime tried to improve agriculture without changing the social structure of the country or providing adequate resources for its modernization, a tactic that was doomed to failure. Tsar

---

[33] Cathy Frierson, "Razdel: The Peasant Family Divided," *Russian Review*, vol. 46, no. 1 (January 1987), pp. 35–51.
[34] Cathy Frierson, *All of Russia is Burning!* (Seattle: University of Washington Press, 2002).

Alexander II emancipated the serfs in 1861, but through a series of stipulations concerning the distribution, tenure, and ownership of the land and membership in existing peasant communes, the peasant remained tied to the land and his village and to small, narrow parcels of land. He had to repay the government loans provided to him to acquire the land that he thought he should be given outright. This left him constantly in debt. He rarely received larger parcels of land best suited to agriculture, or pastures or forest; the nobility kept those for themselves, and peasants often worked as tenants for them on that land. Indeed, the nobility, imperial family, and other private landowners controlled more than one-third of the nation's land, leaving 75 percent of the population to farm the rest of the land in much smaller holdings. In a word, the emancipation did little to change production practices. World War I and the Russian Revolution immediately altered this situation, with estate farming losing its labor to the army at the beginning of the war and with violent peasant confiscations and the revolution itself completing the process.

The tsarist bureaucracy feared that a class of landless peasants who roamed the countryside might form, so it tied the peasants in a variety of ways to the commune. One such way was to require them to pay for the land they received over forty years. The government made the commune as a whole responsible for debt repayment. Any entrepreneur-peasant would have to gain the elders' permission to leave the commune; this stultified interest in expanding one's holdings or trying new techniques. The creation of a peasant land bank in 1885 that allowed peasants to purchase land, produce a surplus, and use the profit to pay off loans began to break down the ties to communes, for example in the Southern Volga region. In areas such as the Southern Volga what used to look like communal land dominated by three-field agriculture turned into a crazy quilt of individual landholdings, with owners experimenting with new (cash) crops. The rise of the profit motive did not lead owners to build fences. Because the communes had served to establish boundaries, entrepreneurs pushed beyond the communes; they saw fences as limiting. In Western Europe and North America, fences were an important indicator of private property and shaped the landscape in significant ways.[35]

Through local organizations of self-rule called *zemstvos*, created after emancipation of the serfs in 1861, agronomists began to make progress

---

[35] Judith Pallot and Denis Shaw, *Landscape and Settlement in Romanov Russia* (Oxford: Oxford University Press, 1990).

in overcoming peasant resistant to adopt different farming techniques.[36] The zemstvos, although operating with little financial support from the regime, had some impact on environmental issues through their activities. In spite of their weak financial position and limited powers, the zemstvos helped to improve village infrastructure by building roads, bridges, and schools, and also brought doctors and teachers to the countryside. Several zemstvos determined to modernize agriculture by educating peasants, giving agricultural aid, and developing a kind of agricultural extension service. Especially in the years after the Russian Revolution, *kraevedenie*, (literally "local studies") was a kind of refuge of zemstvo science that went beyond agriculture. Practitioners of kraevedenie investigated local areas and studied human–environment interactions. Many of them lacked advanced degrees but remained sincerely interested in their studies and enthusiastically filled free time with their study. They hoped to contribute to understanding between rural and urban residents and understanding of local history, economics, culture, and ecology. This interdisciplinary approach to human–nature interactions may have differed from area to area and region to region, but eventually architects, biologists, ethnographers, historians, and ecologists joined the effort and contributed nationally to the rise of a kind of contemporary environmentalism within studies of local lore.

The increasingly intransigent and conservative government under Tsar Nicholas II found the zemstvo organizations to be too liberal and also thought that the existing social structure in the countryside was the bulwark of the Russian nation. Therefore, beginning in the 1890s, the government cut back on funding for extension services, which were already quite low, and threatened to close the Agricultural Academy after the last student graduated in 1893. Although it established a Ministry of Agriculture and State Domains in 1837, the government generally showed a lack of concern about land use problems. But in the face of widespread opposition, it allowed the academy to remain open as the Moscow Agricultural Institute to train agronomists. All in all, however, modern agricultural science lagged considerably.

A great crop failure in 1891 and subsequent famine encouraged zemstvo organizations to work with the Ministry of Agriculture, itself founded only in 1894, to establish experimental stations and agricultural schools. Because the government offered a mere pittance in support of

---

[36] Kerans, *Mind and Labor*. See also Kerans, "Toward a Wider View of the Agrarian Problem in Russia, 1861–1930," *Kritika*, vol. 1, no. 4 (Fall 2000), pp. 657–678.

these efforts, the zemstvos often were required to hire agronomists themselves. In the absence of extensive data or information about agricultural practices, land under cultivation, production, and so on, the specialists had a hard row to hoe. These agronomists sought "to assist farmers in understanding and applying the best lessons of agronomic science." But they "played no role in taxation or other administration undertakings, such as land surveying, statistical record keeping, or apportionment of duties in kind."[37]

In spite of the presence of a strong scientific tradition that offered the promise of the introduction of modern agricultural practices,[38] reformers remain frustrated. Agronomy had developed by the eve of the Russian Revolution, even though the tsarist government saw little interest in state support for it. A series of leading scholars – including Vasily Vasilievich Dokuchaev (see **Box 2**) and K. A. Timiriazev among them – studied the chemistry, biology, and taxonomy of soils, how to fix nitrogen, phosphorus, and other chemicals in the soils; and set out to develop hybrid plants that better suited Russian climates. They knew that fertilizers alone could raise production two-fold if not more. Klement Timiriazev, a well-known agronomist, criticized the government for failing to support research and get adequate resources into the countryside. The state created no agricultural extension service such as that which functioned in the United States. Timiriazev believed strongly that modern, Western science was the key to lifting backward Russian culture into a modern era.[39]

Throughout, the Russian peasant remained a major obstacle to changes in agricultural practices. Of course, it is wrong to blame peasants in any society for their supposed backwardness and ignorance. Too often, critics of traditional agricultural (and other) practices accuse indigenous people, peasants, and others with little political power or formal education of creating problems through their inefficient use of resources and their deforestation, slash-and-burn, planting, and fertilizing practices. But when political leaders and scientists bring modern techniques to bear on them, the peasants suffer. Their community life is disrupted, and the emphasis on production of cash crops leads many of them into failed

---

[37] Kearns, *Mind and Labor*, pp. 390–395.
[38] On the history of Russian agricultural practices, see Lazar Volin, *A Century of Russian Agriculture* (Cambridge: Harvard University Press, 1970); and David Joravsky, *The Lysenko Affairs* (New York: Harvard University Press, 1970), pp. 18–24.
[39] K. A. Timiriazev, *The Life of a Plant* (London: Longmans, Green and Co., 1912), and *Nauka i Demokratiia* (Moscow: Gosizdat, 1920).

Box 2. Vasily Vasilievich Dokuchaev

Dokuchaev studied at St. Petersburg University during its so-called golden age in the 1870s when chemist Dmitrii Mendeleev and physiologist Ivan Sechenov were in attendance. He was interested in swamp drainage, soils, cartography, and chemical typology of soils. In 1875, the Ministry of the Interior and Government Property asked Dokuchaev to participate in the creation of a map of soils of the European part of Russia. Dokuchaev worked on classification of chernozem (black earth) soils, coming up with eight basic soil types. He worried that until his study peasants, agriculturalists, and biologists had no sense of the diversity of soils or their chemical makeup, which made it difficult to promote scientific agriculture. Dokuchaev carried out extensive research on soils in the summers of 1877 and 1878 that included chemical and geological studies of soils over a huge region.

In a rare moment of cooperative spirit, in response to the 1891 famine, and at Dokuchaev's initiative, the government asked him to conduct expeditions to identify areas for nature preserves, to develop the science and engineering to prevent droughts, and to promote modern agriculture in the steppe regions to prevent famines. In 1888, he founded the Soil Commission of the Free Economic Society, where he delivered talks on his research on the black-earth regions of the steppe and called for a systematic approach to improving the situation in science and agriculture.[40] Dokuchaev's *Russian Black Earth*, his crowning work, was published in the works of the Free Economic Society. Dokuchaev joined an expedition to the Poltava region that resulted in the publication of sixteen volumes of materials; joined a special expedition for the Imperial Forest Department that resulted in eighteen volumes of materials; performed research on the soils of Bessarabia and the Caucasus; and contributed to several other important publications. Dokuchaev commenced study of Kamennaia Steppe soils in 1892 in the aftermath of the famine that suggested to him and other leading scholars the need to put agriculture on a scientific basis.

(*continued*)

---

[40] In 1912, this commission was named after Dokuchaev, and from that time on a number of research institutes have been involved nearly continuously in the study of soils in this steppe. In 1956, the Lenin Academy of Agricultural Sciences established the Dokuchaev Institute of Agriculture in the chernozem (black earth) region.

> **Box 2** *(continued)*
>
> Dokuchaev wrote *Our Steppes Then and Now (Nashi Stepi Prezhde i Teper)* (1892), a popular scientific treatise, in which he proposed to transform the steppe into a rich agricultural area to push public awareness of the problems.[41]

agricultural pursuits. Poverty and underproduction of staple products result, and men are often forced to migrate to the cities in search of jobs with the intention of sending earnings home; in many cases, they fall prey to the vices of the city. As we shall see, Stalin's collectivization effort, with its significant environmental costs, was first and foremost a declaration of war on the backward peasants and destruction of his traditional ways of life, and led to tremendous social dislocation through coercive exile and spontaneous emigration to burgeoning industrial centers.

It is also true that the peasant had little interest in new techniques, rarely strove to develop hybrid varieties, and followed the time-proven – but quite inefficient – grain-fallow method of production. Yields were low and there were frequent famines, growing destitution, increasing reliance on small grains, and the plowing up of meadows and pastures for food crops, which left livestock short of forage. Were the peasant to have adopted diversification of crops, complex rotations, intensive cropping, and modern implements, production would have improved significantly. As agricultural historian Lazar Volin notes, "The peasant naturally saw his difficulties in the shortage of land, his only salvation in an increase of land."[42] This meant expansion of the same inefficient kind of farming over larger amounts of land and into forests, meadows, and pastures, with significant environmental costs.

Ideas about conservation grew up along with a burgeoning effort to study natural resources to conquer problems of famine, draught, and profligate use that had plagued the nation. Serious droughts in 1873 and 1875, the famine of 1891–1892,[43] and a cholera epidemic that accompanied it shocked even the most complacent bureaucrat. In 1891, in several areas of the steppe there was no rain all spring or summer. The drought

---

[41] On Dokuchaev, see B. B. Polnov, *Vasilii Vasil'evich Dokuchaev* (Moscow: Academy of Sciences, 1956) and I. A. Krupenikov, *Vasilii Vasil'evich Dokuchaev* (Moscow: Molodaia Gvardiia, 1949).

[42] Volin, *A Century of Russian Agriculture*, pp. 69–70.

[43] Richard Robbins, *Famine in Russia, 1891–1892* (New York: Columbia University Press, 1975).

enveloped twenty provinces in which 35 million persons lived, including in Ukraine and the Central Black Earth Regions. This was a great embarrassment to the government, which considered itself to be a modern, European nation. Whatever grain harvest the peasants managed meant there would be no seed the next year; the situation with potatoes and vegetables was even worse.

In the aftermath of the 1891–1892 famine, Dokuchaev actively encouraged fellow scientists to be involved in the improvement of agricultural conditions. Like his progressive-era counterparts in the United States, Dokuchaev was convinced that humans ought to take direct, systematic, and active action on nature to fight drought, storm, and erosion through science and state institutions.[44] He believed that state-sponsored irrigation projects could be a hedge against natural and human disaster. He called for greater attention to the influence of poor agricultural practices on forest degradation, including those that promoted erosion. He called for protecting forest growing along rivers and other bodies of water and for the planting of forest defense belts. All of this would slow spring thaw runoff and erosion. Dokuchaev warned that migration, resettlement, and expansion of agriculture would lead to the disappearance of the biodiversity in the steppes. To promote study of this danger, he founded three zapovedniks in 1892 in the Luganskie steppe.[45] Several of the zapovedniks in whose formation he was involved survive to this day and preserve wells and other experimental artifacts of his research into the water table and other factors.

Inland Fisheries During the Tsarist Era

The fisheries industry of the Russian Empire – whether inland or coastal fisheries – lagged behind those of England, Norway, and elsewhere. Several major obstacles slowed the growth of this industry. One was the absence of systematic knowledge about fish populations, migrations, mating seasons, water currents and chemistry, and climate generally. Another was the relative poverty of fishermen and the lag in capitalization of their enterprises. Norwegian, Scottish, and English fisherman had acquired larger vessels and powerboats much earlier. Nonetheless,

---

[44] David Moon, *The Environmental History of the Russian Steppes: Vasilii Dokuchaev and the Harvest Failure of 1891* (London: Transactions of the Royal Historical Society, 2005).
[45] Moon, *The Environmental History of the Russian Steppes*.

by the eve for the Russian Revolution in 1917, these resources too were at risk for overfishing in some places and destruction of habitat in many others.

Just as the tsarist government engaged in half-hearted and contradictory efforts to support the charting and conservation of the forest, so too its officials had limited understanding of the positive role they might play in promoting the rational use of such resources as fisheries. Russia's coast, by far the longest in the world, is ice bound for much of the year, so her inland fisheries were more important than ocean fisheries until after World War II. The inland fish include perch, pike, salmon, salmon-trout, char, thirty-five kinds of white fish, carp, and, most important, Caspian herring and sturgeon (the latter, in 2010, under threat of extinction). Before the Russian Revolution, 80 percent of fishing was concentrated in internal bodies of water, where overfishing occurred, often out of season. Other practices destroyed spawning grounds, especially of sturgeon. Hatcheries existed on a small scale, more as a curiosity than a scientific endeavor, and often were connected with pond farming on the lands of individual nobles. As in other countries, the hatcheries were not very successful. On the eve of World War I, there were perhaps 3,500 hectares of pond fisheries in the Russian empire, most of which were located in Ukraine and Belarus. Mostly, however, fish was salted to preserve it, and nascent oceanographic and fishery science studies to promote use of this resource depended on private initiatives.[46]

Poorly equipped individual fishermen scraped out an existence in the few coastal fisheries. Businessmen and entrepreneurs were not interested in technological improvements because of an oversupply of cheap labor.[47] Fishermen used rowboats and small sailing vessels. Only occasionally did they command a steamship. Usually, ten to fifteen small boats fished together within 50 to 100 kilometers of the coast; this limited the quantity and variety of the catch. The fishermen were at the mercy of sudden storms which, on occasion, decimated the working men in an entire fishing village. The men relied on their instincts, not scientific surveys to locate fish. They averaged 9,000 tons of cod in good years – less than the British

---

[46] Frederick Whymper, *Fisheries of the World* (London: C. Cassell and Co., 1884), pp. 333–334. Indicating the limited extent of the government's concern about fisheries, the Russian Court developed the floating fishmonger's shop for the Neva and canals of St. Petersburg, shops fitted with floating tanks to keep fresh fish alive, and others in which fish was kept frozen in ice and snow to ensure availability in the capital city.

[47] A. V. Terent'ev, "Russkii Prioritet v Mekhanizatsii Rybnoi Promyshlennosti," *Rybnoe Khoziaistvo*, no. 11 (1950), pp. 7–10.

and Norwegian fishermen harvested in the very same waters and both of whom already had motorized boats (and motorized fleets by 1920).[48]

Humans make environmental history. In the way that Vasily Dokuchaev represented the effort among geographers and other specialists to bring science to bear on agriculture, the oceanographer Nikolai Mikhailovich Knipovich (1862–1939) strove to indicate how conservation measures held great promise for modern fisheries practices. His difficulties in gaining support for his research efforts to identify ocean fisheries and protect inland ones give a sense of the challenges specialists faced.[49] Like other scientists of the tsarist era, he remained convinced of the power of science to solve the nation's problems. (On Knipovich, see **Box 3**.)

Knipovich's research in oceanography had tremendous applied importance, for it showed the great potential for trawl fishing in open seas instead of the craft industry that predominated on Russia's shores. Knipovich's expeditions along the Murmansk shores and Barents Sea from 1898 to 1902 astounded many traditional Russian scientists, who saw little need for applied scientific research that might support industry. Knipovich rejected their attitude; he was disturbed by the Russian lag behind European scholars in such fields as biology, hydrology, and physical geography. The governments of Great Britain, Norway, and Germany had already recognized the importance of this work and provided funding to support research and trawling; only in Soviet time did Russia develop a powerful trawling fleet.

Knipovich shared the ideas of such individuals around the world interested in conservation as Gifford Pinchot, who believed that science was apolitical and capable of establishing the objective truth. He wrote: "Only in-depth, universal knowledge of the nature of fishery waters will it be possible to give a firm foundation for the full and rational utilization of their natural wealth." Knipovich knew that fish reserves were extensive, but finding them required study. He had not the slightest doubt that their economic significance demanded "putting them on rational ground." Yet, fisheries were not inexhaustible. Knipovich warned of "bitter experience

---

[48] A. Kiselev and A. I. Krasnobaev, *Istoriia Murmanskogo Tralovogo Flota, 1920–1970 gg.* (Murmansk: Murmanskoe Knizhnoe Izdatel'stvo, 1973), pp. 18–25. If 30,000 Norwegians were fishing in the Lofoten Islands, only 4,000 Russians worked the Murmansk region.

[49] This biography of Knipovich is taken from P. I. Usachev, ed., *Sbornik, Posviashchennyi Nauchnoi Deiatel'nosti Pochetnogo Chlena Akademii nauk SSSR, Zasluzhennogo Deiatelia Nauki i Tekhniki, Nikolaia Mikhailovicha Knipovicha* (Moscow-Leningrad: Pishchepromizdat, 1939), pp. 5–12.

> **Box 3. Nikolai Knipovich**
>
> Knipovich spent his youth near the sea and never lost his interest in it. After graduating from the physics-mathematics department of Petersburg University in 1881, Knipovich turned to hydrobiology and the study of structure and development of conchs, taking a position in the university's zoology department. In 1885, the twenty-three-year-old Knipovich began his first research on fisheries in an expedition to the lower Volga to study herring under Professor O. A. Grimm, who supported the development of fish farming. Grimm's activities gave impetus to the creation of the Imperial Society of Fishing and Fish Farming in 1881, with its own journal, *The Herald of the Fisheries Industry* (*Vestnik Rybopromyshlennosti*).[50] Other journals soon appeared that reflected growing interest in the development of fisheries and rising concern over overfishing.[51]
>
> After studying with Grimm, Knipovich became infatuated with the Russian north in the late 1880s, and spent the remainder of his life studying fisheries of the Far North, the Volga River, and the Caspian Sea. He assembled hydrobiological and zoological data of the White Sea at the Solovetsk Biological Station. The Petersburg Society of Naturalists underwrote his research, enabling him to complete his master's thesis. In 1892, he returned to Petersburg University, where he served as a conservator in a zoology museum, then as a senior zoologist, and from 1911 to 1930 as a professor of zoology and general biology at Petersburg Medical Institute.
>
> A freak storm in 1894 destroyed twenty-five ships with a great loss of life, leading to the formation of an emergency aid committee with some funds both for the families of the sailors and for a research expedition to prevent a recurrence. Knipovich nearly single-handedly organized and carried out the scientific fisheries expedition that drew scholars from Norway, Denmark, and Germany, for it was one of the first attempts to combine the study of fish with those of oceanographic conditions. In 1897, while abroad he secured equipment for a research

---

[50] O. A. Grimm, *Nikol'skii Rybovodnyi Zavod* (St. Petersburg: P. P. Soikin, 1902). Grimm later published *Besedy o Prudovom Khoziaistve* (Discussions of Pond Farming).

[51] The journals included *Rybnoe Delo* out of Astrakhan, the center of rich sturgeon fishing and caviar production, an illustrated monthly journal published in Moscow, *Rybolov-Liubitel'*, and the journal of the Fisheries Commission, *Rybopromyshlennost'*.

vessel, the *Andrei Pervozvannyi*. The need to turn to private funding indicated the unwillingness of tsarist officials to recognize the crucial role of modern science in resource management. Knipovich produced detailed maps of the Barents Sea, including its currents, and identified much of the marine life that filled its waters. A three-volume study and a series of articles resulted.[52]

with the illusion of unlimited natural resources." Another novelty of Knipovich's research was its multifaceted scope, its focus on the complex interaction between flora and fauna, not only on fish, and on the physical chemistry of the water. He recognized a food chain, although he did not describe it as such: "We see that the entirety of living matter which inhabits these waters comprises one complex biological whole." Finally, Knipovich insisted on the necessity of carrying observations over time and in comparison with data collected in other countries.[53]

"Environmental" Concerns in Imperial Russia and the Rise of Nature Preserves

Along with these specialists on forests and soils, agriculture and fisheries, many members of the intelligentsia pondered Russia's special relationship with the environment. By the mid-nineteenth century, specialists had conducted extensive research on Russian natural resources – its flora, fauna, and mineral wealth. In the 1830s, these included leading university and academic thinkers who participated in expeditions that were not only cartographic, but also taxonomic in the Linnaean tradition, and included soil and geobotanical surveys. This led to the establishment of extensive botanical gardens not only as curiosities, but also for study (see the work of A. N. Krasnov, who founded the Batumi, Georgia, Botanical Gardens in 1880). In "The Life of Organisms in Relation to External Conditions" (*Zhizn' Zhivotnykh po Otnosheniiu k Vneshnim Usloviiam*), K. M. Rul'e set forth ideas connected to modern ecology in a specialty he called zoobiology, which is the science of organisms and their influence on the environment. Some specialists drew on the work

---

[52] P. I. Usachov, ed., *Sbornik, Posviashchennyi Nauchnoi Deiatel'nosti Pochetnogo Chlena Akademii Nauk, Zasluzhennogo Deiatelia Nauki i Tekhniki, Nikolai Mikhailovich Knipovicha (1885–1939)* (Moscow: Pishchepromizdat, 1939).

[53] N. M. Knipovich, *Ekspeditsiia Nauchnopromyslovykh Issledovanii u Beregov Murmana*, I (St. Petersburg: Khudozhestvennaia Pechat', 1902), pp. 1–6.

of the forerunners of ecological thinking in Europe and North America; A. I. Voeikov introduced and developed ideas of George Marsh, whose *Man and Nature* (1864) argued, among other things, that civilizations fell because of environmental degradation. L. I. Mechnikov wrote about the role of geography in a hydraulic history of civilizations.[54] Such members of the intelligentsia as Peter Lavrov and N. V. Shelgunov wrote about the importance of "progress" to advance Russian civilization. The nobleman and chemist Alexander Engel'gardt was a Darwinian acutely aware of the intertwined nature of natural processes. He sought to improve the agricultural situation on his estates and beyond through the application of modern science. During political exile to his estate, he wrote annual "letters from the countryside" to the *Otechestvennye Zapiski* (The Great Fatherland Notes), a major literary journal, in which he described in detail the ecology of peasant agriculture and its backwardness.[55]

One of the leading figures of modern environmental thought was Vladimir Vernadsky. A biogeochemist and a member of the Academy, the university, and several scientific societies, Vernadsky experienced all of the challenges of research and public service that his colleagues did. He is important to us because of his notion of the noosphere, a new state of the biosphere in which humans play an active role in change that is based on man and woman's recognition of the interconnectedness of nature. The tendency in Russia had been for scientists to study narrowly, to be closely connected with the empirical, in part to avoid wrestling with broader questions of science and social utility that might upset the bureaucrats.

Vernadsky began his career as a soil scientist, in the 1880s participating in several of Dokuchaev's expeditions, and also studied with Dmitri Mendeleev. From his study of the mineral composition of different soils, Vernadsky became deeply interested in mineralogy and crystallography. He undertook research expeditions to Crimea, the Kerch and Taman Peninsulas, the Caucasus, Tambov and Poltava provinces, and traveled to Germany, Denmark, the Netherlands, France, Scotland, and France, where he met with his counterparts and developed his ideas further. Like Knipovich, he rejected the view that science was apolitical, consisting only of facts. Rather, scientists were an integral part of any

---

[54] L. I. Mechnikov, *La Civilisation et les Grands Fleuves Historiques* (Paris: Hachette et cie, 1889).

[55] Alexander Engelgardt, *Letters from the Countryside*, trans. and intro. by Cathy Frierson (New York: Oxford University Press, 1993).

modern society. The task of science was to achieve the common good, and this was to be accomplished by organizing the pursuit of knowledge to discover laws of nature and apply them for the betterment of humankind. What was the best path to modernization? Progress, according to Vernadsky, was not inevitable, nor was the onward advance of civilization guaranteed by the laws of history. Rather, enlightened specialists needed to pursue knowledge to spread a scientific outlook throughout society, and Russia in particular required extensive contacts with western science and culture, indeed ought to pursue secular Western science.[56]

Vernadsky participated in the First General Congress of the zemstvos, held in Petersburg during the 1905 revolution to discuss how best to pressure the government to be more responsive to the needs of the masses; became a member of the Constitutional Democratic Party (KD); and served in parliament, resigning to protest the tsar's proroguing of the Duma. He served as professor and later as vice rector of Moscow University, from which he also resigned in 1911 in protest over the government's reactionary policies. At about this time, he organized the first expeditions in the Russian Empire to locate radioactive minerals (radium) for research, and organized and headed the permanently acting radium expedition of the Academy of Sciences. From 1915 on he was chairman of KEPS (*Kommissiia po Izucheniiu Estestvennykh Proizvoditel'nykh Silakh*, the Committee for the Study of the Productive Forces), and after the February revolution served on several commissions of agriculture and education of the provisional government, including as assistant minister of education. He moved to Kiev in 1918, where he was the first president of the Ukrainian Academy of Sciences. In all of these activities he strove to promote the dissemination of modern science in the Russian Empire to promote the well-being of the masses, including of their environment.

On the more conservative or mystical side of understanding of the relationship between humans and nature were the *pochveniki* (from *pochva*, meaning soil, ground, or earth). Like their successors, members of the genre of literature called "village prose" in the late Soviet period (see Chapter 5), the pochveniki were nationalistic and conservative, believing strongly in the bonds of the soil to the Russian, particularly the peasant. They considered man and woman to have been spoiled by the city and by contact with non-Russians, in particular Jews. For them, the simple world of agrarian toil was far more appropriate for the Russian nation.

---

[56] Kendall E. Bailes, *Science and Russian Culture in an Age of Revolutions* (Bloomington: Indiana University Press, 1990).

The representatives of the pochveniki included the well-known novelist Fyodor M. Dostoevskii, as well as Gleb Uspenskii, N. F. Fedorov, V. S. Solov'ev, S. N. Bulgakov, and Nikolai Berdiaev.[57]

Of course, the "environmentalism" that existed at the beginning of the twentieth century differed substantially from that that has existed since the 1970s. Experts did not have fully developed concepts of biodiversity; there were no advanced degrees in the ecological sciences and no specialized journals. Yet, many scientists recognized the importance of being actively engaged in studies of the environment to understand, for example, the often deleterious impact of such human activities as mining, forestry, settlement, industrialization, and so forth on a world that appeared to be increasingly sensitive to these changes. Few could ignore the destruction of inland fisheries that accompanied damming of rivers willy-nilly or the impact of smoke and soot from factories in industrial centers on the surrounding landscape.

But in several respects, environmental consciousness grew in the last decades of the tsarist era. Authors and artists may have been the first to note the negative impact of industrialization on nature and public health, later to be joined by scientists and naturalists. Charles Dickens's *Tale of Two Cities* centers on the poverty and destitution of the industrial revolution. Henry David Thoreau and John Muir worried about the assault on nature of the capitalist impulse and called for preservation of wilderness. So too in Russia, even before the revolution, several individuals addressed the costs of what they perceived to be rampant deforestation and of industrialization. Writers and other members of the intelligentsia were determined to defend nature.[58] They advocated the preservation of nature, historical sites, and the cultural legacy of the nation, with the first zapovednik established in 1898 at Askaniia Nova under F. E. Fal'ts-Fein.[59]

Vladimir Arseniev (1872–1930), an explorer of the Far East, recorded the great wealth of flora and fauna of that region and determined that it must be protected. Although also interested in the economic development of Siberian resources, he pursued the goal of preservation of the nation's heritage even in fiction. In his *Dersu Uzala* (1923; there is

---

[57] Wayne Dowler, *Dostoevsky, Grigor'ev and Native Soil Conservatism* (Toronto: University of Toronto Press, 1982).
[58] Costlow, "Imaginations of Destruction"; and Christopher Ely, *This Meager Nature: Landscape and national Identity in Imperial Russia* (DeKalb: Northern Illinois University Press, 2002).
[59] V. V. Dezhkin, V. V Snakin, *Zapovednoe Delo: Tolkovyi Terminologicheskii Slovar-Spravochnik s Kommentariiami* (Moscow: NIA-Priroda, 2003).

also a 1977 Akira Kurosawa film by the same name), which is based on Arseniev's own expeditions, a tsarist officer struggles to chart the great forest and mineral wealth of the Russian Far East. He and his men are bemused when they encounter a Nanai native of the forest, Dersu Uzala, an indigenous man whose clothes, seemingly backward behavior, and views of nature rooted in shamanism are foreign to them. Yet, Dersu's knowledge of the wealth of Siberia is in many ways more complete than that which the promoters of Western science and industry could ever obtain. One of the messages of the book and film is that these men – scientists like Dokuchaev, Knipovich, Vernadsky, and others mentioned here – seek to control and change nature, to extract its riches, not really to understand it.

As Douglas Weiner and others have pointed out, scientists were, of course, mistaken in believing that nature preserves could be inviolable, as if ecosystems had closed borders. Yet, the importance of zapovedniks as symbols of conservationist and preservationist thought cannot be ignored. On the eve of World War I, scientific and political forces joined to create the first nature preserves in Russia. Grigorii Kozhevnikov, a professor of invertebrate zoology at Moscow University, had visited the United States and Germany in 1907, where conservation and preservation efforts left him uninspired. He saw the American concept of national parks as too anthropocentric and realized that German nature monuments were tiny islands in a vast, human-dominated European ocean. Kozhevnikov advanced an ecocentric principle of absolute inviolability of large, uninhabited natural spaces, in part motivated by what appeared to be vast pristine areas of imperial Russia. Supported by other prominent Russian scientists, Kozhevnikov conceived of zapovedniks as protected areas in which the unspoiled natural world could ebb and flow following its own rhythm, forever free of human interference. To the founders of the preservation movement, this was akin to a sacred proscription – a zapoved – to relinquish voluntarily mankind's self-imposed "King of the World" title and surrender to nature's boundless and sacred wisdom. Granted, such lofty aspirations were pushed into the deep background of concerns under Soviet industrialization, collectivization, resource development, and nature-enhancement policies. After 1930, the zapovedniks's founding principle of inviolability disappeared from legal documents on the protection of the environment. Yet, the ideal was kept alive and propagated through the teachings and examples of Russian environmentalists and ardent disciples of the founders' philosophy.

At the twelfth Congress of Russian Naturalists and Physicians in December 1909, before a distinguished crowd of biologists, Kozhevnikov proposed the creation of nature preserves.⁶⁰ The participants recognized the important role they might play in saving flora and fauna in the face of rapid expansion of agriculture and settlements especially along the black earth steppe. The efforts to this time to create zapovedniki had succeeded or fallen based on private initiatives. Now, the Academy of Sciences also got involved, and even the Imperial Russian Geographical Society, which had mining and forestry departments, sponsored a Permanent Conservation Commission. Nongovernmental conservation groups emerged throughout the empire from Kazan to Simferopol to Orel to Kharkiv.⁶¹ Natalia Danilina argues that when the minister of agriculture presented a decree establishing a hunting nature preserve in the Zabaikal region of the empire in December 1916, the state-run system of preserves truly began. This preserve was first proposed in 1914 by Franz Shillinger, who established almost twenty other zapovedniks, including the Altai, Pechoro-Ilych, and Kondo-Sosvinsk.⁶²

The Weak Institutional Foundation of Ecology

Most of the activities concerning environmental thought and action took place among individuals, not in institutions, although professional societies, universities, and the Imperial Academy of Sciences had begun to serve as a forum for exploration of these ideas. The Free Economic Society, the Moscow Society for the Admirers of Nature, and several other organizations, biologists, zoologists, geologists, and chemists discussed issues of continuity, change, and makeup of nature that we would consider environmental questions today. A weakness in their approaches was that the societies were national in membership but truly centered in St. Petersburg and Moscow, so that scholars in the provinces often felt left out of national policy discussions, and the leading scientists in the center often ignored their work. Tsarist ministers disregarded the work of these societies except after the outbreak of World War I (e.g., the Russian Physical Chemical Society established a commission to study poison gas).

Things were little better in the universities. Laboratories lagged far behind their European counterparts in equipment. Leading Russian scholars preferred to do research and even to get their doctorates abroad.

---

⁶⁰ See, in particular, Feliks Shtilmark, *The History of Russian Zapovedniks, 1895–1995*, trans. G. H. Harper (Edinburgh: Russian Nature Press, 2003).
⁶¹ Weiner, *Models of Nature*, pp. 9–13, 15.
⁶² Natalia Danilina, "The Zapovedniks of Russia," *The George Wright Forum*, vol. 18, no. 1 (2001): pp. 48–49.

According to scientists who studied there, Moscow State University had weak teachers to go along with equipment shortfalls. The chemist Vladimir Ipatieff was happy for a miserly allocation of 200 rubles for chemical apparatuses and reagents, and often had to turn to his own pockets. To make matters worse, the government constantly attacked university autonomy. After the assassination of Alexander II in 1881, the government established quotas to limit entry by various minority students and the poor, slowed curriculum reform, and showed more interest in supporting philology, theology, and the classics. It centralized an already overbureaucratic administration. The Ministry of Education gained power to veto foreign travel of such illustrious scientists as Vernadsky. The ministry had to be consulted on awarding of prizes, degrees, dissertation defenses, and even class schedules. Things reached a head in 1910 when a member of the parliament (Duma), V. M. Purishkevich, the leader of the anti-Semitic Black Hundred group, called for the minister to purge the universities of "undesirable" elements. Fully one-third of the faculty resigned in support of student demonstrations.

Until the eve of World War I, members of the Imperial Russian Academy of Sciences, the leading such body in the nation, had little interest in "environmental questions." It was strong in two areas of the natural sciences, astronomy and seismology, but not so in the biological sciences. The conservative scholars who dominated its membership were largely specialists in the humanities. As noted, the Tsarist government avoided close cooperation with doctors, scientists, and engineers. Its officials feared the growing political power of the middle class, which included these individuals. As a result, they rarely turned to experts for systematic advice except when crisis demanded crucial input from specialists, for example, after the Crimean War, after the Revolution of 1905, and during World War I.

Precisely during World War I, government and science came together in KEPS. KEPS had the potential to contribute to resource exploration and management questions. Russia had relied heavily on European nations, especially Germany, for manufactured goods and strategic materials made from natural resources (e.g., explosives). The onset of war left the nation isolated and at risk. Russian specialists and policy makers recognized that the nation could not survive without Western technology. The fact that the Black Sea ports were cut off in 1915 when Ottoman Turkey joined the war forced the nation to accelerate construction of a railroad line to Murmansk that contributed ultimately to the development of Arctic resources.

Specialists in the Academy, who remained patriots even if they disliked the Tsarist regime for its conservatism and closed-mindedness, offered assistance to the nation in this time of need through KEPS. They began to chart the extent of natural resources and to offer research, development, and manufacturing solutions to the sudden cutoff of trade from the west. The work of KEPS moved slowly until 1916 because of logistical and funding problems that to a small degree reflected the government's reluctance to engage the scientists even in a time of crisis. Ipatieff expressed shock over the failure of his government to support KEPS in the effort to produce explosives and other war materiel.[63] Ultimately, KEPS surveys and inventories were crucial to postwar and Bolshevik plans for electrification of Russia (known by the acronym GOELRO), the Soviet successor of KEPS called SOPS (the Council for the Study of the Productive Forces), and even for the Soviet state planning agency, Gosplan. In essence, KEPS was the first stage in "scientifically based" planning that was central to Soviet economic life and was supposed to prevent irrational resource use that, Soviet planners believed, was an essential factor in capitalism. KEPS and its successor organizations were filled with individuals who believed that humans ought to tame and exploit natural resources; some were conservationists, but most saw raw economic value in nature.

Environmental Concerns After the Russian Revolution

World War I, the Russian Revolution, and civil war led to unimaginable human and environmental costs to the Russian Empire over the period of 1914 to 1921. Millions of people died from battle, starvation, and disease. The natural environment fared little better. World War I led to the degradation of lands along the western front and ultimately to the "perdition" of such species as the European bison. The Provisional Government assumed power after the abdication of Tsar Nicholas II in February 1917. Many specialists saw in the new government a chance for cooperation on a variety of scientific, educational, and cultural projects. Conservation activities accelerated, including the founding and expansion of a series of scientific and other societies throughout the spring and summer. Many of those of a conservationist mind set determined to work with the Provisional Government. Among them, many also rejected petit bourgeois attitudes about the sanctity of private property and recognized the need for a strong central government to ensure nature protection. Some of them,

---

[63] Vladimir Ipatieff, *The Life of a Chemist* (Stanford: Stanford University Press, 1948).

for example. A. P. Semenov-Tian-Shanskii, spoke in favor of expropriating private lands in the interest of expanding protected areas. Scholars sought to increase the network of zapovedniks manyfold with the creation of perhaps forty-six reserves. Speaking at the founding congress of the Association of Russian Naturalists and Physicians in the summer of 1917, G. A. Kozhevnikov said that "it is difficult to imagine more unfavorable conditions for a discussion of conservation matters than those of the present time we meet."[64] The participants nonetheless determined to cooperate with the new regime as part of the three-stage evolution of the relationship between man and nature. As Kozhevnikov saw it, Russia would move inevitably from a primitive to a growth-oriented (i.e., modernization) stage and eventually toward a conservation phase that relied on the knowledge and experience of Russian conservationists.

But after months of effort to work with the Provisional Government, everything fell to naught with the October Revolution, the seizure of power by the Bolsheviks, and the need for scientists to establish a new relationship with a new government. This was especially difficult at first because of the unknown shape of Bolshevik policies toward agriculture, let alone toward protected territories and resources. Yet, the effort to use science, engineering, and industry to understand how nature operated and to improve on it entered a new phase with the establishment of Soviet power under Vladimir Lenin and the Bolsheviks in 1917. The Bolsheviks welcomed the application of science to the control of nature, an attitude with frightening consequences for humans and nature under Joseph Stalin.

When the Bolsheviks took power, they immediately set out to nationalize industry, establish workers' control of major enterprises, and give land to the peasants. The government's decree "On Land" two days after taking power nationalized all forests, waters, and subsoil minerals with the goal of rational use but encouraged anarchy. Nationalization indeed followed, although the peasants in many regions had already simply confiscated land from the gentry; and anarchy, confusion, and Bolshevik terror also led to the confiscation of private property in the cities. The cities were emptied of population; residents poached wood in the cities and the countryside, destroying parks and botanical gardens.

Lenin's government approved "On Forests" in May 1918 to create a central administration to ensure proper management, protection, and reforestation. A degree in 1919 on hunting to establish seasons and

[64] Kozhevnikov as cited in Weiner, *Models of Nature*, p. 21.

protections followed. These and other measures, however, served as a background for a fourteen-year battle between the Commissariat of Enlightenment (Narkompros) and the Commissariat of Agriculture (Narkomzem) over control and direction of conservation, with the former more in concert with those who favored conservation, and those of the latter seeing nature for its contribution to the economy.[65]

Simply put, the Russian Revolution and resulting civil war and anarchy put "nature" and such cultural monuments as zapovedniks at great risk. The Academy of Science journal *Priroda* (Nature) reported on wanton destruction of parks, gardens, forests, and manor houses. Peasants seized tracts of virgin steppe. Several preserves were destroyed, and only in rare instances were scientists able to organize successfully to prevent irreversible damage elsewhere, for example, the establishment by Vladimir Vernadsky in 1918 of a Nature Protection Society in the Poltava province, Ukraine. However, the hopes of several founders of the nascent environmental movement to combine the revolutionary enthusiasm of Bolshevism with scientism and rational use of nature did not play out.

As the Bolsheviks struggled to hold onto power, they turned to War Communism (1918–1921). War Communism involved strict, even violent military measures to maintain control of the economy, especially the so-called commanding heights of industry. In the absence of money, direct exchange of goods, theft, and confiscation were the rule. Under Lev Trotsky, the Red Army carried out war against monarchists and other "white" oppositionists and against the peasantry as they used armed detachments to confiscate grain for the cities. Confiscation of grain (*prodrazverstka*) was unforgiving and led to the stripping of all resources from the countryside. In this environment of political and economic turmoil, agricultural and industrial production fell to pre-1913 levels; and war, social displacement, and scavenging left scars across the landscape. Soldiers, workers, peasants, and their families descended on zapovedniks to rob them of anything that might be used as food or fuel. We cannot overstate the ecological consequences – anarchy in the countryside, confiscation of land, burning out of landowners – and of *prodrazverstka*. Paradoxically, the sharp decline in economic production meant that in other ways, the natural environment was not under assault from industry.

Consider the impact of the Russian Revolution on fishery research and Nikolai Knipovich. World War I, the revolution, and civil war

[65] Weiner, *Models of Nature*, pp. 20–26.

interrupted his field research but gave Knipovich the chance to work through the extensive material he had gathered with great energy. He produced several outstanding works in this time, including *Hydrological Research in the Caspian Sea* (1921). Surprisingly, the Russian Revolution enabled the establishment of modern oceanographic and inland fishery research. This had a lot to do with personal contacts. Lenin knew of Knipovich; they had met in the 1890s, and Lenin himself signed off on government support of fishery research. The Bolsheviks supported Knipovich's inland fisheries along the Volga River and the Azov, Black, and Caspian Seas.

Yet, the revolution was no blessing for the fisheries themselves. The war with Germany, growing anarchy throughout the summer of 1917 and beyond, revolution and civil war led to vast migrations and displacement from the cities. Famine broke out. Starving migrants fished wherever they could. In the absence of authority – no game wardens or police dared to interfere – only immediate rivals for a fishing spot might stop fishing. The anadromous fish industry – bream, sazan (a kind of freshwater carp), and sheatfish – was nearly destroyed. A few wealthy fishing firms usurped the local fishing rights of residents and fished illegally, even in zapovedniks. As the civil war escalated, Red and White deserters alternated fishing and overfishing with impunity. Many zapovedniks were destroyed and took years to recover. As troops came and went, confiscating what they could, the fishermen overfished to stay afloat.[66]

Similarly, the sturgeon and caviar industries declined rapidly. In Russia, the caviar industry had been a state monopoly since Peter the Great in the eighteenth century, but by the early 1900s most of the enterprise had fallen into private hands. After the Bolshevik Revolution, the Soviet state took control of the sturgeon fisheries and saw in this trade a way to earn hard currency. This industry suffered from the civil war and from the inattention of the government and poor understanding of inland fisheries. Later, because of industrial pollution and the construction of hydroelectric power stations along the length of the Volga and other European rivers, the breeding areas of sturgeon were destroyed. Catches fell precipitously,[67] and to this day they remain in a precarious state.

---

[66] N. M. Knipovich, *Kaspiiskoe More i Ego Promysly* (Berlin: Z. I. Grzhebin, 1923), pp. 82–83.
[67] http://natzoo.si.edu/Publications/ZooGoer/2001/3/sturgeonandcaviar.cfm.

The Caspian experience was a sad reminder of the limited knowledge scientists and officials possessed of Russia's natural resources and how fragile they are. Knipovich wrote:

> We poorly know our own motherland. Its diverse nature is still far from sufficiently studied. Those natural resources which are spread abundantly to all ends of Russia, from the desolate cold north to the fertile heat of the south, and on the surface of the land and in its bowels, and in the depths of the waters, are all poorly understood. And that which we know, we don't use rationally and wisely. The huge sums of money that are paid out by other governments for that which we could and ought to produce ourselves places us in a controlling economic dependence on them. The difficult experience in the last war shows clearly how deep and significant this dependence is. And that which we already have studied, researched at the present time, that which we know, remains primarily the province of scholarly specialists, and the broad masses of people as before remain poorly informed.[68]

Knipovich believed that humans might, indeed ought, to control nature. He concluded, "The only way for him to be the sovereign of the earth – is knowledge, science. Science is the basis of all human welfare, from her, first of all and more than anything else depends the possibility of radical improvement of the lives of humans."[69]

During the turmoil of civil war and War Communism, the Bolshevik government managed to support scientists, including some working on issues of environmental concern. And scientists, with this support, expanded their environmental activities. They tried to protect and expand the network of zapovedniks, and many spread the notion of conservationism. They gained government support for these activities through the Main Scientific Administration (Glavnauka) of Narkompros, which funded a variety of scientific societies whose numbers proliferated between 1918 and 1925, especially in Moscow and Leningrad. The members of many of these organizations were concerned with the life sciences, including the safety and expansion of a system of zapovedniks. The Il'menskii nature preserve, the first state-funded such reserve in the world set up entirely for scientific research, was founded in 1920 with the support of Lenin.[70] Glavnauka also supported the establishment of new centers of biological research whose work was important in the development of the nascent science of ecology. Through various economic

---

[68] Knipovich, *Kaspiiskoe More*, pp. 72–77.
[69] Ibid.
[70] Weiner, *Models of Nature*, p. 29.

commissariats, the government supported the study of fish, forest, and other natural resources to feed and house the population.[71]

During the New Economic Policy (NEP, 1921–1927), a return to small-scale capitalism that saw the nation's economy recover, a series of conservation measures was enacted. In 1921, Narkompros gained responsibility for nature conservation after Lenin's decree "On the Protection of Monuments of Nature, Gardens, and Parks." However, insufficient funding and tardy approval of charters handicapped its efforts. This meant that the zapovedniks struggled to survive in face of bandits and economic demands on their resources. Meadows were used as pasture in some cases. In addition, existing laws were too flexible or nonexistent, and uneven and unbalanced exploitation resulted, including state-sponsored sales of forest and fur for export. Diminishing populations of red deer, sable, wolverine, greylag goose, wildcat, and so on resulted, and some went extinct.

Still, several well-respected representatives of the natural sciences and fine arts participated in the work of Narkompros, not only as employees but also as organizers of public commissions and associations in the scientific and cultural spheres. Through the 1920s, a series of initiatives connected with nature protection advanced in connection with the sprouting of civil society, a phenomenon that would be curtailed in the Stalin era. The government approved the establishment of four state zapovedniks in the Narkompros system (the Il'menskii, Crimean, Kosino, and Caucasus) and several local ones by 1924. Furthermore, several important organizations increased their activities. In the Tsentral'noe Biuro Kraevedeniia (TsBK or the Central Bureau for the Study of Local Lore), established in 1922 under the auspices of the Academy of Sciences, a number of activists pushed conservation. TsBK was not an armchair scientific organization, but rather a uniquely mass organization. By the late 1920s, 2,270 branches of the TsBK embraced roughly 60,000 members, including not only academically based naturalists but representatives of the provincial intelligentsia as well. The Vserossiiskoe Obshchestvo Okhrany Prirody (VOOP, the All-Russian Society for the Protection of Nature, 1924), the most influential voluntary society devoted mainly to nature protection, flowered into a mass organization, carried out work on mass education

---

[71] Ibid.; Paul Josephson, "Science Policy in the Soviet Union, 1917–1927," *Minerva*, vol. XXVI, no. 3 (Autumn 1988), pp. 342–369.

and scientific research, and, most importantly, established a new journal, *Okhrana prirody* (Conservation) in 1928.[72]

The network of zapovedniks grew from nine state and fifteen local reserves in 1925 and to a total of sixty-one, with a total area of 3,934,428 hectares (up from a slightly more than 1 million hectares) in 1929.[73] The zapovedniks grew on a foundation of ecological studies, which had been expanding in Russia since the nineteenth century through the work of such phytosociologists as P. N. Krylov, A. N. Krasno, and S. I. Korzhinskii. Community ecology then found full expression in the forest system of G. F. Morozov, who viewed the forest community not as a mechanical aggregation of trees but as a complex organism. By the end of the NEP, the zapovedniks of Narkompros were "fast being transformed into centers for the study of ecological communities," with Askania-Nova, in the Ukrainian steppes, the locus for the most important studies under the direction of Vladimir Stanchinskii, whose concept of food webs became a way of identifying the boundaries of communities in nature.[74]

From Tsarism to Bolshevism to Stalinism – and the Environment

In the late imperial Russia and early Bolshevik periods, environmental ideas were largely shaped by the growing influence of a scientific community whose members believed that natural resources should be managed "better" to improve the people's lives. The limited, yet important results of ecological thought before 1917 included the advancement of ideas for the scientific management of fish, forest, and water resources, and the establishment of the nation's first nature preserves. However, a series of obstacles remained. World War I halted cooperation between tsarist officials and scientists in areas of conservation. Still, scientists joined the tsarist regime in an effort to save the nation in war by ensuring Russia's self-sufficiency in raw materials and developing military material.

The Russian Revolution disrupted scientific, including environmental, activities. Economic production plummeted, to recover only in the mid-1920s after the introduction of the NEP. In spite of economic and social turmoil, scientists managed to establish dozens of new research institutes with state support and were permitted to found scores of professional organizations to promote their research and social programs. These

---

[72] Weiner, *Models of Nature*, pp. 31–53.
[73] Ibid., pp. 60–62.
[74] Ibid., pp. 64, 70, 78–80.

included nascent efforts in environmental science and policy: the expansion of the network of nature preserves; the study of natural resources, from surveys to analysis of data in support of efforts at scientific management; and the development of ecological ideas unique to the Soviet setting, for example, Vladimir Vernadsky's concept of the noosphere.

A tension over the role of specialists in the new socialist society played out between the scientists and Bolshevik officials from the first days of the revolution that had significance for environmental concerns. Although many scientists rejected the Bolsheviks as infuriating interlopers, they also welcomed a government that professed interest in supporting modern science and technology. They had advanced water melioration, reclamation, and afforestation projects to improve on nature since well before Soviet power. What they lacked in financial support from the tsarist government or well-appointed research institutes they attempted to overcome in vision. Agriculture, forests, and fisheries would be the main beneficiaries of these visions. Scientists welcomed Bolshevik interest in their projects. The scientists argued that they knew best what research topics merited attention and that the state ought to fund their new institutes and research programs without interference. They believed that the determination of the disposition of natural and mineral resources was an empirical task, not one of any ideological content. Many of them continued to believe that a "backward" peasantry interfered with the improvement of nature for the betterment of humankind.

In contrast to the autonomy scientists believed was crucial to their success, Bolshevik administrators wanted scientists to be accountable to the state for their funding. Reasonably, these officials believed that no money should come without strings. They sought to see scientific results translate into benefits for society, for example, increased harvests of lumber, greater catches of fish, and improvements in agricultural production. During the 1920s, the tensions between autonomy and accountability, and between the desire of scientists to conduct basic research and the desire of bureaucrats to emphasize applied science, grew more pronounced. On the one hand, the scientific enterprise grew in numbers of scientists, institutes, and professional organizations and in numbers of publications and foreign contacts. Biologists were able to secure support to expand the network of zapovedniks as inviolable preserves, even as economic pressures to find value in nature grew. Leading scholars developed concepts in population biology, genetics, and the biosphere that serve to this day as the foundations of modern ecology. On the other hand, for many Bolsheviks, especially impatient, radical young Communists who wished to transform not only society during this revolution but also nature, the scientists' approach amounted to "ivory tower" reasoning of little value to the proletariat. They believed that they best understood which directions of research were most important to the new state. When they gained the upper hand under Stalin, nascent ecology would experience stultifying political, economic, and ideological pressure.

The rise of Stalinist policies in the economy, and the cultural revolution in science, education, and the arts, meant a series of setbacks for the conservation movement in the Soviet Union. Marxists insisted on utopian projects for the transformation of society and nature, which stood in contrast to immutable limitations in nature that many conservationists knew would prevent great advances in productivity. The theoreticians of the cultural revolution, according to Weiner, "wanted science to contract to the narrow limits of handmaiden to technology, eschewing broader social or political questions."[75] Practice – economic growth – would be the final judge of the success – and victory – of proletarian science over bourgeois science. As part of the cultural revolution, mirroring what happened to other scientific organizations, conservation groups faced the pressure to contribute to the socialist reconstruction of society.[76]

The result was a series of attacks on the professoriat, their societies and associations with the accusations that they were divorced from youth, the masses, and practice. VOOP, TsBK, and the Moscow Society of Admirers of Nature were purged. Conservationists sought to protect their achievements through the "protective coloration" of rhetoric – *Okhrana prirody* was renamed *Priroda i Sotsialisticheskoe Khoziaistvo* (Nature and the Socialist Economy), whereas VOOP became the All Russian Society for Conservation and the Promotion of the Growth of Natural Resources.[77]

The program of rapid industrialization and collectivization of agriculture adopted under Stalin's self-proclaimed Great Break included the language of war or conquest of nature and the people in it. Rapid industrialization and collectivization of agriculture were accompanied by nature transformation projects fulfilled by massive construction trusts determined to over-fulfill target plans. Thousands of factories, smelters, boilers, and power stations were built in a few years whose production practices resulted in vast environmental degradation. Because production was paramount, any slackening in tempos was seen as an effort to wreck

---

[75] Ibid., p. 125.
[76] Ibid., pp. 85–92.
[77] Ibid., pp. 142–146.

plans. Armed with rudimentary equipment – or with big machines not known for their sophisticated handling of material – laborers extracted ore, coal, oil, and so on from the earth with little regard for the scars left behind. They moved through the forests cutting rapidly, leaving valuable lumber to rot on the forest floor, wasting much of the trees in the mill. They dumped pollution into rivers, lakes, and ponds, or buried it just below the surface, where it might leach into ground water. In a word, the environmental history of the Soviet Union took a great turn for the worse under Joseph Stalin in the late 1920s.

Even in arid regions or those with low rainfall the result would be a harvest, a "Stalinist plenty" of fruits and vegetables – and a happy nation.

Under Stalin the government began extensive efforts to grow grain and cotton in arid Soviet Central Asia a program that continued until the fall of the USSR in 1991.

2

Stalinism

Creating the Socialist Industrial, Urban, and Agricultural Environment

During the New Economic Policy of the mid-1920s, the state focused more on economic recovery than on large-scale mining, metallurgy, forestry, fisheries, and geological engineering projects, although it insisted on controlling the "commanding heights" of industry. By the late 1920s, in a period of relative economic stability, if growing political intrigues involving who would succeed Lenin, the Bolsheviks were prepared to subject nature itself to their plans, no less than industry or labor. Stalin's self-proclaimed Great Break with past social, political, and economic policies involved – among other things – programs for rapid industrialization and collectivization of agriculture that had long-term extensive environmental impacts. Engineers who suggested a circumspect approach to construction projects, dams, forestry enterprises, and so on risked facing charges of subversion or wrecking. Stalinism was therefore not only a polity and economic program, but also a transformationist doctrine that would rebuild nature and the people in it for the "socialist reconstruction" of the nation.

During the Stalin era, party officials, economic planners, and engineers joined in the effort to master the empire's extensive natural resources toward the end of economic self-sufficiency and military strength. At their order, armies of workers began the process of constructing giant dams and reservoirs on major European rivers – the Don, Dniepr, and Volga. Irrigation systems spread across vast arid and semiarid areas of Central Asia, although never to the scale that planners hoped. The workers erected massive chemical combines, metal smelters, and oil refineries in both European and Siberian parts of the country, paying little attention to the pollution they produced. They put up entire cities to house the laborers, whom they exhorted to meet plans and targets irrespective of the environmental costs and the risks to the workers' own health and safety. A major aspect of the labor enterprise was slave labor camps of the gulag system, which had their own environmental – and, of course, human costs.

Many of the achievements of Stalin's five-year plans for industrialization were destroyed after Hitler's armies invaded in June 1941. World War II involved complete ruination of vast regions of the country, its agricultural and forest lands, on top of the deaths of more than 20 million people. But rather than focus on investment and policies to rebuild human lives, homes, and agriculture, in the postwar years Stalin redoubled industrialization efforts in the fourth Five-Year Plan. Millions of people lived in rubble or barracks, and at least 1 million people starved. Stalin then advanced a bold Plan for the Transformation of Nature, promulgated in 1948 to turn the European Soviet Union into a well-functioning machine based on massive reclamation, afforestation, irrigation, and other projects. On top of this, the Soviet Union and the United States engaged in the Cold War, with its significant environmental costs of the production of nuclear, chemical, and biological weapons.

When Joseph Stalin (1878–1953) established his position as unquestioned leader of the Communist Party and the Soviet Union, he immediately commenced a rapid program of industrialization and collectivization of agriculture. Stalin and his followers had grown impatient with the New Economic Policy (NEP) passed under Vladimir Lenin in 1921 to rejuvenate the economy. They believed that the NEP relied too much on market mechanisms and had permitted a class of petit bourgeoisie ("the NEP-men") to prosper. They also believed that to wait another generation or even decade to industrialize would expose the Soviet Union to dangers in the international arena in the event of inevitable war. They determined to pursue a socialist economy at breakneck speed through a process of socialist reconstruction. They established state control of the entire means of production and emphasized heavy industry and the military sector at the expense of light industry, housing, medicine, and food. They transformed society, too. They eliminated elements they deemed to be hostile to Soviet socialism, including the NEP-men, but also the somewhat wealthier peasants (the so-called kulaks) and specialists allegedly beholden to capitalist ideas and power. This economic, social, and cultural revolution from above, which Stalin proclaimed to be a "Great Break" with past institutions and ideas, resulted in great human and environmental costs and in a new, almost entirely utilitarian way of looking at resources and nature itself.

In three major ways, the Stalinist Great Break had both short-term and immediate, as well as long-term and extensive, negative impacts on the environment. First, in seeking confidently to industrialize in one or two five-year plans, Communist leaders wished to leap to the level of the advanced capitalist nations, and especially the United States, in the production of iron and steel, the extraction of coal and ore, the pumping of oil and gas, the felling of timber, and the creation and expansion of networks of roads and railroads. They created entire production cities dedicated to iron, concrete, asbestos, and mining (e.g., "Asbest," founded in the Urals region in 1933). In the process, they refused to tolerate deviation from plans, even though the plans were in reality far-fetched; targets did not reflect real possibilities, but symbolized determination to transform the nation into an industrial powerhouse.

Granted, targets for two-, three- and fourfold increases in production generated significant growth in output, although never the amounts specified, leading to sectoral imbalances and bottlenecks. The political leaders emphasized production to such an extent that safety of workers and provision of pollution control equipment was an afterthought. Party officials attacked as "wreckers" or otherwise anti-Soviet those individuals who worried about waste of resources in the national campaign to industrialize, who identified the bottlenecks of supply that arose in the feverish construction and production activity or who indicated the human costs associated with the effort. In any event, if the Soviet Union managed before World War II to become a major industrial power, then this economic development was extremely costly in terms of the environment and likely far exceeded the human and environmental costs of industrialization in the capitalist West, although the worker was supposed to benefit from the many advantages of the socialist economy.

Second, as in the case of industry, so in the pursuit of collectivized agriculture, Soviet planners and political officials impatiently established targets for the destruction of individual peasant holdings and the creation of larger cooperative farms that resulted in great human and environmental costs. In the effort to bring the modern machinery of tractors, combines, and harvesters, as well as fertilizers, to the countryside, they promoted profligate use of land, accelerated erosion, and poisoned the soil. Millions of peasants were forced out of familiar ways of life and driven from their homes and villages into the farms, often at gunpoint. Millions of others were simply exiled to Siberia or the Far North as "kulaks," allegedly wealthier peasants who exploited their neighbors, but likely village strongmen. And perhaps as many as 12 million more simply fled collectivization or hoped for a better life in the burgeoning industrial cities. In any event, the authorities and the peasants themselves clearly labeled people who were not kulaks as kulaks. Many resisted passively; many peasants resisted actively. The peasants slaughtered perhaps one-half of the nation's livestock rather than see the animals go into the farms. Although officials proudly announced that 90 percent of small holdings had been collectivized in short order, the cost was a famine that led to the deaths of millions of individuals in Kazakhstan in 1931 and of perhaps another 3 million peasants, mostly in Ukraine, in 1932.

Third, Stalin's revolution from above required a new view of nature as an object to be manipulated according to plan. Of course, in other nations in the twentieth century, engineers joined forces to develop natural resources, tame rivers, produce electricity, and so on. In the Soviet Union, the engineers adopted an even more aggressive attitude, while Party officials shelved, weakened, or isolated nascent programs in ecology. Although the network of nature preserves expanded somewhat during the Stalin era, the political authorities saw them as economic units. Ecologists now had to tread very carefully so as not to be seen as wreckers of glorious plans to control nature. Nature itself, in many ways, came to be viewed as an enemy to be subjugated to the positive forces of the Communist Party to reshape it in a socialist fashion. Rivers and forests would operate according to plan; ore would be mined with minimal concern over pollution. Officials pushed Enlightenment notions of the desirability – and engineering ability – to transform nature to serve society's ends to the extreme. The attitude of nature as a capricious, irrational entity under capitalism gave way to one of a socialist nature that served the whim of planners. The epitome of this attitude was the promulgation of the Stalinist Plan to Transform Nature (1948). In this chapter, we also consider the environmental consequences of World War II, of the postwar plan to rebuild the nation's economy, and of the burgeoning Cold War defense industry.

An Evaluation of the Environmental Costs of Stalin's Plan for Rapid Industrialization

During the first Five-Year Plan, Soviet leaders determined to create a military and industrial superpower that, according to the slogans of the era, "reached and surpassed" the might of the capitalist West, and especially that of the United States. Toward that end, the authorities dedicated all investment programs toward the construction of new, basic industries. These industries – mining and metallurgy, aluminum, construction, asbestos, and so on – are among the most potentially polluting in any economic system. Of course, the construction of 9,000 major industrial enterprises over the 1930s had direct impact on air, water, and soil quality. Millions of peasants flooded the growing work sites, creating sewage, water, and other challenges. No one thought about any kind of filters or pollution control equipment or about worker safety.

Not only massive dislocation, but also revolutionary changes in the nature and structure of the labor force accompanied the industrialization and collectivization campaigns. Militant young Communists poured into cultural, educational, and other institutions determined to make good on the Great Break. These included newly trained specialists dedicated to production. The Central Committee of the Communist Party ordered the mobilization of thousands of young communists to study engineering. They had a narrow view of their engineering, by which they understood technical training toward the ends of industrial growth. Industrial safety, pollution, and management of ecosystems seemed to them to be "remnants" of bourgeois society. In all fields, in addition to studying the classics of Marxism-Leninism-Stalinism and Party history, they promoted the study of highly specialized subjects to make routine the transformation of natural resources into production valuable to the state. Such specializations included a PhD in ball bearings. These narrowly and rapidly trained militant young communists participated in the creation of a 1,700-page optimum variant of the "Five-Year Plan for National Economic Construction" that envisaged the doubling of the industry's fixed capital stock between 1928–1929 and 1932–1933. They forecast a tripling in pig iron output from 3.3 to 10 million tons per year, coal from 35.4 to 75 million tons, and iron ore from 5.7 to 19 million tons. Light industry would expand by 70 percent, national income by 103 percent, agricultural production by 55 percent, and labor productivity by 110 percent. Workers would labor out of enthusiasm for the cause and not for personal gain. (And this was the result, for real income and wages declined in the 1930s.) Given the tempo and scale of the national endeavor, significant waste of resources resulted, and managers of projects made no effort to protect local or regional ecosystems, nor had they the resources to do so.

Rapid industrialization in no way permitted a place for the environment as something to be preserved or defended, but only as an object to reach production and modernization targets, that is, almost exclusively as a resource for industry. The focus of industrialization was both existing industrial regions and the new republics surrounding the empire that had to be brought into the Stalinist industrialization scheme. This activity found reflection even in literature. Ilf and Petrov, who wrote of their incredulity at American life in *One-Storey America*, had earlier written about the experiences of their NEP-man, Ostap Bender, on the Turksib (Turkestan-Siberia) Railroad, an epic Stalinist construction project, in their satirical *Golden Calf*. In the 1930s through the 1950s in the genre of "socialist realism," authors glorified production in production novels whose plots essentially centered on the nearly single-minded dedication of the manager or loyal party member to the industrial ethos.

Already by 1926–1927, three giant new projects were approved as part of the doubling of investment in new enterprises; large-scale, centralized projects became the standard approach to economic development. The Volga-Don Canal, the Dnieprostroi hydroelectric complex in Ukraine, and the Turksib railway line to link Turkestan's cotton fields to Siberia's grain and timber regions were early signature events. Leaders assumed these projects were the most efficient way to accomplish modernization, but, as we have noted, they turned out to be quite costly in terms of resource waste – including humans. When the authorities declared these and other projects and the first Five-Year Plan complete in December 1932, they had not in fact met any major targets. However, they had dramatically transformed the nation and the economy. In the Ural Mountains, the Kuznetsk basin (*Kuzbass*, a rich coal mining region in southwestern Siberia), the Volga district, and Ukraine, hundreds of mining, engineering, and metallurgical enterprises were up and running or under construction. More than half of machine tools onstream in the Soviet Union by 1932 were fabricated or installed after 1928. Many were built from scratch, such as the Magnitogorsk combine (part of the Ural-Kuzbass iron and steel complex); the Turksib railway that opened in 1930; and Dnieprostroi (Dniepr Hydroelectric Power Station), whose turbines began to spin in 1932.

In the second Five-Year Plans, many of the projects started under the first plan – the metallurgical enterprises of Tula, Zaporozhe, and Lipetsk, and the Azovstal combine on the Black Sea, began operations. All of these enterprises contributed to extensive environmental change and pollution. For example, Azovstal accelerated the destruction of the Azov Sea fisheries with extensive heavy metal pollution and profligate water use that led to increased salinity. (At the turn of the twentieth century, Azovstal was using 2.5 million cubic meters of sea water daily for the cooling of its equipment.)

What explains Stalin's determination to demonstrate a "Great Break" with the past in the economy and especially industry? Stalin and the Bolsheviks needed to pursue industrialization that had been interrupted by World War I, the revolution, and civil war. The NEP pulled the nation out of a deep economic crisis, but the Bolsheviks, in the words of Trotsky, demanded "rapid tempos of industrialization and collectivization" to ensure the transition to large-scale collectivist production.[1] In April 1927, at a plenary session of the Central Committee of the Communist Party, Stalin seemed skeptical that breakneck industrialization was the correct path. He initially disputed that Dnieprostroi was the appropriate vehicle for modernization. He claimed that the Dnieprostroi project would be "for us the same as the peasant buying a record player rather than a cow." Lenin and Trotsky seemed to argue that modern technology – the tractor, combine, printing press, film project, and so on – would modernize the peasant, change his world view and bring him into the orbit of Soviet power, although Lenin thought this process would take a generation longer. At this point, Stalin attacked the so-called Rightists for their fantastic industrialization plans on the backs of the peasant. But as soon as he took power, he adopted the Rightists' program with even greater vigor and total impatience toward any slacking in tempo. He may have assumed that eventually Communism would spread throughout the world. But to save the Soviet Union, he pursued "socialism in one country" through unbelievably high tempos of economic growth. This meant high levels of investment in heavy industry and extraction of capital from the countryside. Hence, industrialization was achieved on the backs of the masses, both the workers and the peasants, and through the stripping of resources from them. They were forced to become modern at Turksib, Magnitogorsk, and Dnieprostroi.

Stalin, the Great Break and Nature

If climate, geography, resources, and inhabitants suggested that different approaches be applied in Soviet efforts to build new cities, establish mining, smelting, and forestry operations, and expand road and railroad networks, then the centralized campaigns of the Stalin era inhibited recognition of these differences. Major engineering firms in Moscow, Leningrad, and Kiev established baseline designs for all industrial, transport, and agricultural operations. This meant that frequently machinery

[1] Leon Trotsky, *Revolution Betrayed* (New York: Pathfinder Press, 1972).

and equipment appropriate in one field or area of application might be applied willy-nilly across the landscape. Major construction organizations grew rapidly around one operation – for example, subway construction in Moscow or dam construction on the Volga River. As the organization grew and work neared completion, the authorities would hive off thousands of employees to create a new organization that they sent to another river basin or ore deposit to build another dam or establish another mine. In a word, Stalinist construction and engineering firms acquired great momentum to pursue other construction projects. All of this worked against innovations of efficiency, safety, and environmental soundness, because plan fulfillment was the major indicator of success, the construction firm forever sought new projects, and few individuals recognized the need to introduce new processes or reforms of safety because what had worked before would likely work again. In the market system, generally speaking, once a project was completed workers left to find other jobs, meaning that projects did not merely happen but required the coming together of political, economic, and engineering interests over time.

The Turksib Railway had both direct and indirect impacts on local residents and on ecosystems. All railroad construction provokes changes in ecosystems and across large swaths of land. With an average minimum right-of-way of 20 meters for one track and 30 meters for two tracks, the excavation, grading, graveling, laying of timbers, and so on inevitably alters the earth, leads to erosion, and cuts through habitat. (Granted, railroads have a significantly narrower right-of-way than roads and highways per lane.) On top of this, the construction of bridges, drilling of tunnels, use of explosions to clear ledges and hills, and so on also contribute to environmental changes. The construction of a new trans-Siberian railway, BAM (the Baikal-Amur Magistral) in the 1970s and 1980s was noteworthy precisely for its technological failings and significant environmental impact. No less, the Turksib Railroad, at 1,400 kilometers from Tashkent to Novosibirsk, across fragile desert and through mountainous terrain, caused immediate and long-term degradation.

The Turksib contributed to environmental changes throughout the region by facilitating unchecked industrial development and inefficient agricultural consolidation. Turksib was connected with agriculture, irrigation, meat-packing, and other manufacture. Turksib was intended to serve *smychka*, the all-important link between the city and the countryside, to accelerate the expansion of cotton culture, improve trade and transport, and accelerate collectivization of agriculture in Central Asia.

Like such other railroad transport nexuses built in the 1920s and 1930s, Turksib was a vital project of socialist reconstruction[2] – in the southern Urals to connect the south with the industrial areas to the north; from Petropavlovsk to Balkhash in Siberia and Kazakhstan to open the Karaganda Coal Basin and the Balkhash copper deposits; and a direct line from Moscow to the Urals via Kazan.

Human suffering also accompanied the construction of Turksib, not the least of which was the exploitation of laborers on the project. Planners anticipated that products from Kazakh animal farms would be shipped to Russia through an improved transport system connected with the building of the Turksib, along with attendant meat-packing and canning plants. Collectivization and de-kulakization ruined their plans. Hundreds of thousands of dekulakized peasants were sent to Kazakhstan to push the process of "de-nomadization" of the local people by settling heretofore pastoral lands. As elsewhere, peasants slaughtered livestock rather than give it up or saw it die in the harsh collectivization process; livestock disappeared, and Kazakh herds would be unable to provide meat until 1937. Between 1928 and 1934, the percentage of Kazak livestock of the total Soviet herds dropped from 18 percent to 4.5 percent. Families were left with nothing. Famine spread through Kazakhstan a year earlier than in Ukraine – already in the autumn of 1931. Large numbers of people died in epidemics of typhus, scurvy, and smallpox after suffering from malnutrition. Cannibalism spread. People ate dogs, cats, rodents, carrion, refuge, and roots.[3]

Engineers understood that the development of coal and metallurgical industries in Ukraine required generation of copious amounts of electricity. This led them to design the Dnieper hydroelectric power station ("Dnieprostroi," one of Stalin's first "hero" projects of socialist reconstruction). As part of the State Electrification Plan (known as GOELRO, according to its Russian acronym), they pursued hydroelectricity at a variety of sites on rivers of the European Soviet Union. Some of the engineers worried about the feasibility of the projects, the low "head" (the distance

---

[2] Edward Ames, "A Century of Russian Railroad Construction: 1837–1936," *American Slavic and East European Review*, vol. 6, no. 3/4 (December 1947), pp. 57–74.
[3] Sarah Cameron, "The Hungry Steppe: Soviet Kazakhstan and the Kazakh Famine, 1921–1934," PhD dissertation, Yale University, New Haven, CT, 2010. See also Matthew Payne, *Stalin's Railroad: Turksib and Building Socialism* (Pittsburgh: University of Pittsburgh Press, 2001). See also Niccolo Pianciola and Susan Finnel, "Famine in the Steppe: The Collectivization of Agriculture and the Kazak Herdsmen, 1928–1934," *Cahiers du Monde russe*, vol. 45, no. 1/2 (January-June 2004), pp. 137–191.

water fell downstream to power turbines) and volume of the relatively slow-flowing rivers. They spoke out about the need to consider simultaneously housing and work conditions for the laborers. One of them, Peter Palchinsky, paid for his circumspection with his life and career. His arrest and execution was intended to convey a message to Soviet engineers that Communist Party dictates, not engineering considerations, should be the design basis for the hero projects of Stalinism. But Party officials believed that the Dnieper River, with its rapids that interrupted transport and its spring floods that covered farmland, ought to be transformed into a socialist river that worked for the proletariat.[4] Rebuild the river they did, at significant cost to soils, flora, and fauna, in particular fish, with a long-term impact on fisheries in the Black Sea and Sea of Azov into which the Dnieper flowed. The Dnieper, the third-longest river in Europe, flowed smoothly through a broad alluvial plain with regular spring inundations. But from 1934 to 1975, Soviet engineers built a cascade of six reservoirs and dams from the confluence of the Pripiat River (not far from where they built the Chernobyl nuclear power station) to its delta at the Black Sea. The reservoirs were shallow but inundated rapids, and with locks this permitted passage of large boats.[5] (See **Box 1**.)

The worldwide experience for power generation based on hydroelectricity since the 1920s has been founded on the determination to build ever larger reservoirs and dams on rivers. This has led everywhere to rapid depletion of water resources, serious pollution, and basin-wide ecological catastrophes, including heavy losses in fisheries on rivers and coastal waters, especially for anadromous (migrating) fish like salmon and sturgeon. Whether the work of the Bonneville Power Administration and the Columbia River in Washington State, the Tennessee Valley Authority, various ministries and power consortiums in Brazil, or ministries of electrification in the Soviet Union, years of experience confirmed what some engineers and biologists knew from the start: the benefit of power generation would mean a great loss for biodiversity, and the benefits of flood control have rarely lived up to the promises. It has been well known for

---

[4] On the construction of the Dniepr Hydroeletric Power Station, see Anne Rassweiler, *The Generation of Power* (Oxford: Oxford University Press, 1988). On Peter Palchinsky, see Loren Graham, *Ghost of the Executed Engineer* (Cambridge: Harvard University Press, 1993).

[5] On the impact of Dnieprostroi on macrophyte biomass, see Štepán Husák and Valerij P. Gorbik, "Macrophyte Biomass in the Kiev Reservoir (Ukraine)," *Folia Geobotanica & Phytotaxonomica*, vol. 25, no. 3 (1990), pp. 265–273. On the early days of Dnieperstroi, see "Dedicate World's Largest Power Plant in Russia," *Science News Letter*, vol. 22, no. 594 (August 27, 1932), p. 137.

> Box 1. Fisheries and Hydroelectricity
>
> About 40,000 large dams at more than fifteen meters in height impound more than two-thirds (approximately 10,000 cubic kilometers) of the freshwater flowing into the oceans. The dams provide water for municipal, agricultural, and industrial purposes and can produce electricity and service flood control. Yet, they also affect riparian ecosystems. Three-quarters of the 139 rivers in the United States, Canada, Europe, and the former Soviet Union that exceed 350 meters per second have succumbed to these dams. The dams obstruct migration pathways for fish, while their reservoirs trap waterborne sediment. They have an impact on flood extent and the nutrient load. Inundation leads to loss of habitat and organisms and introduces environmental problems. In many cases, including in the Soviet Union, land was inundated before being cleared of vegetation – tens of millions of cubic meters of timber. When that vegetation decomposes, it releases greenhouse gases and other substances that can be harmful to fauna. Downstream effects of large dams include increased water use and evaporation losses. This can reduce groundwater recharge in the riparian zone. Dams also change geomorphologic processes such as sediment cycling. Waterlogging and soil salinization are common in many semiarid and arid areas where dams are built to provide water for irrigation. As noted, the Aral Sea in Kazakhstan and Uzbekistan is a prime example of devastation from altered river hydrology after large amounts of water were withdrawn from the Amu Darya and Syr Darya Rivers. This led to an 80 percent decrease of water volume of the Aral Sea in less than forty years, and the sea is now saltier than the ocean.[6]
>
> Impoundments on the Danube, Dnieper, and Dniester rivers immediately reduced fish yields in the Black Sea and the Sea of Azov. Over the Soviet period, the discharge of the Volga River into the Caspian Sea was reduced by almost 70 percent; that of the Dniester, Dnieper and Don into the Black and Azov seas, by about 50 percent. Because the decline in freshwater, the salinity in the estuaries of these rivers increased by up to fourfold and that in their deltas by up to tenfold.
>
> *(continued)*

---

[6] Christer Nilsson and Kajsa Berggren, "Alterations of Riparian Ecosystems Caused by River Regulation," *BioScience*, vol. 50, no. 9, Hydrological Alterations (September 2000), pp. 783–792.

> Box 1 (*continued*)
>
> The most valuable commercial fisheries were reduced by 90 percent to 98 percent.[7]
>
> Another negative impact of human activity was the eutrophication (oxygen depletion) of coastal waters and shelf zones. This resulted from such nutrients as fertilizers, animal waste, and untreated urban sewage entering the water in high concentrations in the Soviet Union, particularly in the northwestern Caspian Sea, where the Danube, Dniester, and Dnieper flow into it. The increase in nutrients triggered the mass development of phytoplankon and an expansion in the area of its blooms. Sturgeon catches in the Caspian Sea are only 1 percent to 2 percent of historical levels and have been totally eradicated in the northwestern Black Sea and Sea of Azov (a northeastern appendage of the Black Sea), with combined economic losses to the fishing industries of the Black, Azov, and Caspian in the decade between 1977 and 1987 estimated at $35 billion dollars.[8]

decades that dams and other engineering modifications of rivers have a direct, significant, and lasting impact on the physical, chemical, and biological processes of the marine environment, including changes in temperature and speed of the water that lead to negative impacts on migration patterns, spawning habitat, species diversity, water quality, and distribution and production of lower trophic levels.[9]

Soviet studies quietly confirmed the impact on inland fisheries over the years.[10] Ukrainian and Russian specialists knew quite well even before

---

[7] David Tolmazin, "Black Sea – Or Dead Sea?" *New Scientist*, vol. 84, no. 1184 (December 6, 1979), pp. 767–769. See also Tolmazin, "Soviet Environmental Practices," *Science*, vol. 221 (September 16, 1983), p. 1136, in which he reiterates the "abysmal state of fresh and brackish water resources in the most populous regions."

[8] http://www.internationalrivers.org/en/node/1640. See also Vadim Birstein, "Sturgeons and Paddlefishes: Threatened Fishes in Need of Conservation," *Conservation Biology*, vol. 7, no. 4 (December 1993), pp. 773–787.

[9] Kenneth F. Drinkwater and Kenneth T. Frank, "Effects of River Regulation and Diversion on Marine Fish and Invertebrates," *Aquatic Conservation: Marine and Freshwater Ecosystems*, vol. 4, no. 2, (1994), pp. 135–151.

[10] On the community of species in Central Asian rivers before extensive Soviet geoengineering that destroyed the fishers, see G. V. Nikolski, "On the Influence of the Rate of Flow on the Fish Fauna of the Rivers of Central Asia," *Journal of Animal Ecology*, vol. 2, no. 2 (November 1933), pp. 266–281. See also S. G. Lazalumi, "The Fish Fauna of the Lower Reaches of the Dnieper," *Journal of Ichthyology* (English translation of *Voprosy Ikhtiologii*), vol. 12 (1970), pp. 249–259.

Dnieprostroi opened that the station would change the regime of the river completely, altering the makeup of soils and flora in the downstream flood plain. The European rivers flowed relatively slowly because of the general flatness of the land. The spring thaw resulted in extensive flooding across the lands. Before the station was constructed, the river rose significantly, and a broad lake nearly ten kilometers across covered the river valley with soil, seeds, and other material left behind that contributed to fertility and biodiversity.[11] This flooding ended with Dnieprostroi. One of the aspects of hydroelectricity in the Soviet Union that remains to be studied in greater depth is how various ministries and bureaucracies concerned with different aspects of the natural resources resolved their disputes, for example, those concerned with electrification who pursued dams, and those concerned with inland fisheries who worried about the impact of dams on fisheries.

Many of the new industrial cities, especially in the Arctic and Siberian regions, were built by gulag prisoners and were part of a huge national network of prison camp mining, construction, and other operations. One such city, Norilsk, not far from the Arctic Ocean was the center for Norillag – a mining and metallurgical operation that continues to function to this day. Established in 1935, true industrial production began in 1939, when prisoners were put to work mining on some of the largest nickel deposits in the world. They also extracted copper, cobalt, platinum, palladium, and coal with rudimentary equipment and built a railroad across the fragile tundra to carry the ore and ingots to the central industrial regions of the country. Extensive and still not-fully-evaluated extremely high levels of pollution resulted.[12]

Work sites were beehives of activity. Political commissars pushed the workers to higher and higher targets, almost oblivious to the dangers of construction, manufacture, smelting, and other processes. Contributing to the feverish and confused production process was the fact

---

[11] S. Illichevsky, "The River as a Factor of Plant Distribution," *Journal of Ecology*, vol. 21, no. 2 (August 1933), pp. 436–441.

[12] A. R. Bond, "Air Pollution in Norilsk: A Soviet Worst Case," *Soviet Geography*, vol. 25 (1984), 665–680; I. J. Kirtsideli, N. I. Vorobeva, and O. M. Tereshenkova, "Influence of Industrial Pollution on the Communities of Soil Micromycetes of Forest Tundra in Taimyr Peninsula," *Mikologiia i Fitopatologiia*, vol. 29, no. 4 (1995), pp. 12–19; B. A. Revich, "Public Health and Ambient Air-Pollution in Arctic and Sub-Arctic Cities of Russia," *Science of the Total Environment*, vol. 161 (January 1995), pp. 585–592; and A. A. Vinogradova and V. A. Egorov, "Contributions of Industrial Areas of the Northern Hemisphere to Air Pollution in the Russian Arctic," *Izvestiia Akademii Nauk Fizika Atmosfery i Okean*, vol. 33, no. 6 (November–December 1997), pp. 750–757.

that many of the workers were illiterate, and many of them were also first-generation workers who had recently left the countryside, often because they had been forced to leave the farm. This meant that the production was even more dangerous than it might have been, for raw labor recruits were literally forced into the battle of raising factories out of the mud with inadequate equipment and tools, weak understandings of production and safety issues, and miserable clothing and food. As the American engineer John Scott chronicled at Magnitogorsk, work sites quickly transformed into garbage dumps and waste heaps, sites of deep freezes in the winter and muddy swamps in the summer, where inexperienced young workers lost limbs and lives as an everyday occurrence. He wrote, "... the Arctic winter broke suddenly into spring.... By May 1 the ground had thawed and the city was swimming in mud. Leaving our barrack we had literally to wade. Garbage heaps and outdoor latrines near the barracks thawed and the winter's accumulation had to be removed immediately to avoid contamination."[13] Scott noted how a disastrous explosion destroyed furnace number 2 in 1934, blew the roof of the cast house, and seriously injured everyone who was nearby.[14]

Of course, the creation of heavy industry, the advancement of the notion of socialism in one country, and the defense of the nation from hostile capitalist encirclement and the preservation of territory – all of these things, whether slogans or real programs, both contributed to and required the creation of a powerful military state. Toward that end, the authorities hoped to acquire in one way or another the world's most advanced technology. They invited Western specialists to assist in installing the machinery at the same time the country engaged a massive literacy campaign that also involved training engineers; by the breakup of the Soviet Union one-third of the world's engineers. The Soviet Union imported some 300,000 machine tools from 1929 to 1940,[15] and roughly one-quarter of the equipment brought into operation was imported. The entire Magnitogorsk steel combine was modeled on the Gary, Indiana, U.S. Steel Bessemer mills, considered among the most modern in the world. It was designed with the help of American engineers, run with their help, and used Western equipment.[16] (See **Box 2**.)

[13] John Scott, *Behind the Urals* (Bloomington: Indiana University Press, 1973, originally published in 1942), p. 65.
[14] Scott, *Behind the Urals*, p. 140.
[15] Anthony Sutton, *Western Technology and Soviet Economic Development, 1917 to 1930* (Stanford: Stanford University Press, 1968), and *Western Technology and Soviet Economi Development, 1930 to to 1945* (Stanford: Stanford University Press, 1971).
[16] Scott, *Behind the Urals*.

> **Box 2. Magnitogorsk's Failure to Become a Modern Industrial Center**
>
> "Magnitogorsk," the major industrial center on the Siberian side of the Ural Mountains, literally means "the city by the magnetic mountain" and refers to the Magnitnaia mountain – a geological anomaly that once consisted almost completely of iron. When urban planning started in 1929, everything seemed to be going right: Magnitogorsk was copied from two of the most advanced steel-producing cities in the United States at that time: Gary, Indiana, and Pittsburgh, Pennsylvania. Hundreds of foreign experts arrived to carry out and supervise the project, including German architect Ernst May, who had successfully built worker housing in Frankfurt. A linear city design was planned, with the steel works on one end and housing blocks for the workers on the other, separated by a greenbelt that was supposed to absorb noise and pollution. This was a good plan, but Ernst May didn't know that some parts of the steel works and some housing complexes had already been completed. This forced May to abandon his original plan and come up with patchwork solutions. After a considerable delay, construction finally started in 1932, when 100,000 workers were already living in makeshift accommodations. Things got worse when building materials meant for housing were used for the steel mill – the project that always came first. Disgusted, May left in 1933 after being blamed by the Soviet government for all that went wrong with city planning. Rushing to build the Soviet Union's model steel town according to Stalin's Five-Year Plans of the 1930s, the steel mills started operations before emission controls and plant security were even considered. Housing was eventually located downwind from the plant, subjecting the town's residents and workers unnecessarily to air pollution. Bronchitis, asthma, and lung cancers followed, as well as an entry into the Dirty Thirty – the world's most polluted cities.[17]

To summarize the tremendous economic impact of Stalin's first three five-year plans: between 1929 and the Nazi invasion of June 1941, 9,000 new enterprises were established.[18] The chemical, aeronautics, farm equipment, and other industries grew at a rapid pace. According

---

[17] S. Preuss, "Magnitogorsk: Once Stalin's Model Town, Now a Polluted Hell-Hole," http://www.environmentalgraffiti.com/featured/magnitogorsk-once-stalin-model-town-polluted/16708.

[18] *Narodnoe Khoziaistvo SSSR za 70 Let* (Moscow: Gosstatizdat, 1987).

TABLE I. *Change in Per Capita Production of Output, Soviet Union, in Comparison with Other Countries, 1913–1940*

Country	1913	1928	1932	1937	1940
United States	0.24	0.19	0.27	0.32	0.29
Great Britain	0.30	0.29	0.31	0.40	0.36
France	0.45	0.35	0.49	0.53	0.45
Germany	0.46	0.39	0.49	0.53	0.45
Italy	0.58	0.51	0.53	0.73	0.70
Japan	1.13	0.78	0.82	1.08	0.87

Source: Davies et al., 1994.

to official statistics, the volume of industrial production grew annually over the 1930s at 18 percent; according to Davies and others the figures are somewhat lower but still impressive at perhaps 9 percent to 16 percent. Table 1 indicates that the Soviet Union far exceeds most other nations in per capita production of output.[19] Industry and construction grew from 9 percent to 27 percent of employment, whereas agriculture declined from 80 percent to 48 percent between 1928 and 1950. The contribution of industry and construction to the national economy grew from 29 percent to 64 percent over the same period, while that of agriculture declined from 45 percent to 22 percent.[20] Stalin both summarized the achievements of the first Five-Year Plan and forecast future results when, in 1933, he declared, "We didn't have ferrous metallurgy.... Now we do. We didn't have an automobile industry. We have that now, too. We didn't have machine building. We have this, too."[21]

In addition to these tremendous industrial feats, the Soviet Union engaged in a series of technological displays intended to prove the verve of the young nation: flights across the North Pole, assimilation of the Northern Sea Route, the saving of the steamship *Cheliuskin*, and wintering and drifting in the ice.[22] All of this was propaganda to show the brilliant achievements of the economy while hiding the tremendous human and environmental costs.

---

[19] R. W. Davies, M. Harrison and S. G. Wheatcroft, *The Economic Transformation of the Soviet Union, 1913–1945* (Cambridge: Cambridge University Press, 1994), and *Evropa i Rossiia. Opyt Ekonomicheskikh Preobrazovanii* (Moscow: Nauka, 1996).

[20] *Trud v SSSR*, 1988; Vainshtein, 1969.

[21] Joseph Stalin, "Itogi Pervoi Piatiletki," in *Sochinenie*, vol. 13 (Moscow: Gosizdatpolit, 1955).

[22] John McCannon, *Red Arctic* (New York: Oxford, 1998).

Urbanization in the Stalin Era

Rapid urbanization naturally accompanied the Five-Year Plans. Soviet leaders, economists, urban planners, and visionaries understood that "socialist reconstruction of the nation" under Stalin would include remaking old cities and building new ones, and required critical thinking about the relationship between the labor, the work site, and the home. Socialist cities, so the utopian version went, would be greener than capitalist cities, with more open space and parks. Indeed, Soviet cities have extensive green spaces. Second, public transportation would be more extensive, less expensive, and calming for the worker. The subway, tram, and bus would whisk the worker quietly to the factory and return him or her home each evening rested, whereas in capitalist cities, the daily commute, workers believed, was a noisy, intrusive, and violent process. Third, Soviet cities would, of course, be cleaner, safer, and quieter than capitalist cities.[23] (See **Box 3**.)

Although there may be some uncertainty about the course and extent of Soviet urbanization in the 1930s because of the purges and efforts to hide human losses, the 1937 national census and other sources indicate massive in-migration and urban growth. From 1929 through 1933, urban population increased by 12 to 13 million people, and over the period from 1929 to 1939 by 27 to 28 million individuals. The percentage of urban residents in the nation over the 1920s and 1930s grew from 18.4 percent to 31 percent (and reached 50 percent with the 1959 census). The number of cities with more than 100,000 inhabitants grew from 24 in 1926 to 89 in 1939 (of which seven were the result of the expansion of the Soviet Union in 1939, but fifty-one were the result of in-migration).[24] Industrial centers mushroomed at an astonishing rate, expanding on average at a rate of some 50,000 inhabitants a week between 1928 and 1932. In the words of the commissar of heavy industry, Sergo Ordzhonikidze, the Soviet Union had become one huge "nomadic Gypsy camp."[25]

Unfortunately, the construction of housing, stores, schools, hospitals, and other social overhead capital lagged significantly behind the influx of new residents. Although from 1926 through 1939 the urban population grew by more than twofold, the housing fund grew by only 68 percent.

---

[23] For a discussion about how some of these ideas played out, see Blair Ruble, *Leningrad: Shaping a Soviet City* (Berkeley: University of California Press, 1990).

[24] A. Vishnevskii, *Serp i Rubl'* (Moscow: OGI, 1998).

[25] Sergo Ordzhonikidze, quoted in Moshe Lewin, *The Making of the Soviet System* (New York: Pantheon, 1994), p. 221.

> **Box 3. The City and Environmental History**
>
> As Joel Tarr and other urban historians have pointed out, the "relationship between the city and the natural environment has actually been circular, with cities having massive effects on the natural environment, while the natural environment, in turn, has profoundly shaped urban configurations." The natural environment required urban dwellers to respond to a series of problems in their built environments: floods, fires, pests, hurricanes, and earthquakes. Planners have shaped the environment to ensure proper transport, water supply, waste disposal, and energy production. They constructed a built environment of paved streets, malls, houses, factories, office buildings, and churches. In the process, they altered urban biological ecosystems for their own purposes, killing off animal populations, eliminating native species of flora and fauna, and introducing new and foreign species. They created a local micro-climate with different temperature, wind patterns and precipitation from the surrounding countryside. City mothers, fathers, and industrialists advanced projects to shape rivers to support factories and power generation, and thereby destroyed habitat and interrupted patterns of life of flora and fauna. They have become energy and resource sinks that consume resources of water, food, and so on, and they produce heat, waste, and pollution that often end up being piled up or disposed of within the urban environment.[26] In all of these ways, Soviet officials dealt with the circularity of "natural" and urban interactions. For example,

---

[26] For a general discussion, see Joel Tarr, "The City and the Natural Environment," http://www.gdrc.org/uem/doc-tarr.html; and Martin V. Melosi, "The Place of the City in Environmental History," *Environmental History Review* (Spring 1993), pp. 1–23. See also Nelson Blake, *Water for the Cities: A History of the Urban Water Supply Problem in the United States* (Syracuse, NY: Syracuse University Press, 1956); Willian Cronon, *Nature's Metropolis: Chicago and the Great West* (New York: W. W. Norton, 1991); Margaret Davis, *Rivers in the Desert: William Mulholland and the Inventing of Los Angeles* (New York: Harper Collins, 1993); Hugh Gorman, "Manufacturing Brownfields: The Case of Neville Island, Pennsylvania," *Technology and Culture*, vol. 38 (July 1997), pp. 539–574; Suellen Hoy, *Chasing Dirt: The American Pursuit of Cleanliness* (New York: Oxford University Press, 1995); Martin Melosi, *Garbage in the Cities: Refuse, Reform, and the Environment, 1880–1980* (College Station, TX: Texas A and M University Press, 1981); Theodore Steinberg, *Nature Incorporated: Industrialization and the Waters of New England* (New York: Cambridge University Press, 1991); Joel Tarr, *The Search for the Ultimate Sink: Urban Pollution in Historical Perspective* (Akron, OH: University of Akron Press, 1996).

> Leningrad, built on a swamp in the Neva River delta, experiences floods regularly, with catastrophic ones (with water 300 centimeters and higher) in 1777, 1824, and 1924. And for decades engineers and city fathers have sought some kind of hydrotechnical solution, with engineers from Gidroproekt proposing a series of designs to this day.[27]

This created a crisis that persisted in one way or another, especially after the devastation of World War II, until the end of the Soviet Union. People lived in barracks, like sardines in small rooms, and even in dugout underground huts. In Stalin's own Moscow, tens of thousands of people lived in overcrowded factory barracks up to 200 meters long and 4.5 meters wide, containing as many as 500 narrow, uncomfortable beds covered with mattresses stuffed with straw or dried leaves. There were no pillows or blankets or screens or walls to give any privacy. The barracks were breeding grounds for various intestinal, pulmonary, and other ailments. Smoky wood or coal stoves provided limited heating; toilets were on the street and threatened water supply. Kerosene was also widely used. Air, water, and soil pollution resulted. Living standards fell continuously between 1928 and 1933, stabilized in 1934, rose throughout the "three good years," and fell back after 1936, but at no point did they recover to the late NEP levels.

There are no data available to assess the level of environmental pollution in Soviet cities in the Stalin period, although we know that pollution and other costs of industrialization are present in every society and political system. We can indirectly gauge the extent of the problem, however, through a series of Soviet publications. During the late 1920s and 1930s, the Soviet authorities were already attempting to control pollution, as indicated in the first official Soviet report on air pollution control that was issued in 1973.[28] In 1929–1930, the Soviet hygiene and epidemiology network gained responsibility for investigation and control of air pollution and played a very important role in the period. (See **Box 4**.) Public health officials focused, in the first place, on the most important sources of air pollution in large industrial centers, including Moscow, Leningrad, and cities in Ukraine and the Urals. They also dealt with the threat of

---

[27] See A. P. Kosinskii et al., "Flood Control Structures of Leningrad," *Power Technology and Engineering*, vol. 22, no 2 (February 1988), pp. 92–96.

[28] N. F. Izmerov, *Control of Air Pollution in the USSR* (Geneva: World Health Organization, 1973).

> **Box 4. Stalin's Epidemiology Service**
>
> K. F. Meyer wrote that
>
> cholera, typhus, typhoid fever, smallpox, dysentery, malaria, and sometimes plague broke out in large epidemics and the morbidity and mortality rates were exceptionally high in Soviet Union in the post revolution years. The typhus morbidity reached 20 to 30 million between 1917 and 1922. Conservative estimates of the incidence of relapsing fever from 1918 to 1922 exceed 15 million cases. Malaria (mortality rate, 3 to 20 percent) became the most important disease in Russia and spread during the famine [of] 1921–1922 throughout European Russia and as far north as Arkhangelsk. In 1921 there were 205,000 cases of cholera, always endemic in southeastern Russia, particularly in the area of famine in the Volga Basin; and the case mortality rate for all Russia was 50 percent. Cases of typhoid fever doubled and trebled, and in 1920 the reported figure was 424,487 cases. Over 166,000 cases of smallpox, with a mortality rate of 10 to 20 percent, were reported in 1919.[29]
>
> Lenin's slogan "Either lice conquer socialism or socialism conquers lice" appeared in 1919.
>
> The Soviet authorities recognized the need to establish public health and industrial safety organizations to accompany the stormy industrialization process and attendant migration and urbanization. They both established new institutes and drew on a series of pre-revolutionary establishments. To combat epidemics, the first sanitary and epidemiological center was established in 1922. In 1931, a series of sanitary and epidemiological centers gained permanent status, and in the following year the government determined to set up a national network. By the end of 1940, there were 1,958 centers in the Soviet Union, with a staff of 11,121 sanitary physicians and other university-trained specialists. In 1921, scientists founded the Moscow-based F. F. Erisman Institute of Hygiene (named after a nineteenth-century pioneer in sanitary research), and in 1923 they established the Institute for the Study of Occupational Diseases, later the Work Hygiene and Occupational Diseases Institute of the Academy of Medical Sciences of the Soviet Union. In 1931, the Gamaleia Scientific Research Institute of Epidemiology and Microbiology was set up in Moscow on the base of the Chemical-Bacteriological Institute (itself founded in 1891). Together, these institutes and specialists succeeded in eradicating

---

[29] K. F. Meyer, "Some Observations on Infective Diseases in Russia," *American Journal of Public Health*, vol. 47 (September 1957), pp. 1083–1092.

> epidemic typhus quickly, with cholera eliminated by the end of the 1920s. Compulsory vaccination, introduced in 1920, gradually reduced the annual morbidity of smallpox to the extent that by the mid-1930s, only a few cases were reported.[30] Public bathhouses and strict disinfestation measures contributed to the rapid improvement of all measures of public health. These achievements served as a propaganda victory that enabled Soviet leaders to claim that industrialization and urbanization in the Soviet Union served the proletariat, not the capitalist bosses, and had minimal public health costs.[31]

epidemics that accompany mass migrations, urbanization, and other similar phenomena.

In April 1935, specialists gathered for the first All-Union Conference on Air Pollution Control in Kharkiv, Ukraine. They discussed such topics as clean air in towns, methods of studying the atmosphere, and organization of measurement stations for air pollution control. During 1935–1936, a special corps of urban sanitary inspectors with responsibility for air pollution control was established in a number of cities. The All-Union Institute of Community Sanitation and Public Health of the People's Commissariat of Health of the Soviet Union began investigations of the problems of increased atmospheric pollution caused by automobile and truck exhaust. Yet the authorities issued a national standard that set five levels or classes of the threat that industrial enterprises posed to public health based on the hazards they presented and the technological processes they used only after the war in 1947. The standards indicated modest exclusion zones around such enterprises reaching no more than two kilometers, and usually 100 meters, and these prohibitions were not followed.[32]

In 1948, the Council of Ministers of the Soviet Union issued a decree to regulate air pollution in Moscow. This instructed the Ministry of Health of the Soviet Union to work out atmospheric standards for urban areas in short order. On the basis of an analysis and summary of toxicological research, clinical statistics, and so on, a special committee under

---

[30] Meyer, ibid.
[31] Laurie Garrett, *Betrayal of Trust: The Collapse of Global Public Health* (New York: Oxford University Press, 2001); Meyer, ibid.; and Lev Razgon, *Plen v Svoem Otechestve* (Juriya Kuvaldina "Knizhnyi Sad," 1994).
[32] Izmerov, *Control of Air Pollution in the USSR*.

the chairmanship of Professor V. A. Riazanov, a specialist concerned with exposures to heavy metals and a founder of the field in the Soviet Union, worked out standards for maximum permissible concentrations of the substances most frequently encountered in the air of towns: sulfur dioxide, hydrogen sulfide, carbon bisulfide, carbon monoxide, nitrogen oxides, chlorine, mercury, lead, dust, and soot. These standards, set forth in 1951, were among the world's first.[33]

A series of debates among leading Bolshevik administrators as well as leading figures in Soviet science, culture, civil engineering, hygiene, journalism, fine arts, architecture, and urbanism took place at the end of the 1920s over the nature and structure of the socialist city as distinct from the capitalist city. The participants included Lenin's widow, Nadezhda Krupskaia; the commissar of enlightenment, Anatolii Lunacharskii, Gleb Krzhizhanovskii who headed the State Electrification Program; Nikolai Semashko, the commissar of public health; and Stanislav Strumilin, a leading economic planner. Le Corbusier and several other prominent European architects and urban planners joined in the discussions. Students of Soviet architectural institutes took part in the drafting of Green City models. The discussion had widespread press and radio coverage. Both the "urbanists," supporters of maximum urban concentration, and the "de-urbanists," adherents of dispersed housing to promote a bridge between city and countryside, participated in the discussion, and both sides saw the number and distribution of green spaces and greenbelts as a key issue of future Soviet cities. Both advanced the hope that workers would live in proximity to nature.

This reflected an ongoing discussion among urban environment conservationists and urban environment futurists. The former had been influenced by theorists engaged in the study of local lore and customs and later by Ebenezer Howard's utopian vision of a green city and by his followers. Although the relationship between man and nature was the central point of the discussion, in essence it had much wider aims: the development of the ways and means of the reconstruction of Soviet cities in support of intensive and rapid socialist industrialization and urbanization. Architecture and town planning were considered as a powerful means for the change of mass psychology, and in the end for the creation of a new woman and new man of the socialist future.

In the broadest sense, the "Green City" played the role of antithesis of the densely populated and polluted capitalist city, with its lack of open

[33] Izmerov, *Control of Air Pollution in the USSR*.

spaces and reflected avant-garde concepts in urban design and new planning attitudes. In one project, planners considered the full transformation of Moscow into a giant park with the preservation of small sites of old Moscow. In this and other projects, the "old city" (downtown) would become a kind of historical-cultural reserve. In 1930, a questionnaire concerning the future of the Moscow capital was disseminated among the participants of the discussion as well as among some workers of Moscow plants. The questionnaire indicated that Soviet urban planners were at the vanguard of the European urban thought. Moreover, this vanguard had influenced European urban thought for decades. The trend of utopian thought ended with two resolutions of the Central Committee of the Communist Party, "O Rabote po Perestroike Byta" (On Work for the Reconstruction of Daily Life, 1930) and on "O Razrabotke General'nogo Plana Moskvy" (On the Development of the Moscow Master Plan, 1935). Ultimately, however, all visionary plans fell by the wayside as tightness of resources and the desideratum of maximum speed in factory construction precluded significant "greening of the Soviet city." As in other policy areas, a manhunt for left- and right-wing deviationists commenced that put visionary projects out of consideration – and their supporters often in prison.

Nevertheless, the very ideas of man–nature relationships that matured in these discussions became basic in Soviet urban planning for decades, and some conceptual principles were implemented in the process of construction of new Soviet neighborhoods and worker settlements that turned out, in fact, to be far greener than their capitalist counterparts, even if the neighborhoods struggled under the weight of air and water pollution and unfinished infrastructure. In the postwar years, as a manifestation of the preservation of civic culture in spite of all the pressures to subjugate one's feelings to the state, activists established the All-Russian Society for the Promotion and Protection of Urban Green Planting (VOSSOGZN) in 1947. Representing both the implementation of the ideas of the "Green City" and current needs of urban development, the society became an important augmentation to the environmental movement.

War on the Agriculture: Soviet Agriculture

If in other countries the destitution of the masses was neither planned nor centralized, and if their exposure to dangerous processes and noxious smells and chemicals was an expected outcome of industrialization, then in the Soviet Union this was the result of a planned, highly centralized process based on violence and coercion against the citizenry as a kind of government-sanctioned terrorism, especially against the peasantry. If before the revolution more than half of the national income was created in the agricultural sector, and by the end of the 1920s a little less than half, then it becomes understandable why the political authorities determined that the countryside was the source of capital for the industrialization effort. They organized the collectivization campaign both to secure this capital for investment and to establish regular grain deliveries to burgeoning cities. They named an entire class of peasants, the kulaks, as "enemies of the people" to justify the violent expropriation of all of their property, down to their clothing. De-kulakization served several interrelated goals: collectivization; formation of investment capital; establishment of modern, technologically based agriculture; and creation of a cheap source of labor. The state sold grain and even works of art abroad to finance the further impoverishment of the masses.[34] The violence extended rapidly from the kulak to the rural working class, middle peasants, and even poor peasants.

Archival documents indicate the great human and other costs of the collectivization campaign. Stalin and the Communist Party leadership intended for collectivization to end once and for all the threat of an independent class of wealthy peasants who were hostile to Bolshevism. They believed generally that the peasants were hostile to Bolshevism. Lenin had ultimately come to the view that forcing the peasant to collectivize prematurely would fail. The peasant had to be warmed to the idea through modern technology. Lenin believed that such a technology was the tractor, which was vitally important as a symbol of the unification of industry with agriculture. Lenin saw more in the tractor than just an implement for tilling the soil; he saw it as a vehicle for luring the peasantry to Communism. On March 23, 1919, he told the delegates to the eighth Party Congress that tractors would create Communism: "If we could give 100,000 first class tractors tomorrow, provide them with gasoline, with mechanics (you all know quite well this is a fantasy), then the middle peasant would say, 'I am for communism.'" The peasant, Lenin continued, "needs the industry of the city, without it he cannot live."[35] Lenin concluded that it would be a generation or longer before the peasant came

---

[34] V. I. Danilov-Danil'ian et al., *Okruzhaiushaia Sreda Mezhdu Proshlym i Budushim: Mir i Rossiia* (Moscow: s.n., 1994).
[35] V. I. Lenin, *Sochinenie*, XXIV (Moscow: PartizdatTsKVKP(b), 1935), vol. XXIV, p. 170.

over to the side of the Communists and that coercion would not work to change his essentially conservative world view.

Crucial to the rise of modern agriculture was the destruction of traditional peasant land ownership and management practices through technology and Bolshevism. In many villages, tiny holdings for one peasant family were spread quite a distance from one another. The peasants and their draught animals spent a great deal of energy inefficiently moving from one field to another. On top of this, the countryside was poorly mechanized. Very few tractors, let alone combines or harvesters, had made their way to the farms. As noted in Chapter 1, peasants used the three-field system and poorly engaged fertilizing and crop rotation practices to improve production. This would change overnight with Stalin's policy of collectivization of agriculture at breakneck speed. Collectivization involved the move from "economically irrational, small-scale, tradition farming to the modern economically rational, large-scale farming that lent itself to mechanization."[36] The peasant would become the rural proletarian, working fields that stretched to the horizon with modern machines, controlled and surveilled through Machine Tractor Stations (MTS) that were not only central distribution points of equipment, but also the Communist Party's eyes and ears of political control.

The legendary American tractor became of interest to mechanize agriculture. The Soviets established the Amtorg Trading Corporation in May 1924 to procure agricultural and other machinery and opened offices in several American cities. Amtorg engaged in small-scale acquisitions until the Soviets switched over from concessions to technical assistance contracts in 1928. They hired hundreds of engineers for employment in the Soviet Union who worked in the Cheliabinsk Tractor Plant, the Tractor Plant Construction Trust, the Stalingrad Tractor Plant, and the Kharkiv Tractor Plant. These engineers were involved primarily with agricultural machinery industry and to a lesser extent with the fertilizer industry and in irrigation. The Detroit firm of Albert Kahn, which designed the Ford plant, also provided the blueprints for the Stalingrad Tractor Factory, Leon Swajian, who worked on construction of River Rouge, was involved in the Kharkiv Tractor Factory and later at Cheliabinsk.[37] In absolute numbers and symbolically, the tractor was the *technology de résistance*.

---

[36] Sheila Fitzpatrick, *Stalin's Peasants* (New York: Oxford University Press, 1994), pp. 104–105.
[37] Dana G. Dalrymple, "American Technology and Soviet Agricultural Development, 1924–1933," *Agricultural History*, vol. 40, no. 3. (July 1966), pp. 187–206.

Dalrymple writes, "Seldom has a major agricultural technology been adopted so quickly and on such a vast scale as was the tractor in Soviet Russia." Until Soviet factories were capable of meeting domestic need, the nation imported huge numbers of tractors toward the end of modernization of agriculture, primarily from the United States. From 1924 to 1933, 86,377 American tractors were shipped to the Soviet Union, nearly 23,000 in each of the peak years of 1930 and 1931.[38] By the end of 1933, there were 2,446 MTS with almost 75,000 tractors that served 50 percent of the sown area according to official, if doubtful statistics. Still, many kolkhoz workers preferred draught power without MTS because of high cost. At least tractors did not fall prey to lack of fodder and diseases that plagued animals throughout the 1930s.[39] In a decade, the number of tractors increased from 1,000 to more than 200,000, a fact "all the more remarkable because it took place in a rural economy which had, from a technological point of view, changed little from the Middle Ages."[40]

The poor quality of the tractors was accentuated by their misuse, even though initially American agricultural specialists were engaged in training Russian peasants in how to use the tractors properly. The Russians lacked mechanical backgrounds. They took little interest in proper operation because the tractors were owned by the state, not the operators. Maurice Hindus, after a tour through his home area, spoke of "the reckless treatment of machinery on all the socialized lands." He continued: "The resulting breakage is colossal. Fleets of disabled tractors dot the Russian landscape ... machines are left with no cover over them in yards and in far away fields, exposed to the devastation of wind, rain and sun."[41]

To a smaller extent, the Soviets worked with American companies to expand the fertilizer industry, in particular with Du Pont of Wilmington, Delaware, and Nitrogen Engineering Corporation of New York. Americans were also involved with two substantial irrigation projects: the Central Asian Water Trust (Sredazvodkhoz) and the Transcaucasian

---

[38] Dalrymple, "American Technology."
[39] Fitzpatrick, *Stalin's Peasants*, pp. 306–307.
[40] Dalrymple, "American Technology." American specialists reported that Soviet tractors did not meet American standards of quality, and that maintenance was still a severe problem because of substandard raw materials and severe production difficulties. The problems included leaking radiators, poorly cast cylinder heads, loose bearings, broken valve springs, unsatisfactory threading on spark plugs, and so on.
[41] Dalrymple, "The American Tractor," and Maurice Hindus, *Red Bread* (New York: J. Cape and H. Smith, 1931), pp. 357–358.

Water Trust (Zakvodhoz). For example, Arthur Powell Davis, former director of the United States Reclamation Service, served as chief consulting engineer with Sredazvodkhoz.[42] Tractors and irrigation projects accompanied collectivization, with significant environmental costs.[43] (See Box 5.)

Collectivization was more than mechanization of agriculture and the creation of a modern agronomy to feed burgeoning cities. It was violent, brutal, and murderous coercion, a revolution of totality, rapidity, and violence. Collectivization meant to do away with the individual, to make agricultural property, tools, land, and livestock, as well as labor, collective. Between 1929 and 1934, the number of peasants living in collective farms grew from 4 percent to 71.4 percent. Rural Russian became a world of nationalized land whose livestock, grain, and tools belonged to the state, where peasants no longer scheduled their own labor, where all decisions that affected their lives came down from above. About 10 million peasants were displaced from their villages and deported into exile in Siberia or the Far North for resisting collectivization, being labeled kulaks, or being in the wrong place. About 4 million people were imprisoned, about 200,000 were shot, and as many as 3 million more died in the famine of 1932–1933. A major reason was the extraction of grain and seed grain from the countryside for the cities and for export. But this was a war directed against the Soviet citizen.

On top of this staggering loss of human life, livestock was killed too. Of 34 million horses in 1929, 16.6 million remained in 1933; of 147 million sheep and goats in 1929, fewer than 51 million remained in 1933; of 21 million hogs in 1929, only 12 million remained in 1933. The situation with draught animals will give a sense of the anarchy and devastation in the countryside. If in 1928 peasants owned 34 million horses, then at the end of 1932, when 60 percent of peasants had joined collective farms, only 12 million horses were in collective farms. By 1934, there were only 15 million horses in the entire Soviet Union. Tractors replaced them, but not in sufficient numbers to compensate for the loss of draught power.[44] Davies and Wheatcroft meticulously point out how overoptimistic sowing, harvest, and procurement plans failed in the face of active and passive peasant resistance, drought, famine, and the flight of peasants to escape the famine. While officials debated what to do, and

---

[42] Dalrymple, "American Technology."
[43] Dalrymple, "The American Tractor."
[44] Fitzpatrick, *Stalin's Peasants*, pp. 66, 136–137.

> ## Box 5. Irrigation and Agriculture in Central Asia
>
> A major irrigation project on the Golodnaia steppe accompanied Turksib. Although in some cases machinery replaced mass labor and canals were lined with concrete, basically the irrigated area followed the development of a canal system developed by tsarist authorities that relied on human labor and outmoded technology. Irrigation was intended to move beyond wheat, barley, and millet that grew in twenty to fifty millimeters of annual rainfall to enable dense rural populations in the Fergana Valley, along the Zaravshan River and the Qashqa Darya near the Amu Darya, to expand cotton production. Irrigation expansion commenced in 1924 at the Pakhta-Aral state farm and reached some 68,000 hectares for cotton production in 1928–1929, 76,000 hectares by 1934, and some 95,000 hectares by 1939 with water coming from the Syr Darya and moved by railroad from Spasskoe to Taskhent. By 1932, 75 percent of Uzbek cotton came from collective farms, although collectivization saw production plummet.[45]
>
> Ultimately, these projects had disastrous impact on rivers and on the Aral Sea. One specialist characterized "the most serious problems of contemporary Soviet economic geography" to be the expansion of agricultural output. Toward that end, the Soviets pushed to increase the area under irrigation in moisture-deficient lands with intensive land reclamation at a "magnitude and tempo never paralleled." Yet, this program did not advance "as smoothly as planned." Central Asia was the principal site of concentration of irrigated agriculture in the Soviet Union, but most rivers were used to their maximum capacity by the 1950s, resulting in critical water shortages and well-below-average runoff. Technical and economic problems prevented the Soviets from utilizing the Amu Darya River, the largest in the region.[46] Prospects

---

[45] Ian Matley, "The Golodnaya Steppe: A Russian Irrigation Venture in Central Asia," *Geographical Review*, vol. 60, no. 3 (July 1970), pp. 328–346. On pre-revolutionary Russian efforts to pursue agriculture in Central Asia, see George Anderson, "Agriculture in the Undrained Basin of Asia," *Agricultural History*, vol. 22, no. 4 (October, 1948), pp. 233–238.

[46] For an earlier evaluation of the significance of Soviet development programs for Amu Darya riverine ecology, see Neil Field, "The Amu Darya: A Study in Resource Geography," *Geographical Review*, vol. 44, no. 4 (October 1954), pp. 528–542.

> for further expansion looked bleak. By the 1970s, its sources – the Amu Darya and Syr Darya Rivers – essentially gone, the Aral Sea had begun to dry up. This had "catastrophic consequences" for both the inhabitants of the region and for the environment.[47]

in some cases relaxed several policies, targets continued to fall short of plans, and mismanagement of resources was added to the mix of violence, political crisis, and poor weather. As supplies of grain to towns fell and peasants were forced to steal grain to survive, Stalin imposed a ten-year exile or death sentence for "theft of socialist property."[48]

Granted, the agricultural sector experienced a tremendous industrial revolution with the appearance of tractors and combines on the collective farms and the creation of a national farm equipment industry. Yet, the Soviet Union achieved this revolution at great cost to people and land. Collectivization was accompanied by droughts and famine, the latter caused by the Communist Party. The government radically reduced the plan for grain deliveries only in 1936.[49]

The Gulag on the Frontier of the Soviet Empire[50]

The gulag was a major feature of the Stalinist economic system, of course, an immoral exploitation of human life and a contributor to environmental problems.[51] To speed up the tempo of modernization and extract more capital, the authorities created a system of slave labor camps that went beyond the initial reason for their establishment – to "reeducate" political prisoners and other opponents. The gulag immediately became a system to incarcerate any suspected enemy among hardened criminals, and next

---

[47] Norman Precoda, "Requiem for the Aral Sea," *Ambio*, vol. 20, no. 3/4 (May 1991), pp. 109–114.
[48] Robert Conquest, *Harvest of Sorrow*; Fitzpatrick, *Stalin's Peasants*; and R. W. Davies and Stephen G. Wheatcroft, *The Years of Hunger: Soviet Agriculture, 1931–1933* (New York: Palgrave Macmillan, 2004).
[49] N. Dronin and E. Bellinger, *Climate Dependence and Food Problems in Russia 1900–1990* (Budapest: CEU Press, 2005).
[50] For a thorough, engaging study, see Anne Applebaum, *Gulag: A History* (New York: Anchor Books, 2004). See also Paul R. Gregory and Valery Lazarev, eds., *The Economics of Forced Labor: The Soviet Gulag* (Stanford: Hoover Institution Press, 2003); and Oleg Khlevniuk, *The History of the Gulag* (New Haven: Yale, 2004).
[51] Applebaum and Khlevniuk.

a source of cheap labor. By autumn 1929, roughly 64,000 prisoners had been transferred from prisons to timber operations, and the total number of inmates had grown to 200,000 people. The size of labor camps expanded rapidly along with the promulgation of the first Five-Year Plan. Planners requisitioned prisoners to meet production targets, particularly in the field of construction, and in the exploitation of natural resources in the remote regions of the enormous country. The demand for compulsory labor in turn contributed to the number of arrests and deportations carried out by the security police. Western experts estimate the number of prisoners at between 3 and 13 million people and mortality at 30 percent annually.[52]

By the mid-1930s, the NKVD (*Narodnyi Kommissariat Vnutrenykh Del*, or People's Commissariat of Internal Affairs) became not only a security police with its own army (including artillery and air force units), but also a growing industrial and construction operation. The NKVD adopted strangely capitalistic practices of "hiring" as many workers as it could and paying them as little as possible in terms of food, shelter, or clothing. Then, its military leadership "hired out" the slaves to other commissariats. By the eve of World War II, more than two-thirds of gulag forced labor was likely concentrated in construction (3.5 million prisoners), non-ferrous metallurgy and mining, including gold and platinum (1 million prisoners), and lumbering (400,000 prisoners), or a total of nearly 2 percent of the total population and 8 percent of the total labor force). One gulag camp, Igarka, 400 miles inland on the Enisei River, focused on forest resources. It was glorified by Soviet writers including Maxim Gorky as contributing to political education; Gorky did not mention the human costs, and glorified the achievements in subjugating nature. Igarka grew rapidly to 14,000 people, including 4,000 kulaks exiled in 1930 and 1931 during collectivization.[53]

Construction entailed a wide variety of activities that had a direct impact on the environment: the construction of roads, railroads, and canals, reclamation of swamps, straightening of rivers, and dredging of waterways, and farming. NKVD activity in the third Five-Year Plan (1937, interrupted by war) forecast the construction of 7,000 to 8,000 miles of new roads, the extension of navigable inland waterways by some 80,000 miles (including the Moscow-Volga canal), and about

---

[52] Chris Ward, *Stalin's Russia* (London, New York: Edward Arnold, 1993).
[53] H. P. Smolka, "Soviet Development of the Arctic New Industries and Strategical Possibilities," *International Affairs*, vol. 16, no. 4 (July 1937), pp. 564–578.

45,000 miles of new railway lines, many in remote regions – for instance, the Kotlas–Vorkuta line in north central Russia to serve a burgeoning mining industry in the Komi Republic. A Baikal-Amur labor camp (Bamlag) was set up for the construction of the Baikal-Amur Mainline (BAM) across southern Siberia from Taishet to Komsomolsk. The project was stopped with Stalin's death in 1953 and resumed again only in the Brezhnev era (1972), when it was considered an honorary project of the Komsomol (the Communist Youth League) and called "the construction project of the century." BAM was finished only in 1991, itself with significant environmental degradation.

Mining was a second major area of gulag labor activity, with a very large slave labor force and irreversible impact on the environment. The infamous Dalstroi organization in the Far East used hundreds of thousands of slaves in gold, platinum, and molybdenum operations, especially in Magadan and Kolyma, the latter immortalized by Varlam Shalamov in *Kolyma Tales*. Gulag camps also focused on mining gold in Bodaybo, Aldan, and Vilyuy in the Lena River basin; the territory between Zeya and Burey in the Amur River basin; and along the Upper and Middle Tunguska rivers in the Enisei basin. A full sense of the economic – and environmental – role of the NKVD may be gathered by noting that in 1941, the NKVD was the second-largest transportation organization in the country (840 million tons per kilometer) after the Commissariat of Supply (with 1,120 million tons per kilometer). Whether in Nikel, Norilsk, Vorkuta, or any of the other scores of arctic and subarctic cities, the extensive environmental degradation that has resulted from heavy metal and sulfur dioxide pollution and its impacts on public health and indigenous cultures are beyond question and remain a major problem into the twenty-first century.[54]

---

[54] M. L. Gytarskii et al. "Monitoring of Forest Ecosystems in the Russian Subarctic – Effects of Industrial Pollution," *Science of the Total Environment*, vol. 164, no. 1 (March 1995), pp. 57–68; B. A. Revich, "Public Health and Ambient Air Pollution in Arctic and Subarctic Cities of Russia," *Science of the Total Environment*, vol. 161 (1995), pp. 585–592; N. V. Vasiliev et al., "The Role of Migration Processes in Oncological Epidemiology of Siberia and the Far East," *Vestniia Rossiiskoi Akademii Meditsinskikh Nauk*, no. 7 (1994), pp. 34–39; A. R. Bond, "Air Pollution in Norilsk – A Soviet Worst Case," *Soviet Geography*, vol. 15, no. 9 (1984), pp. 665–669; A. A. Vinogradova and V. E. Egorov, "Contributions of Industrial Areas of the Northern Hemisphere to Air Pollution in the Russian Arctic," *Izvestiia Akademii Nauk. Fizika Atmosfery i Okeana*, vol. 33, no. 6 (November–December 1997), pp. 750–757; D. R. Klein, "Arctic Grazing Systems and Industrial Development: Can We Minimize Conflicts?" *Polar Research*, vol. 19, no. 1 (2000), pp. 91–98; V. Shevchenko et al., "Heavy Metals in Aerosols Over the Seas of

The camps were, in fact, forms of technology: human machines with limited operational capacity and a short, if lifetime, guarantee. By the early 1930s, the secret police had determined to use hundreds of thousands of prisoners in such projects as the White Sea-Baltic Canal, and also in forestry, agriculture, and such other prison camp operations as Sevdvinlag, Kargopol'lag, Mekhren'lag, and Iagrinlag (the Northern Dvina, Kargopol, Mekhren, and Iagry Island camps). The camps served virtually all operations – including those of the Glavsevmorput (Administration of the Northern [Arctic] Sea Route) empire, for example, mines on the Arctic coast at Amderma that were tunneled deep into the permafrost and produced spar for the cement industry, fluorite, and rare metals.[55]

The slave laborers often arrived in the middle of nowhere and were told to commence labor. On the one hand, their impact on the environment was limited by heavy reliance on hand tools and rudimentary equipment rather than on bulldozers, trucks, skidders, and harvesters. They worked only as hard as necessary to avoid the wrath of guards and to conserve energy given their inadequate diets. They could accomplish only so much given their tools and the low level of their physical power. On the other hand, they were usually sent to "virgin" regions to begin new operations. They were more concerned about survival than anything else. So, they trampled the grass in the fields and sprouts in the forest. They extended the reach of the nation's industry, roads, and forestry operations. Unfortunately, no one has written systematically about the environmental costs of the gulag system of labor. But we can extrapolate from the experiences of prisoners in a series of case studies of specific sectors of the gulag what the interactions between humans and nature likely entailed.

Take, for example, the forestry industry and the gulag colonies located in relatively inaccessible regions in the north of European Russia, the Urals, and Siberia. On the one hand, owing to limited access to modern technology, it is unlikely that clear-cutting practices were employed. The taiga was nearly impenetrable, and access to logging camps was possible only by narrow-gauge railways and temporary roads (or rivers) that were not extensive. The Upper Kama region was important as a trade nexus

---

the Russian Arctic," *Science of the Total Environment*, vol. 306, nos. 1–3 (May 2003), pp. 11–25; and S. M. Allen-Gil et al., "Heavy Metal Contamination in the Taimyr Peninsula, Siberian Arctic," *Science of the Total Environment*, vol. 301, nos. 1–3 (January 2003), pp. 119–138.

[55] GAAO (State Archive of Archangelsk Region) Otdel DSPI, F. 296, op. 2, d. 1391, ll. pp. 7–14. See Territorial'noe Otdelenie Arkhiva Nenetskogo Avtonomogo Okrug (hereafter TO ANAO), F. 289 (entire), on Glavsevmorput's "Amderma Expedition," 1929–1953, and F. 263 (entire) on the Amderma district Ispolkom.

and for forestry and potash products before the revolution, although the region was sparsely populated on the eve of the revolution, and inhabitants eked out an existence in felling and floating of lumber. But Soviet power changed the region into an industrial zone, where the industrial city of Berezniki on the Kama River contributes chemical (potassium, potash, and sodium) and heavy metal (titanium and magnesium) wastes to the environment.

Slave laborers cut wantonly because of the seemingly inexhaustible timber resources. They fed a pulp and paper combine in Vizhaika, itself a gulag organization. E. P. Berzin, who later commanded labor camps in Kolyma, supervised 18,000 prisoners to build the Vizhaika combine and supply it with lumber. Forestry operations of the Cherdynskii *leskhoz* (forest collective farm) covered 1 million hectares of largely first growth stands. The leskhoz "had responsibility for the conservation and management of forests in the region, including replanting, conservation, fire safety and the surveying of sections for harvesting." The prisoners built barracks and ancillary settlements, forest roads, and other infrastructure. The frontier of lumbering activities was pushed farther and farther north as rapacious harvests unfolded, and laborers spent more and more time on the road. They lived in unheated barracks, tolerating food and forage shortages and terrible labor conditions that led to high accident and mortality rates. From these beginnings, the authorities established such other labor camps as Usollag to build a sulphite plant in Solikamsk. In general, the camps spread along rivers – the Kama, Kolva, Vishera Berezovaya, Visherka, and their tributaries. Within two months of its foundation, Usollag had 10,749 inmates, and by February 1939, 34,403. During World War II, prisoners and others filled the Perm region gulag camps to provide labor for mines, timber, and hydroelectricity construction, a total of 80,000 prisoners at war's end.[56]

One geographer points out that "it will be hard to differentiate between what environmental impacts were caused by Gulag activity specifically, as opposed to felling/logging operations conducted by free or semifree labor (conscripted peasants, etc.)." He asks if forced labor had a greater impact than the labor of ministerial enterprises. It seems likely, given the greater access to tractors and other equipment moving on tank treads

---

[56] Judith Pallot, "Forced Labour for Forestry: The Twentieth Century History of Colonisation and Settlement in the North of Perm' Oblast'," *Europe-Asia Studies*, vol. 54, no. 7 (November 2002), pp. 1055–1083. See also Dominque Moran, "Lesniki and Leskhozy: Life and Work in Russia's Northern Forests," *Environment and History*, vol. 10, no. 1 (February 2004), pp. 83–105.

that, in one pass, compress soils and crush flora and fauna, that the "public" forestry operations had a greater impact than gulag operations that relied on manual labor. Yet he notes that "large swathes of Soviet forest territory wouldn't have been touched at all had it not been for the use of forced labor, nor would many industrial sites and cities have been constructed." On the other hand, his research notes that "the Karelian regional boss Edvard Gylling, who in the 1920s had been very aware of and vocal about the need to ensure sustainability, was forced during the first Five-Year Plan to shut up and put up with" clear cutting. In his archival work in Karelia, he saw "no evidence that during the 1930s those in the NKVD gulag administration, centrally, and in the regional camp systems, were any more or less contemptuous of the local and wider environmental impact of their operations than civilian authorities."[57]

To give another example, the Arctic city of Vorkuta, founded in 1932 as the site of a large Soviet forced-labor camp to develop Pechora basin coal, grew rapidly without consideration of pollution issues because of its contributions to defense industries. Before the war, only one mine had opened. During the war, ten more mines opened, a power station came into operation, and the North Pechora railway connected Vorkuta with the nation's rail network. Pechora coal was crucial during the war since the Don basin had been lost to Nazi armies; it fed blockaded Leningrad, the northwest and the Baltic and Northern naval fleets. Even with the gulag system, and with the creation of the gulag construction operation, Vorkutstroi, in 1938, the dispatch of supplies to Vorkuta and raw materials back was often delayed. Delivery over a sixty-four-kilometer-long narrow gauge railroad from Vorkuta to Rudnik took up to ten hours even when the railroad operated at full capacity. Engineers, managers and prisoners had no idea how to build roadways, structures or mines in the permafrost. Power-generating equipment was in complete collapse; buildings needed repair the moment they were finished; bridges were untrustworthy; and housing, stores, schools, and hospitals lagged in construction.[58]

Another example is the city of Norilsk on the Taimyr Peninsula, which grew out of the Norgulag that from 1935 to 1956 employed tens of

---

[57] E-mail, Nick Baron, School of History, University of Nottingham, UK, July 20, 2010, to Paul Josephson.

[58] P. I. Negretov, "How Vorkuta Began," *Soviet Studies*, vol. 29, no. 4 (October 1977), pp. 565–575. Camp prisoner uprisings occurred in the 1950s, and there is evidence that some of the camps in the region remained open until the 1980s. A bleak, concrete town of 200,000 individuals in 2005, Vorkuta still boasts five of the original thirteen mines.

thousands of prisoners to extract ore from the largest nickel-copper-palladium deposits in the world underneath the permafrost. Thousands died at forced labor, especially during the war and immediately after because of short supplies of food. Production of the metals commenced in 1939. Until World War II, the Norilsk combine was small scale, with a temporary power station, three open-cast coal pits, three mines, and others under construction, sandstone and limestone quarries, a railway, an airport, and a port at Dudinka on the Enisei River. By 1953, the combine produced 35 percent of the Soviet Union's total nickel output, 12 percent of its copper, 30 percent of its cobalt, and 90 percent of its platinum group metals.

The Soviets not only turned to the north for metals and timber, but even dreamed of turning the empty land into agricultural production. The great obstacles to assimilation – climate and desolation – did not deter the effort. Before railroads and roads were built, planners intended to use rivers for transport during the three months of the year they were thawed. They focused on building icebreakers. By the mid-1930s, they had established a chain of fifty-seven radio stations along the coast that operated year round and provide meteorological and other information. "Ice-watch" airplanes also assisted the icebreakers. The goal was to open the Arctic route from Murmansk to Arkhangelsk, to the Ob, Enisei, and Lena deltas, and then to the Kolyma, Indirgika, and Anadyr Rivers of the Far East.[59]

The Arctic region served the state through its rich natural resources. It includes ice-covered areas of the ocean, islands covered by continental glaciers, cold tundra with permafrost, and parts of the northern taiga. Over fifty years, the Soviet leadership established a series of facilities to exploit the raw materials found in the southern Arctic: Pechenganikel and Severonikel, mining and metallurgical combines in Apatity, Vorkuta, Olenegorsk, Kovdor, Lovozero, and Norilsk. Exploitation of oil and gas on the Yamal Peninsula accelerated in the late Soviet period. About 10 million people now live in this network of industrial facilities across the north, from Zapoliarnyi, Nikel, Murmansk, Monchegorsk, Apatity, and Kirovsk, to Vorkuta, and to the surrounding miners' settlements at Labytnanga, Salekhard, Dudinka, Tiski, and Anadry – many of them descendants of political prisoners and kulaks. All of these facilities served the Soviet military establishment directly. Officials built hundreds of thermoelectric power stations and steam boilers to support

[59] Smolka, "Soviet Development."

this population. A typical settlement of 10,000 inhabitants has at least ten heating plants discharging sulphur dioxide, nickel, dust, and other pollutants.

According to Vasily Kriuchkov of the Institute of Economic Problems of the Kola Science Center, "the northern ecosystems cannot sustain such impacts and are being destroyed." The major impacts are from sulfur fallout and heavy metals that have stripped taiga ecosystems of vegetation, leading to "industrially created wasteland." As deforestation accelerated, winds became more intense; snow was compacted; rivers, soil, and lakes froze to greater depth; and summer and winter air and soil temperatures decreased.[60]

Environmentalists and the Nature Protection Movement under Stalin

Surprisingly, environmentalists of various stripes managed to continue their activities in a limited fashion and with great care under Stalin. The politics of the succession struggle from Lenin to Stalin played out on the conservation movement as the state-party apparatus forced the scientists and activists of local lore studies to tie their activities to Soviet economic and political programs. In practice, this meant that any sense of local memory or homeland would be destroyed and replaced by one of feelings for the communal good – and glorious empire under Stalin. The authorities achieved this violently in several ways: first, by means of mass resettlement, mobilization, arrests, and exile, with the unspoken goal of eliminating feelings for locality and substituting feelings for nation; second, by the coercive involvement of peasants in the construction of industry and collective farms; and third, by destroying Russia's pre-revolutionary intellectual elite, who had been custodians of the historical memory of the country. Stalin and his followers had determined to destroy not only remnants of capitalism in the NEP, but also any potential seeds of civil society outside party control. They rewrote history, including the history of conservation, celebrating big projects and ridiculing small ones. Over the next years, they either precluded or subjugated to party organizations local amateur activity concerning both the environment and local lore. As part of the process, party officials dismissed many of the leaders of conservation organizations and arrested others.

[60] Vasiliy Kriuchkov, "Extreme Anthropogenic Loads and the Northern Ecosystem Condition," *Ecological Applications*, vol. 3, no. 4 (1993), pp. 622–630.

Before Stalin cemented power, a number of organizations interested in conservation, preservation, and the environment had formed. Russian and Soviet scientists were involved in community ecology research and teaching activities from the early years of the field. They considered the dynamics, structure, organization, and functioning of biotic communities or parts of them, the interactions of the constituent species among themselves and with the habitat, and methods for their investigation. Their earliest ecological studies were largely botanical, and many of the first Russian studies were conducted by foreign botanists traveling through Russia. By the 1850s, a number of accounts of Russian plant geography appeared. By the eve of the Russian Revolution, a number of animal ecology studies appeared. V. V. Alpatov may have first recognized the biotic community concept in Russia in hydrobiology (in 1923). A number of the Soviet scholars of the 1920s studied abroad, frequently in America, with grants from the Rockefeller Foundation's International Education Board. Upon his return from the United States, D. N. Kashkarov, who, along with Alpatov and V. V. Stanchinskii, was among the first to offer courses on ecology in universities, founded the *Journal of Ecology and Biocenology* that later become *Problems of Ecology and Biocenology*. Kashkarov, who published an article on Soviet zooecology in *Priroda* (Nature) in 1937, asserted that, based on an examination of the *Proceedings* of the Congresses of Zoologists in 1922, 1925, 1928, and 1930, the number of papers on ecology grew throughout the decade. Ecological topics were also clearly set forth at the national Botanical Congresses of 1926 and 1928, the All-Ukrainian Zoological Congress in 1931, and the Hydrobiological Congress in 1930.[61] By the mid-1930s, institutes of VASKhNiL (the All-Union Academy of Agricultural Sciences) and the Soviet Academy of Sciences had several programs focused on ecology.[62]

Beyond these strictly scientific achievements, as part of a brief Soviet flirtation with civic institutions independent of party control, a variety of scientific and other public actors pushed civil initiatives and established voluntary associations, amateur societies and movements aimed at nature and culture protection. They achieved major successes, published broadly, and established several important nature preserves. VOOP

---

[61] J. Richard Carpenter, "Recent Russian Work on Community Ecology," *Journal of Animal Ecology*, vol. 8, no. 2 (November 1939), pp. 354–358. On community ecology, see Weiner, *Models of Nature*, pp. 64–85.

[62] Carpenter, "Recent Russian Work on Community Ecology," pp. 358–361. Carpenter's article includes a nineteen-page bibliography of Russian articles, many in translation, on ecology.

(the All-Russian Society for the Protection of Nature) and TsBK (the Central Bureau for the Study of Local Lore) (see Chapter 1) were the most influential, expanding activities during the NEP with minimal government and party interference. Officials granted VOOP the right to engage in a wide range of activities, including dissemination of environmental information and the organization of nature protection in cities and towns. Its members believed it was crucial to attract laypeople, including the illiterate and poorly educated, to VOOP activities, where they might gain firsthand knowledge from prominent conservationists. In this way, the conservationists viewed laypeople as ignorant, with little to contribute to conservationism. In the late 1920s, the leaders of VOOP and TsBK established close working ties, in part facilitated by overlapping leadership. TsBK was much larger than VOOP in all dimensions – in numbers of members and followers, annual budget, and variety of structure of local affiliates. But the distinguishing feature was that TsBK "united those most interested in learning about their native region in all of its fullness: history, folkways, art, architecture, archeology, and natural history. In many cases, this love of region was only an aspect of love of homeland. Sometimes, in fact, the society seemed to be one of the last legal havens for the expression of an aesthetically colored patriotic sensibility."[63] VOOP's publication of *Okhrana prirody* (Nature Conservation) became a window onto Europe and beyond.

Yet strict top-down control of all institutions, organizations, and bureaucracies accompanied Stalin's rise to power. Such previously independent professional or social organizations as VOOP, TsBK, and dozens of others, from physicists to chemists, were either disbanded or subjugated to Party or ministerial groups. The authorities equated independent organizations with a variety of anti-Soviet tendencies, including technocracy, anarchism, and national separatism, or thought they were front organizations for foreign influences. Occasionally, the authorities permitted bottom-up initiatives, but only with strict ideological monitoring. But essentially civil organizations disappeared, and Soviet organizations based on mass meetings and guided by the Party, trade unions, or the Communist Youth League took their place. An independent intelligentsia as a public source of ideas and as an epistemic and politically influential community largely ceased to exist. Except for those who directly served the state and the Party organs, the intelligentsia, who had been treated

---

[63] Douglas Weiner, *Models of Nature* (Bloomington: Indiana University Press, 1988), pp. 45–46.

as apolitical specialists but "natural" materialists under Lenin, came to be seen as potential technocratic enemies and were obliged to prove their loyalty to the communist ideas. They were forced to play the roles of petitioners, supplicants, and intercessors. From the point of view of ecology, specialists had to be careful not to insist that pristine nature or wilderness were valuable to humanity in and of themselves, because this suggested distance from the needs of the proletarian for socialist reconstruction. They also had to give the sense that their research had immediate applications, not some long-term and indeterminate function. They learned these desiderata firsthand through a series of show trials of engineers in the late 1920s and early 1930s that consumed such people as Peter Palchinsky for imagined crimes against the state, and also through their own experiences.

When delegates to the First All-Russian Congress for Conservation gathered in September 1929, the Soviet Union was quite a different place than in 1917. The industrialization and collectivization campaigns were in full swing. This necessarily meant a conflict between nature as something to be preserved or studied and as a resource for economic growth. A debate at the congress unfolded that reflected these two major ideological positions. At the one extreme stood the Young Naturalists, who spoke against "the naked idea of preservation" and in favor of active utilization of natural resources in the process of the socialist construction, urbanization, and industrialization. Their opponents were old intelligentsia, to whom many people referred as "tsarist remnants," "bourgeois specialists," or by other insults. Sociologically speaking, this represented the public flowering of the latent struggle between the old, widely educated intelligentsia and the ill-educated, narrow-minded, but ambitious new Soviet "working class intelligentsia."[64]

In view of the debate, the results of the congress carried great weight, although the congress did not lead to the complete emasculation of the old specialists. The congress resolutions somehow permitted a nature protection lobby to survive whose members succeeded in saving zapovedniks (nature preserves) that were still under control of the more liberal Commissariat of the Enlightenment from the designs of industrial commissariats. The congress had enabled all Soviet activists in nature protection and local lore to meet, compare their experiences, and establish personal ties. And the congress enabled young activists to meet with the founders of local lore and nature protection movements, as well as with leading

[64] Weiner, *Models of Nature*, pp. 194–199.

figures in natural and social sciences, so that the former gained some sense of the value of conservation and preservation.[65]

Not surprisingly, conditions became more and more hostile to the environmental movement as industrialization and collectivization accelerated. Stalin strove for total transformation not only of the social and political landscapes, but also of the natural landscape in his drive to control the entire society, economy, and polity. The growing divergence between Stalin's aims and everyday policy of Soviet bureaucracy and that of the environmental movement became more pronounced. Conservation organizations were closely surveilled, policed, reorganized, or liquidated. Young, militant workers who moved rapidly upward through Party, economic, cultural, and scientific institutions of all sorts increasingly replaced the old intelligentsia, in spite of their often modest skills and understanding in fields from sociology to mathematics, and from physics and biology to ecology.

Yet VOOP managed to keep active throughout the 1930s. Its members believed that the nature reserves had greater importance under forced industrialization and with the rapid growth of cities. They were needed for research. According to Weiner, Zalygin, and others, zapovedniks became more than laboratories of scientific research; they were "islands of freedom"[66] that maintained the social and professional identity of scientists during the very difficult Stalin era. VOOP members succeeded by the end of the 1930s in gaining approval to form twenty zapovedniks, in part by using the "protective" coloring of Stalinist language to protect their offspring – the preserves. The first charter of nature preserves (1934) referred to their role as the "preservation and expansion of especially valuable genetic natural funds in the economic and scientific sense, and the development of means to enrich, improve and utilize natural resources."[67] This approach persisted over the next decades and contributed to several important, surprising successes in conservation, for

---

[65] Weiner, *Models of Nature*, pp. 200–210. See also Weiner's *A Little Corner of Freedom* (Berkeley: University of California Press, 1999), for a discussion of how the debates played out in the 1940s and 1950s.

[66] Weiner, *A Little Corner of Freedom*. See also S. P. Zalygin, "Povorot," *Novyi Mir*, no. 1 (1987). Zalygin, a hydrologist, became editor of the liberal *Novyi Mir* in the Gorbachev era and wrote a series of articles to attack such nature transformation projects as the plan to "redistribute" Siberian river water to Central Asia through massive transfer canals at great financial and environmental costs.

[67] V. V. Dezhkin, V. V. Snakin, *Zapovednoe Delo: Tolkovyi Terminologicheskii Slovar-Spravochnik s Kommentariiami* (Moscow: NIA-Priroda, 2003).

example, regarding endangered species. When the government approved the establishment of the Sikhote-Alinskii zapovednik in 1936, only 30 to 40 Siberian tigers still prowled the Russian Far East, and their numbers were dwindling. The population now numbers 300 to 400 tigers, although it took the effort of the administration of Russian President Medvedev in 2007, after the repeating urgings of local nongovernmental organizations (NGOs) and the World Wildlife Federation, to establish a 200,000-acre preservation with a huge watershed, the "Roar of the Tiger" (Zov Tigra) preserve, to ensure stability of the critically endangered 500 remaining individuals.[68]

During the 1940 and 1950s, the leaders of VOOP struggled to keep the movement alive within the limits of Stalinist political and ideological strictures. They drafted model decrees, established permanent interpersonal contacts with bureaucrats, sought to move proposals through government bureaucracies, and conducted research and published reports in the struggle to acquaint bureaucrats and Party members with the idea that the reserves as the models of nature were crucial for economic development. To achieve success in their efforts to "educate" the leadership, one-to-one contacts between VOOP leaders and top Party and state officials were of paramount importance. Moreover, such contacts served for gradual shaping of the ecological discourse. Soviet environmentalists had continued their routine activity organizing such holidays as Bird and Arbor Days, expeditions and libraries, and issuing leaflets and brochures. Even during World War II, environmentalists somehow managed to save rare mammals and birds from the zoos, organizing their evacuation to safe sites – along with the millions of people, machinery, and equipment also hastily evacuated by the authorities. They had also set up classes for growing vegetables and other efforts to assist in daily survival. Just after the war, they turned to restoration of zapovedniks destroyed during the fascist invasion, and embarked on a surprisingly successful campaign to save bison in the European Soviet Union.

Unfortunately, the state was diabolically inventive in developing tools to emasculate the environmental movement. Arrests, interrogation, and torture to compel false confessions and false testimony accompanied accusations of espionage, subversion, and slander of the Soviet Union among those, including scientists, who seemed to oppose Stalinist

---

[68] J. R. Pegg, "Efforts to Save Endangered Tigers are Working," ENS, November 21, 2002, and "Russia Establishes National Park for Endangered Siberian Tigers," ENS, June 8, 2007, earthhopenetwork.net/Russia_Establishes_National_Park.

programs. Exile from the center, from capitals and large cities to the provinces, with prohibitions against teaching, public lectures, or organizational activities; dismissal from work in reserves; the confiscation of movable and real property from the environmental organizations to local administrations and state bureaucracy – VOOP and its members lived through all of these things. These practices amounted to unbearable pressure that, like in so many other areas of Soviet life, made independent and reasonable activities nearly impossible.

How could the movement survive under such severe conditions? On the one hand, party-state control was severe but not all encompassing. Both local administrators and individuals in branches of the federal government had tried to protect reserves and support them with the understanding that the destruction of zapovedniks would lead to irreversible *economic* loses. In addition, a number of environmentalists and scientists-turned-Party-bureaucrats developed close working relations. The latter saw representatives of the environmental sciences as relatively innocuous. The environmentalists, for their part, adopted protective coloration: they promoted their own aims while rhetorically demonstrating loyalty to the regime. They presented themselves as part and parcel of Soviet scientific public opinion. But this was not yet enough. Therefore, activists used every opportunity to gain institutional protection of every kind so that they were essentially built into the system. In this way, in addition to protecting zapovedniks, as Weiner describes in *A Little Corner of Freedom*, they protected small islands of pre-revolutionary civic society. These islands, like much of the European landscape, were destroyed during World War II.

Costs of World War II

The Soviet people and country suffered miserably during World War II. At least 20 million citizens perished, and according to several estimates, as many as 30 million people died. In spite of the obvious environmental costs of war – fires, explosions, shells, haphazard and intentional disposal of chemicals, fuels, ordnance, blowing up of factories and oil fields, spraying chemicals, and so on – little has been written in a systematic fashion about these direct costs in terms of environmental history. We may get a sense of these costs by considering the size of the armies that moved across the German-Soviet front, the scale and number of tank, air, and other battles, and the nature of the Soviet evacuation effort. In a word, millions of soldiers engaged in pillaging and plunder, in scorched-earth policies and pitched battles, and they engaged in this devastatingly violent conflict in the middle of the most productive agricultural lands of the Soviet Union and in some of the centers of some of the most vital industry. Farmland and forest, rivers and lakes, and industrial and agricultural objects in those ecosystems were destroyed. (See **Box 6**.)

The Nazi armies that invaded the Soviet Union comprised "the greatest army ever assembled." They included 148 divisions, 67,000 German Norwegian garrison troops, 50,000 Finns, and 150,000 Romanians. More than 3 million men, 7,184 artillery pieces, 3,350 tanks, 2,770 aircraft, 600,000 vehicles, and 625,000 horses joined in the Blitzkrieg toward Leningrad, Minsk, Smolensk and Moscow, central Ukraine, and the Caucasus. More than 2 million Soviet soldiers in 132 divisions, including 34 armored divisions with 10,000 tanks and 2,300 planes, opposed them. The invasion on June 22, 1941, caught no one but Stalin by surprise. Before the end of the third week of the attack, "300,000 Russians had been captured, 2,500 tanks, 1,400 artillery guns and 250 aircraft captured or destroyed." Hitler's armies moved so rapidly through the countryside that they in fact compromised their own supply and communication lines by the early fall. This led the German armies to madly strip the countryside of anything that remained behind in the wake of Stalin's scorched-earth policy.[69]

Stalin and the military leaders were determined to prevent any resources from falling into Nazi hands. They ordered the evacuation of all resources: industrial and military, rolling stock, engines and wagons, fuel and grain, farm animals, and any other raw materials. What they could not withdraw, they destroyed with explosives or fire, including buildings, public records, monuments, homes, dams, locks, and factories. For example, of the original power-generation capacity of roughly 2.5 million kilowatts in the occupied eastern territories, or approximately one-fourth of total prewar Soviet generating capacity, including the Dniepr hydroelectric power station, the largest in Europe at that time, less than 300,000 kilowatts remained intact. Soviet demolition was so effective that it prevented an increase of capacity to 630,000 kilowatts until more than two years later. Basic industry – coal, iron ore, steel, electricity, and cement – was for all practical purposes totally destroyed. Compared with prewar

---

[69] http://www.worldwar2database.com/html/barbarossa.htm; and http://www.infoukes.com/history/ww2/.

> **Box 6. War and the Environment**
>
> Twentieth-century wars have had a growing impact on the environment because of the scale of battles and the nature of weapons. Bombardment from airplanes and tanks, chemical weapons, and nuclear weapons all have had a significant impact on the environment. Chemical weapons lead to despoliation, defoliation, and toxic pollution.[70] In addition, the determination to see civilian casualties as unfortunate collateral damage, if not real targets, as the aerial bombardment of Tokyo, Hamburg, Dresden, Hiroshima, Nagasaki, and many other cities shows, innocent civilians will die in modern battles. The bombardment of the urban infrastructure led to the displacement of millions of people through Europe and the Soviet Union, including Warsaw, Berlin, Hamburg, Dresden, Dusseldorf, Boulogne, Le Havre, Rouen, Budapest, Leningrad, Stalingrad, Kiev, and Krakow.
>
> During World War II, millions of civilians died; their cities became rubble and ash; and the industrial, chemical, and other materials left behind poisoned the environment. Armed forces targeted forests and other ecosystems in order to deprive enemy troops of cover, shelter, and food. On top of this, mass dislocations and other disruptions caused by armed conflict depleted nearby sources of timber and wildlife. People stripped the cities and countryside of anything they could use, including all animals. By the end of World War II, nearly 50 million people were refugees, and in many places, housing stock disappeared. In the Soviet Union alone, 25 million people had no shelter. In Japan, things were no better, with sixty-six cities suffering major damage, and with food shortages, hunger, and malnutrition after the war.[71]
>
> Like elsewhere in World War II, the invasion of the Soviet Union led to the destruction of forests, farms, transport systems, and irrigation

---

[70] See Edmund Russell, *War and Nature* (Cambridge: Cambridge University Press, 2001) for a discussion of the use of chemicals on insects and humans from War War I until the publication of Rachel Carson's *Silent Spring* in 1962.

[71] Jennifer Leaning, "Environment and Health: 5. Impact of War," *Canadian Medical Association Journal*, vol. 163, no. 9 (October 2000), pp. 1157–1161; Walter Laquer, *Europe since Hitler* (New York: Penguin Books, 1984), pp. 15–16; and J. W. Dower, *Embracing Defeat: Japan in the wake of World War II* (New York: W.W. Norton, 1999), pp. 33–120, both as cited in Leaning. See also http://www.pollutionissues.com/Ve-Z/War.html#ixzzou9UcVYoZ.

systems. Dams were destroyed; levees and embankments were dynamited; and ports "were clogged with unexploded ordnance and sunken ships." A lunar landscape remained. Dead bodies and dead animals decomposed in the open sun. Mines and other unexploded ordnance remained a threat for years. According to Jennifer Leaning, land mines accelerate environmental damage through fear of mines; this fear leads them to avoid mined-land with abundant natural resources and arable land and to move into marginal areas to avoid them, which speeds depletion of biodiversity.[72]

Military preparations before war also lead to profligate use and destruction of land. Military bases require large parcels of land for troops, military exercises, tank maneuvers, bombing, missile tests, and so on. The operation of military bases tends to lead to the destruction of flora and fauna that interfere with operations – for example, birds congregating on runways. Before Soviet troops withdrew from former East Germany in 1992, they destroyed 1.5 million tons of ammunition. According to some reports, the ammunition was burned, without filters, in the open air, and nitrogen oxides, highly toxic chemical dioxides, and heavy metals were released to the atmosphere. Millions of gallons of spent tank and truck oil, and chemical wastes were also left behind with the result of severe pollution.[73]

levels, coal mining averaged 2.4 percent, iron ore production 1.2 percent, crude steel production nothing, electricity 8.8 percent, and cement production 11.6 percent.

According to one source, the Soviet scorched-earth policy included the deportation of millions of men, women, and children. The millions of civilians left behind were both at the mercy of the Nazis and also essentially starved by this policy because they had no food. Before the Nazi invasion, Polish, Ukrainian, and Jewish people along the German-Russian frontier were forced from the borderlands, many of them deported to Siberia. The Soviets destroyed primarily the MTS. As the war burned through the countryside from west to east, almost no cattle or other farm

---

[72] Leaning, "Environment and Health: 5. Impact of War."
[73] http://geocompendium.grid.unep.ch/reference_scheme/final_version/GEO/Geo-1-014.htm.

animals, grains, gasoline, or any other supplies remained. The Soviets busily harvested the fields as they retreated. According to some estimates, the so-called Occupied Eastern Territories produced 43 million tons of grain under Soviet rule in 1940. Under German administration, the harvest in 1941 was perhaps 13 million tons and dropped to 11.7 million tons in 1942.

By November 1941, the German army had occupied territory in which 40 percent of Soviet citizens lived and in which 63 percent of coal, 68 percent of iron, 58 percent of steel, 60 percent of aluminum, 38 percent of grain, and 84 percent of sugar was produced; and where 38 percent of cattle, 60 percent of pigs; and 41 percent of railroads were located. Agricultural production fell by 40 percent.[74] A total of 12 to 13 million people were evacuated to the east, while the Germans enslaved or imprisoned 8.7 million people. Almost 32,000 industrial enterprises were destroyed, although the Soviets succeeded in evacuating some 1,500 key factories to the east, in particular to the Ural, Siberia, Central Asia, and Volga regions.[75]

Another measure of the cost to the land and people of the Soviet Union concerns the Kursk tank battle, the largest battle in its history. For its attack on the Kursk salient, the German military leadership gathered 900,000 soldiers, 10,000 pieces of artillery, 2,700 tanks, and 2,000 aircraft, or about one-third of Germany's remaining military strength. Red Army leaders focused 1.3 million soldiers, 20,000 artillery pieces, 3,600 tanks, and 2,400 planes on the salient. They established networks of anti-tank artillery guns. The soldiers dug trenches and other anti-tank traps and laid 400,000 mines. This equaled 2,400 anti-tank and 2,700 anti-personnel mines every mile – more than at the Battle of Moscow and the Battle of Stalingrad. By June 1943, 300,000 civilians were helping the Russians build defenses around the Kursk salient. What remained of the farmland and forests in the region of the battle were gouged-out craters and huge holes; rivers and streams filled with rock, soil, military material, and bodies.

By the end of the war, one-quarter of the prewar capital stock had been destroyed – in occupied areas, as much as two-thirds of it. Oil, gas, chemicals, and wastes had been burned or dispersed far and wide.

---

[74] N. A. Voznesenskii, *The Economy of the USSR During World War II* (Washington, Public Affairs, 1998).

[75] Mark Harrison, *Accounting for War* (New York: Cambridge University Press, 1996); Denis Brand, *L'Experience Sovietique et sa Remise en Cause* (Paris: Sirey, 1993).

Postwar reconstruction consumed immense resources but was focused on industry, and the fourth Five-Year Plan (1946–1950) that was launched in March 1946 included insufficient funding for cleanup, reclamation, and so on. The first step was resurrection of industry in lands cleansed of the occupiers, in part with help of lend-lease machinery and equipment; perhaps 40 percent of the amount indicated for capital investment (66 billion rubles) went to industrial equipment, with reparations from Germany – and outright theft or removal of capital resources – covering fully the balance.[76] The Soviets carted off two-thirds of the Germany aviation and electronics industries and a major part of rocketry, automobile, construction, military, and other factories.[77]

Nazis soldiers attacked not only nature but also singled out Jews, gypsies, and Slavs for enslavement and genocide. At least 160 Nazi concentration camps, some holding tens of thousands of prisoners, were established in Ukraine alone. The Germans had intended the Eastern Occupied Territories as *Lebensraum* (living space) in which to expand, with German settlers using Slavic populations as slave labor in extensive agricultural settlements that served Germany. According to Dallin, the German administration succeeded in seeding no more than three-quarters of the prewar acreage. Fertilizer was practically unavailable. Yields fell. Compared to the average yields per hectare of approximately 2,200 pounds (fourteen bushels per acre) in the Ukraine in the late 1930s, the Germans managed to obtain just 1,500 pounds (ten bushels per acre) in 1942. Furthermore, the Soviet scorched-earth policy began to show its full effects: the use of seed grains to relieve the worst hunger in the cities, increasing partisan menace, and dearth of personnel and machinery reduced the harvest potential drastically.[78]

Many Ukrainians were also sent to Auschwitz and other death camps in Poland. Ukraine lost more people in World War II than any other European country. According to Pawliczko, "At the end of the war, Ukraine lay in ruins: the population had declined by 25 percent – that is by approximately 10.5 million people; 6.8 million had been killed or died of hunger or disease, and the remainder had been evacuated or deported

---

[76] G. Khanin, *Dinamika Ekonomicheskogo Razvitiia SSSR* (Novosibirsk: Nauka, 1993).
[77] Sutton, *Western Technology and Soviet Economic Development*.
[78] Alexander Dallin, *German Rule in Russia, 1941–1945* (London: MacMillan, 1981). See also Walter Sanning, "Soviet Scorched-Earth Warfare: Facts and Consequences," Paper Presented to the Sixth International Revisionist Conference, http://www.ihr.org/jhr/v06/v06p-91_Sanning.html, accessed September 30, 2009.

to Soviet Asia as political prisoners or had ended up as slave laborers or emigres in Hitler's Germany."[79]

During World War II, the authorities had accelerated the effort to control the Far North – the Arctic regions – as well. The Red Army moved in on those reindeer hunters who had moved north to escape them. They confiscated the deer, depriving the hunters of any means of subsistence, and forced many hunters to join reindeer army units – as an exhibition at a local museum in Lovozero in the Kola Peninsula indicates. Famine resulted, forcing the Komi, Saami, Nenets, and other herders to relocate to villages in the south. The ratio of private to collectively owned reindeer fell from 6 to 1 to 1 to 2, and herd size declined from 143,000 in 1942 to 60,000 in 1949. Literacy remained only 6 percent in 1945; even harsh laws setting fines and punishment brought fewer than half of the children to attend schools. This marked the beginning of a forty-year period during which the authorities attempted several times to reorganize reindeer herders in state and collective farms and fishing plants. But the indigenous people found it difficult to live under Soviet institutions. Stalin's death was crucial to relieving the crisis over the short term, for the state relaxed its attack on the herders. For example, the number of reindeer on Yamal increased by fifty percent in the 1950s and doubled by 1980.[80]

The leaders after Stalin were willing to admit that organizing nomadic reindeer hunters into state farms and sending political prisoners and minority nationalities into the wilderness did nothing to enable the systematic development of the resources of the periphery. They had tried to force indigenous people to become modern and Soviet. They had destroyed their shamanistic practices and other ways of life. They managed to establish huge collective reindeer operations and to increase the number of head manyfold. But the herds also became subject to epidemics of disease, and various industrial and agricultural operations destroyed the fragile tundra, especially the oil, gas, coal, and heavy metal mining operations.

---

[79] Ann Lencyk Pawliczko, *Ukraine and Ukrainians Throughout the World* (Toronto: University of Toronto Press, 1994), p. 62; and Bohdan Wytwycky, *The Other Holocaust: Many Circles of Hell* (Washington, D.C.: Novak Report on the New Ethnicity, 1980), p. 59. The commandant of the notorious Janowska Camp in Lviv, SS officer Gustav Wilhaus, frequently sat on his balcony and used an automatic rifle to pick off inmates in the yard as target practice, even shooting children. The camp had an orchestra and played a tune titled "The Tango of Death" specially composed for it. The orchestra members were murdered when the Germans closed the camp during their retreat.

[80] Andrei V. Golovnev and Gail Osherenko, *Siberian Survival: The Nenets and Their Story* (Ithaca: Cornell University Press, 1999), pp. 90–94.

High Stalinism and the Transformation of Nature

Drought plagued the main agriculture zone in 1946. The drought occupied Ukraine, the Central Black Earth region, and the lower Volga and Don River basins. More than 50 percent of the sown area of the Soviet Union was affected. If the weather conditions were not entirely catastrophic, then the general destruction of Soviet agriculture was the other major cause of the crop failure. As in 1932–1933, the authorities in fact provoked the famine by establishing excessive grain procurement in regions with relatively good harvests – Siberia, the Middle Volga, and Kazakhstan. V. F. Zima estimates that the total number of people who died from the famine reached 1 million. He estimates that 4 million people suffered such famine-related diseases as dysentery and pneumonia, a half million of whom died between 1946–1948 in the last mass famine in the Soviet Union.[81]

Stalin was determined to rebuild the nation in short order and prevent any future drought or famine. This led to a massive reclamation and melioration effort with a fantastic component: the determination to rebuild nature itself. On October 20, 1948, the Communist Party voted unanimously to approve a "Plan for Shelter Belt Plantings, Grass Crop Rotation, and the Construction of Ponds and Reservoirs to Secure High-Yield and Stable Harvests in Steppe and Forest-Steppe Regions of the European Part of the USSR." Ultimately more symbolic than real, the so-called Stalinist Plan for the Transformation of Nature represented prevailing attitudes toward nature and the power of the Bolsheviks to shape it. The audacious plan intended no less than the entire reworking of the European Soviet Union into a vast machine. The plan did not address potential costs of the "geoengineering" of nature, but assumed that the benefits far exceeded any minor costs to ecosystems. It must be mentioned that, in many ways, the plan resembled those advanced by the Bureau of Reclamation and the Army Corps of Engineers in the United States to shape the American West, America's rivers, harbors, and forests, and transform "wasted" land and water into goods and services of benefit to society. The Stalinist Plan had three basic components: the sculpting of rivers to serve industry, agriculture, and cities through hydroelectricity,

---

[81] V. F. Zima, *Golod v SSSR, 1946–47: Proiskhozhdenie i Posledstviia* (Lewiston, NY: Edwin Mellen Press, 1999); Michael Ellman, "The 1947 Soviet Famine and the Entitlement Approach to Famines," *Cambridge Journal of Economics*, vol. 24 (2000), pp. 603–630.

municipal supply, transport, and irrigation; the planting of forest defense belts to protect farmlands from draught and hot dry winds; and the building of ancillary transportation nexuses, including roads, railroads, and dams. In each aspect, gulag construction enterprises played a major role.

Regarding rivers, such large-scale construction organizations as Kuibyshevgesstroi (the Kuibyshev Hydroelectric Construction Trust) were enlisted to build a series of stepped reservoirs and dams to store water for a variety of uses along the Volga, Don, Dniepr, and other major European rivers. Virtually all major rivers were tamed through these projects, at great cost to surrounding agricultural regions, especially because of inundation of productive farmland and the destruction of anadromous fisheries, especially sturgeon on the Volga River. Thirteen major dams were built on the Volga River alone. The Volga, Dnepr, Amu Darya, and Don Rivers would be harnessed for irrigation to serve primarily wheat grains, and to a lesser degree vegetable and fruit farming, cotton, and animal husbandry. The Kuibyshev and Stalingrad and other hydropower stations provided the electricity to pump the water to irrigated agriculture and forest defense belts.

In the upper Volga River basin, annual average rainfall is 300 to 400 millimeters, but in the lower Volga it is often half that high. Hot dry winds raise air temperature to 95 degrees and push the humidity as low as 15 percent. The plan for shelter belts centered on the determination to build (that is, plant) massive forest defense belts as windbreaks nationwide. This project made use of scientific methods, advanced technology, and modem machinery and equipment. But at the same time, it led to increased erosion and essentially destroyed natural steppe. The main target of the plan was to combat droughts in the steppe and wooded steppe zones of European Russia by planting giant shelter belts and constructing large irrigation channels and water reservoirs immodestly over some three or four years. More than 5,000 kilometers of shelter belts were to be planted in the southern part of the country. In fact, from 1948 to 1953, the number of trees planted exceeded that of those planted during the previous 250 years of forestry history.

Each belt itself consisted of two to six belts, 30 to 100 meters wide, standing at a distance from one another of 200 to 300 meters. Planners acknowledged that irrigation and forest belts would affect the climates of a significant part of the country. This did not trouble them; they had studied local impacts and were comfortable making substantial changes in plant cover, soil composition, and moisture level through melioration, irrigation, and planting. They also recognized that "[o]nly under Soviet power under the leadership of the great Stalin" was the implementation of a great plan to transform nature possible. One popularizer of Stalinist gigantomania noted, "Here there will be gardens, vineyards, melon fields, and plantations of technical culture."[82] By 1965, they built eight forest belts on the shores of the Volga, Ural, Don, northern Donets, and other rivers, each from 170 to 1,100 kilometers in length, with a total length of 5,320 kilometers stretching more than 118,000 hectares.

Soviet Forests and the Stalinist Plan to Transform Nature

What was the backdrop to Stalinist plans to plant millions of acres of forest? As noted earlier, in Europe in previous centuries, and increasingly in the United States in the nineteenth century, population pressures, industry, and other factors led to steady encroachment on Russia's vast forests from the late nineteenth century onward; indeed, clear cutting near existing roads became the essential approach of the forestry industry. In the last decade of tsarist power, a movement in favor of protection emerged among foresters, who saw the solution in "the establishment of a single, rationally planned national forest economy, utilizing scientific methodologies and technical knowledge, managed in a consensual association between center and localities... and based on state ownership."[83] Foresters initially saw the collapse of the tsarist regime in February 1917 as the opportunity to realize their plans to introduce modern forestry practices based on universally recognized principles of conservation. The collapse of authority after the revolution resulted, however, in an orgy of tree felling by peasants that continued into the 1920s.

The Bolsheviks responded with a series of laws to secure the forests for economic reasons and protect them from poaching. Yet, their approach was "more facultative than creative" with regard to forest legislation, that is, facilitating, not requiring, fundamental changes. The Basic Law on Forests of 1918 nationalized forests, centralized administration, put scientific planning front and center, and began to establish Bolshevik control over technical experts. The implementation of the law was hampered by the chaos of the civil war of 1918–1921, shortages of resources,

---

[82] L. Ognev and P. Serebriannikov, eds., *Velikie Sooruzheniia Stalinskoi Epokhi* (Moscow: Molodaia gvardia, 1951), pp. 119–120, 206–207.

[83] Brian Bonhomme, *Forests, Peasants, and Revolutionaries: Forest Conservation and Organization in Soviet Russia, 1917–1929* (Boulder, CO: East European Monographs, 2005).

bureaucratic problems, and the continued wide-scale uncontrolled felling. A 1923 "Forest Code" separated nationally planned state forests from locally planned and managed peasant forests. This was a concession to the peasants in line with the NEP of 1921 and the fact that peasants had indeed seized forests after the revolution. On paper, if not in practice, both the 1918 law and the 1923 code sought long-term sustainability by seeking a balance between exploitation and conservation. The two laws differed, however, on the relative balance between central and local interests. The implementation of a long-term plan for the forest industry of the Russian Federation from 1925 to 1928 allowed the industry to increase production to the 1913 level. However, the growth of the industry and the growth of the population increased demand for timber even more rapidly. Consequently, enterprises and managers chose to ignore forestry principles. Bonhomme notes that in 1929, when Stalin's industrialization drive put the emphasis firmly on exploitation rather than on conservation of natural resources, these laws lost any strength.[84] In general, the rule for the Stalin period was that wherever laws existed to protect resources, pressures to meet production targets always led to the laws being pushed aside or ignored.

The authorities and scientists commenced discussions in the early 1930s about forest use and related issues of forest management, forestry, and forest regulation. But principles of conservation turned out in essence to be unacceptable to economic interests, because they contradicted socialist forest economics that saw more rapacious harvest as acceptable. The essence of conservation, harvest rotation, was replaced by age harvesting; this led lumber organizations to clear entire forests. Forest regulation was practically abolished as it was replaced by perfunctory inspection and inventory carried out according to agency directives devoid of any scientific basis. This followed the politburo's decision to transfer authority for the nation's forests from the People's Commissariat of Agriculture to the industrially oriented Supreme Soviet of the Economy (VSNKh). When the damaging results of this decision became clear, the Soviet government put a quick end to this arrangement; less than a year after the decision, the Soviet Union's forests were divided into two zones, a water protection zone and the forest industrial zone, and commercial logging in the former was prohibited.[85]

---

[84] Bonhomme, *Forests, Peasants and Revolutionaries*.
[85] Stephen Brain, "Stalin's Environmentalism," *Russian Review*, vol. 69, no. 1 (January 2010).

This did not occur. Industrial enterprises still pursued forest where accessible and exploited it heavily. M. G. Zdorik, a contemporary observer, described the purpose of the forest in the late 1930s in this way: "The forest, as an object of nature, should serve the goal of building socialism. For this to happen, economic activity had to be organized in the forests. The forest industry, as one branch of the state economy, had to be in complete compliance with the goals of state economic policy. Therefore, any breach of this policy or attempt to preserve the old forms of management was viewed as bourgeois and reactionary." The People's Commissar for Forest Industry, Z. Zh. Lobov, shared this view and its rejection of conservation as anti-Soviet in 1932, stating that "it is imperative to decisively expose opportunistic, kulak-capitalist, damaging theories and practices that stem from the 'principle of sustainability' in forest use, which until recently were ensconced in forest management and science. The main principle of forest exploitation must be concentrated clear cutting."[86]

Yet, in the postwar years, conservationists had convinced officials, including Stalin, of the need to rebuild Russia's forests using methods of conservation. As Stephen Brain points out, the Stalinist Plan for the Transformation of Nature was a bold attempt to reverse human-induced climate change by creating nearly 6 million hectares of new forest in southern Russia in eight massive shelter belts so as to cool and moisten the climate. The plan, like many big projects in Soviet history, had tsarist roots, as we recall Dokuchaev's desires to rework nature to prevent famine such as that in 1891 (see Chapter 1). Protective forest belts may be fantastic at first glance, but are not an anomaly, having been proposed and introduced in the United States, Russia, China, and elsewhere in 1941, with 30 million trees distributed in Wisconsin alone for shelter belts.[87] Like others elsewhere, Soviet conservationists sought to fight draught and hot dry winds with the shelter belts. The promulgation of the plan heightened an ongoing debate among those who advanced conservationist

---

[86] V. K. Teplyakov et al., *A History of Russian Forestry and Its Leaders* (Pullman, WA: Department of Natural Resource Sciences and Cooperative Extension, Washington State University, 1998), p. 7.
[87] Raleigh Barlowe, "Forest Policy in Wisconsin," *Wisconsin Magazine of History*, vol. 26, no. 3 (March 1943), pp. 261–279; Chunyang Li, Jarkko Koskela and Olavi Luukkanen, "Protective Forest Systems in China: Current Status, Problems and Perspectives," *Ambio*, vol. 28, no. 4 (June 1999), pp. 341–345; and Dey Ber Krimgold, "USSR: Conservation Plan for the Steppe and Timber-Steppe Regions," *Land Economics*, vol. 25, no. 4 (November 1949), pp. 336–346.

ideas about how to manage Soviet forests and those utopians who believed nature could be remade in a socialist fashion. The conservationists, whom Brain calls "technocrats," pushed the notion of sustainable yield. Stalin seems to have endorsed their view of the need to reorganize forests into protected and industrial zones.[88] In a word, there was forest conservation under Stalin.

To ensure sustainable harvests, quotas for felling should not exceed annual reproductive growth of forest in a given region. Specialists proposed logging according to this formula in the late 1940s. The authorities adopted it as a theoretical basis for the establishment of felling targets in the early 1950s. The plan was to tie every lumber, paper, and pulp enterprise to a specific area of forest entitled to harvest timber in an amount equal to annual growth of forests on the entire area. In addition, in 1943, the forestry industry determined according to "Principals of Forest Regulation" three categories, or regimes, for the utilization of resources that established in effect three groups or kinds of forests that gave some indication of the effort to conserve them. The first group included forest reserves, water and soil protection zones, and urban parks. These forests were excluded from industrial felling, and only selective felling of mature and sick trees was permitted. These forests comprised about 5 percent of the total forest area. The second group of forests was located in densely populated industrial regions with relatively limited reserves, for example, in the Central and Black Earth regions. There, industrial felling was permitted only if it did not exceed rather limited annual quotas. These forests comprised about 9 percent of the total forest area. The third group of forests had exclusively industrial significance. They were located in the northern regions of the European part of the country (e.g., Arkhangelsk province), the Urals, and Siberia. These forests were to be felled according to plan targets, although belatedly felling quotas were introduced to slow the clear-cutting activities of lumber enterprises.

Trofim Lysenko, a quack biologist with neo-Lamarckian views of inheritance, advanced other ideas for the transformation of nature with the forest defense belts. He believed he could transform the natural environment, and the flora and fauna in it, to the betterment of agriculture. In 1948, after the leading agronomical institution, VASKhNiL, declared that Lysenko and Lysenkoism were supreme in biology and that genetics

---

[88] Stephen Brain, *Song of the Forest: Russian Forestry and Stalin's Environmentalism, 1905–1953* (Pittsburgh: University of Pittsburgh Press, 2011), chap. 6, "Transformation."

was a "bourgeois pseudo-science," the political climate for expressing any environmental concerns based on Darwinian notions grew untenable in many fields. Many officials – and increasingly scientists who fell under the spell of Lysenko – saw nature solely as an economic resource. They also believed firmly in the ability of the Bolsheviks – and Bolshevik scientists – to alter nature for the better, and in so doing to alter plants and animals to serve socialism. They rejected genetics because of their impatience. The plant hybridization specialist Ivan Michurin wrote, "We cannot wait for favors from Nature; we must wrest them from her."[89]

Toward the end of defense belts, the Academy of Sciences established a supervisory commission under Vladimir Nikolaevich Sukachev. Sukachev was the editor of *Botanical Journal* and an opponent of Lysenko, if not of the transformation of nature by other, scientific means. Unfortunately, against Sukachev's recommendations based on analysis of soil-hydrological, silvaculture, and other experimental data, a decision was made to cluster plant oaks for the belts. Lysenko advanced another quack notion of a kind of "floral collectivism," according to which planting trees in clusters helped protect plants against enemies. Lysenko ludicrously proposed that trees (oak, for example) be planted in very compact groups, which he called the "nest" method of planting. Unlike Darwin's views of competition, Lysenko's view held that members of the same species helped each other to fight weeds if planted in clusters or nests and he also urged planting of winter wheat, oats, potatoes, alfalfa, and so on alongside acorns to join in the battle with weeds – which, perhaps symbolically, stood for other enemies of Soviet society. According to his concept, plants belonging to one species did not compete with each other but even collaborated – like workers against the bourgeoisie? – against weeds and other common enemies. He and his followers carried out a violent attack against Sukachev and the commission in such journals as *Socialist Geography*, *Forest Industry*, and *Forest and Steppe*.[90] Because of the urgings of Lysenko to adopt the collectivist planting technique, by the autumn of 1956 only 4.3 percent of the trees were healthy. The failure of the plan cost an enormous amount of money, and after 1953 no mention of the plan was found in Soviet media.[91] The belts failed, and

---

[89] I. V. Michurin, "Results of My Sixty Years' Work and Prospects for the Future," http://www.marxists.org/reference/archive/michurin/works/1930s/results.htm.
[90] S. V. Sonn, *Vladimir Nikolaevich Sukachev* (Moscow: Nauka, 1987).
[91] A. A. Lubishev, *V Zashitu Nauki* (Leningrad: Nauka, 1991).

planners and silvaculturalists resorted to less ambitious, but ultimately as damaging forestry techniques of clear cutting, pesticides, herbicides, and brute-force technology.

Lysenko's success at this time was tied to three factors. First, the party promulgated slogans that touted the conquest of nature. Second, mass mobilization was required for such large-scale projects as forest defense belts and hydroelectric power stations. Armies of workers in mass construction trusts but often with rudimentary equipment were deployed to build socialist nature. This struggle with nature somehow resembled class struggle and might take years, the authorities acknowledged, but would ultimately be successful. Third, Lysenko was of peasant origin and in no way a "tsarist remnant."

According to Douglas Weiner, the goal of increasing cereal crops in those years – when the postwar famine that killed another 1 million people still had not abated – had not only practical but also symbolic importance, for it signaled the attack on anarchic nature and its replacement by planned, socialist nature. If bourgeois specialists worried about natural disasters resulting from human interference and disasters, Soviet biologists would defend grain from predators with trees or something else. (See **Box 7**.) Stalinist climate change would increase the harvest in the steppe and forest-steppe regions. Planners had their eyes on 14 million hectares of the Sarpinskaia and Norgaiskaia steppe, land with rich soil but low rainfall in the lower Volga and northern Caspian regions, as well as other land in Ukraine, Kazakhstan, the North Caucasus, and some land west of the Ural Mountains. (The Norgaiskaia steppe, at 6.5 million hectares, had more land than the Netherlands and Belgium together.)

Of course, Lysenko's methods robbed seedlings of water and, without constant thinning and weeding, let the oak saplings die out. Furthermore, heavy reliance on collective farm labor to plant and tend the belts doomed the project. The peasants were poorly motivated and did not fully understand the projects, and the local authorities did not give resources to the project because they have other things to do. The battle between Lysenko's followers and the conservationists played out until Stalin died, when, because of great costs and for other reasons, the forest belt project withered under the hot winds of de-Stalinization, with professional foresters having successfully built an arsenal of field research data to show the foolhardiness of Lysenko's ways.[92]

---

[92] Brain, *Song of the Forest*.

> **Box 7. William's Theory of Grass Rotation as a Panacea to Prevent Crop Failure**
>
> From the late 1930s, the soil scientist V. R. Villiams was the leading supporter of a grass rotation system in Soviet agriculture. The system was based on an idea that permanent rotation of crops on a given plot would restore soil quality. The main emphasis in the system was to sow perennial grass. A special network of MTS was established to implement the grass rotation system throughout the Soviet Union. This concept of grass rotation was not new or representative of the glories of the Soviet philosophy of science, dialectical materialism, as its supporters claimed, since most Western countries had used similar practices since the nineteenth century. Villiams's concept was dogmatic and put too many hope in improvement in soil structure due to grass rotation. The grass rotation proved to be effective in the wet forest zone of Russia but provided little advantage in dry steppe and forest steppe zones. However, Villiams proclaimed that the climatic (relatively dry) conditions of Russia did not rule out grain yields as high as 100 centners (10 tons) per hectare if soil structure could be improved radically by his grass rotation system. He also offered theoretical reasons for opposing planting of winter wheat. To secure his concept as a national priority, Villiams wrote in *Pravda* in March 7, 1937, that the very idea of the grass rotation system came to him during his night's dream after reading Lenin's major philosophical work, *Materialism and Empiriocriticism*, which, of course, has nothing to do with agriculture.[93]

Another Stalinist project of these "transformationist years" was the construction of a railroad in western Siberia (1947–1953). This uncompleted railway was one of the most shameful projects of the Stalin era, involving death, dislocation, and environmental degradation. The scheme was dropped after Stalin's death in 1953. By that time, fewer than 600 kilometers were in working operation, even though up to 300,000 persons had been involved and about a third of them had perished, while more than 40 billion rubles of capital investment had been wasted. Ghostly labor camps, rusting rolling stock and rails, and hundreds of rickety

---

[93] I. K. Alexandrov, "Snizhenie Energoemkosti Elektrifizirovannyx Razdatchikov," *Mekhanizatsiia i Elektrofikatsiia Sel'skogo Khoziaistvo*, 1982, no. 3.

bridges remain in what has been called "an open air museum of human technology" preserved by nature's refrigerator – the tundra.

The Stalinist plan led officials to increase pressures against members of VOOP for their quiet opposition to the audacious determination to transform nature. Party investigative commissions turned their microscopes on VOOP. They concluded that VOOP publications were dangerously "apolitical" and failed to meet the needs of society, whereas its members were accused of being "bourgeois academics" at a time when such an accusation could lead to loss of a job or worse. The main accusation was that VOOP did nothing to promote the Stalinist plan. Many Party officials and conservationists-turned-bureaucrats argued that Soviet society did not need any middleman like VOOP, but rather that nature must be made to serve state interests directly. As a result, in the last eight years of Stalin (1945–1953), only two zapovedniks were founded, yet both of them, shockingly, in 1945, in the glow of the war.

Conquest of Siberia and the Far North and the Rise of the Modern Defense Industry

World War II waylaid Siberian development. The Soviet leadership had to focus on resources extraction in the struggle to survive. Yet the war required the authorities to order industry and research institutes evacuated to Kazakhstan, the Ural Mountains, and Siberia, a policy that ensured postwar Siberian development, especially so that industry and resources would be distant from the front in the next war with capitalism that Marxist leaders claimed was inevitable in the struggle of two world systems. The authorities recognized the need for a more systematic approach to resource exploitation, because they did not know the extent of the resources available in the empire, nor did they comprehend how to power new industry or provide labor to run the factories. For example, during the war years, Soviet planners and engineers were forced hurriedly to build forty new hydroelectric power stations to power the relocated armaments industry. The speed with which they built them suggested that they could build power stations and factories anywhere quickly and could get away with cursory examination of local geological and climatic conditions. They commenced study of Siberian potential at this time, in particular on the Ob and Angara Rivers. All of this meant that after the war, Siberia and the Far North would come into the direct attention of the planner and builder, and that the environment would suffer the consequences.

In 1947, the first postwar conference on the development of the productive forces of East Siberia was held in Irkutsk, a large Siberian town on the Angara River, seventy kilometers downstream from Lake Baikal. (As discussed in later chapters, Lake Baikal has a special place in this saga of Soviet environmental history.) In 1952, the Committee on the Study of the Productive Forces, the Krzhizhanovskii Energy Institute, and the Ministry of Electrical Power Stations began planning for a series of hydropower stations throughout the region. The Lengidroenergoproekt Engineering and the Zhuk Gidroproekt Institutes, both of which were connected with gulag construction operations, joined the planning effort. The goal was to transform the Amur, Enisei, Angara, Ob, and Irtysh Rivers into planned, rational Soviet waterways – in the same way that European rivers were now planned Soviet rivers. In an unbounded dream of hydroelectricity, one specialist noted that the Amur alone had greater hydropotential than the entire Volga basin at 45,000 megawatts, and suggested plans to build twenty-five stations at a total of 7.6 megawatts.[94] This systematic study led to concrete project proposals and extensive construction after the twentieth Communist Party Congress in 1956, when Khrushchev put Siberian industrial and agricultural on the public agenda. A major conference on "the development of the productive forces in East Siberia" followed in August 1958.[95]

Planners and engineers in Moscow focused their attention on the vast mineral, fossil fuel, water, and forest resources of Siberia. Beyond the general motivation to secure the Soviet Union's long-term economic future well beyond the reach of any future invaders, they had two specific goals. One was the taming of each river through the construction of a series of hydropower stations to power extractive industries – oil, gas, coal, rare metals, and iron – and to facilitate logging. The second was securing water for Central Asian agriculture through a series of grandiose diversion canals. So, they prepared to build massive hydroelectric power stations on the Ob, Irtysh, Angara, and Enisei Rivers,[96] while at the same time identifying the rich oil and gas reserves of Tiumen province in northwest Siberia and preparing to harvest coal in the Kuzbass. Siberian coal was crucial to the future because one-third of Donbass reserves

---

[94] S. V. Klopov, *Gidroenergeticheskie Resursy Basseina Amura* (Blagoveshchensk: Amurskoe knizhnoe izdatel'stvo, 1958).

[95] I. P. Bardin et al., *Razvitie Proizvoditel'nykh Sil Vostochnoi Sibiri* (Moscow: Academy of Sciences, 1960), 13 vols.

[96] A. N. Voznesenskii, ed., *Gidroenergeticheskie Resursy SSSR* (Moscow: Nauka, 1967), p. 16.

had been exhausted and the rest were of poor quality and hard to extract.⁹⁷

The development of Siberian energy resources brought into full relief a significant issue for long-term investment policy in the nation. The vast majority of industry and population remained in the European Soviet Union, whereas energy resources on which to base future industrial growth and consumer well-being were thousands of kilometers away. The cost of transporting them in their primary form in railroad coal cars or pipelines grew rapidly. A total of 30 to 70 percent of all freight transported in the Soviet Union was fossil fuel. One alternative, to build power-generating stations near fuel sources linked by power lines to the European energy grid, was also exceedingly costly, and year by year open spaces were filled with unsightly towers carrying power lines more than 900,000 kilometers in total length. As in Brazil, the major problem was the concentration of population and industrial centers far from this great natural wealth.

The tasks at hand were overwhelming, for Siberia stretched from the Ural Mountains to the Pacific Ocean and to the Arctic Circle from central Kazakhstan, Mongolia, and China. This was a land mass of 12,800,000 square kilometers of tundra, taiga, forest steppe, and steppe, with a population of 40 million, or slightly more than three people per square kilometer. What the Soviet Union lacked in technological sophistication, its engineers and policy makers made up for with their unbridled enthusiasm. Without the impediments of public opposition or the legal requirements of environmental impact statements, they quickly moved to change forever the face of Siberia with serious long-term human, economic, and ecological costs. Soviet plans never lacked enthusiasm for the belief that engineers could improve on nature's gifts or take advantage of the unanticipated payoffs of their hubris.

The largest share of investment in the energy sector went to oil, gas, and coal, especially to new oil and gas fields in the Tiumen region and to the Kuznetsk coal basin. Gas, oil, and coal also comprised the largest share of Soviet energy production, with the latter two filling more railroad cars than any other substance. Yet, hydroelectricity remained important, and the construction crews needed to build them grew to epic

---

⁹⁷ Leslie Dienes and Theodore Shabad, *The Soviet Energy System: Resources Use and Policies* (Washington: 1979), pp. 250, 259; G. D. Bakulev, *Razvitie Ugolnoi Promyshlennosti Donetskogo Basseina* (Moscow: Gosizdatpolit, 1955); and Z. G. Karpenko, *Kuznetskii Ugol'nyi* (Kemerovo: Kemerovskoe Knizhnoe Izdatel'stvo, 1971).

proportions. Angarastroi comprised 70,000 workers; Bratskgesstroi, formed in 1954 to tame the Angara, had 6,000 employees by 1955 and 35,000 by 1961, and the town where most of its workers lived had grown to 51,000. Industroi, Giproshakht, Sredazgiprovodkhlopok, and other organizations responsible for the scientific, design, and construction activities surrounding transformationist projects acquired technological momentum seemingly greater than, for example, the Army Corps of Engineers or Tennessee Valley Authority in the United States.

In one last major – and equally violent – way, Soviet leaders promoted policies that contributed to significant environmental degradation. As in the United States, France, Great Britain, and China, the other four original nuclear powers, the Soviet Union adopted haphazard waste disposal practices in the production of nuclear, chemical, and biological weapons. These Cold War practices destroyed so many ecosystems across such a vast landscape that costly and ongoing remediation in the twenty-first century may never catch up. We discuss the intractable problems associated with the production of weapons of mass destruction in later chapters in greater detail. Suffice it to say here that, under Stalin, the enormous pressure on scientists and engineers to develop and then mass produce these weapons led them to skimp, take shortcuts, and ignore not only common sense but also common decency.

Hiding under the blanket of "national security," scientists and engineers dumped high-level and low-level nuclear and other toxic wastes across the landscape, in lakes and rivers, and even in the oceans. They developed reactors that used "once-through" cooling, so that water from a lake or river that cooled the reactor was piped out back into the lake or river contaminated with radioisotopes and at a higher temperature that destroyed water ecosystems. In developing weapons production and test sites ("nuclear polygons"), they moved people out of their homelands, especially the Nenets from Novaia Zemlia and Kazakhs in Kazakhstan. Yet, like the Bikinians and Polynesians in the South Pacific, forced from their homes by the Americans and French, respectively, they moved them to areas where they might still suffer the consequences of radioactive fallout. In a number of cases (see Chapter 3), the nuclear waste spread far and wide, contaminated thousands of square kilometers of land, and required tens of thousands of people to be evacuated – although with some delay. The regions most affected by Stalinist Cold War weapons of mass destruction were the Southern Urals, the region around Semipalatinsk in Kazakhstan, and the Arctic.

Transformationist Economic and Political Desiderata and the Soviet Environment

During the Stalin era, an ideology of modernization prevailed in which the industrial world was considered the highest form of civilization. This view persisted in the Soviet Union until its collapse – and it persists in most countries of the world to this day. Yet, the Stalinist approach to modernization differed in a number of fundamental ways from those in other nations. First, Stalinist industrialization involved a self-proclaimed break with the past to reach and surpass the countries of Europe and the United States, behind which it lagged. Second, it essentially ignored such regions of the economy such as agriculture that remained on a primitive or poorly funded level. Third, it was carried out with the identification of internal and external enemies, including nature itself, to ensure the concentrated involvement of the nation's citizens in the revolutionary process. Nature and human enemies bore the consequences. Fourth, in conditions of "class struggle," nature was also an enemy of the state and would suffer. Fifth, the Stalinist regime pursued a policy of the impoverishment of the masses to reach state economic goals.

Unlike other regimes, the Soviet Union carried out industrialization and agricultural modernization through terror, state-sponsored famine, falsification, and the widespread establishment of slave labor camp construction operations. Finally, this was not simply industrialization, but military industrialization in which the entire postwar economy was geared to the inevitable war – perhaps nuclear war – with the West. The first nuclear catastrophe of the world, in Kyshtym in the Urals in 1957, was an echo of Stalinist modernization but did not signal a warning to many people. Most were ignorant of the explosion, and those responsible for it justified it as a cost of achieving nuclear parity with the United States. This was the logical outcome of an attempt of the regime to direct all aspects of life, from the economy to nature itself, and to require people and nature to follow a plan. Although touted as a rational plan, as we have seen it was rational only in industrializing the nation. It was supremely irrational in its human and environmental costs. On the side of agriculture, rather than stabilize or increase the harvest and guarantee food for the nation, it destroyed the countryside and the livestock in it. Yet, in spite of a series of reforms triggered during the era of Nikita Khrushchev, who followed Stalin, the Soviet Union continued down the path of destruction of nature in the name of the glory of the proletariat – in another series of great plans to transform nature.

Four major reasons explain the extensive environmental degradation of the Stalin era. The first was the Stalinist imperative to industrialize rapidly and to focus all investment on increasing production. Little effort went to innovation once factories were online, and less still went to pollution or safety control equipment. Second, given the feverish nature of the campaign, workers, machines, and materials were thrown into the effort, with little concern about waste. Third, the concentration of resources at production cities created not savings of economies of scale but battlegrounds of garbage, mud, haphazard disposal of waste, and so on. Finally, the emphasis on mining and metallurgy and other such processes ensured that people and landscapes alike would experience the costs of industrialization even more than the unfortunate exploited worker in the West.

Nikita Khrushchev, leader from 1953–1964, pushed a series of programs with environmental impact, and the same time attacking the utility of nature reserves. Khrushchev promised the achievement of communism by 1980, by which he also meant an improved standard of living.

In the 1950s through the 1980s, under Nikita Khrushchev and Leonid Brezhnev, Soviet economic development programs included significant efforts to open Siberian resources of oil, gas, various ores and timber. "We are opening the pantry of nature!"

3

The Khrushchev Reforms, Environmental Politics, and the Awakening of Environmentalism, 1953–1964

At the end of World War II, the Communist Party approved the fourth Five-Year Plan (1946–1951), which, like the first Five-Year Plans, was dedicated to building – and in this case also rebuilding – the country's heavy industry and military sectors. As in the 1930s, investment income was extracted from the agricultural sector, whereas housing, light industry, and medicine were ignored. A famine in 1946 killed hundreds of thousands of people. Millions of citizens lived in burned-out hovels, rubble, or holes in the ground. The Stalinist Plan for the Transformation of Nature (1948) ensured that nature, too, no less than Soviet citizens, would serve reconstruction purposes through the taming of its resources and through its coal, oil and rivers generating power. The Soviet Union was an industrial economy, yet leaders had superimposed it on a backward agricultural community, and the economy remained dependent on forced labor in such sectors as mining, timber, fossil fuels, and power production. This was the Stalinist centrally planned economy, with its ability to mobilize resources fully and harshly, if not rationally.

When Stalin died in March 1953, a succession struggle broke out between Georgy Malenkov, Viacheslav Molotov, secret police chief Lavrenty Beria, and Nikita Khrushchev. By 1956, Khrushchev had won this struggle and embarked fully on a reform program that included the economy, culture, and society. After the novel *The Thaw* (*Ottopel'*, originally published in the journal *Novyi Mir*) by Ilya Ehrenburg, the Khrushchev era became known as the "thaw" period. De-Stalinization became the official policy, especially after Khrushchev condemned the excesses of Stalinism in a special session at the close of the twentieth Communist Party Congress in February 1956. He described such abuses

The Khrushchev Reforms

of power as the purges, imprisonment and execution of innocent people, the Great Terror, the murder of Sergei Kirov in 1934, the elimination of thousands of Party members, and Stalin's paean to himself, the "cult of personality." Khrushchev's speech led many individuals to expect greater liberalization in Soviet politics society. The younger of these individuals, including many idealistic university students, were called the "children of the twentieth party congress." They embraced socialism but with a human face that they believed could be achieved without the great costs or failed rhetoric of the Stalin era. Yet, many Party officials and government bureaucrats remained dedicated to a closed political system, believed in the importance of maintaining power through the secret police if necessary, and supported economic policy with decisions based on planners', not consumers', preferences. From this time until the breakup of the Soviet Union in 1991, tension between reform and conservatism characterized Soviet politics.

As general secretary, Khrushchev pushed a series of reforms intended to improve the performance of the economy, many of which had significance for the natural environmental. These included efforts to develop Siberian resources, build new irrigation complexes in Central Asia, rejuvenate agriculture, and increase the output of consumer goods, including the food industry with higher production of meat and dairy products. The economy grew rapidly at 8 percent annually. The government also weakened its strict control on publication somewhat, including statistics about the costs of economic development and pollution, although it never hesitated to impose the stamp of secrecy on controversial topics. Khrushchev pursued reforms in politics and society as well. These reforms were more a pastiche of thrusts and parries rather than a systematic effort to change society from top to bottom, but they did open society to a greater extent than ever before to discussion about some of the strengths and weaknesses of the Soviet system.

But reform went only so far, because in no way did leaders intend to abandon the centrally planned economy. Planners continued to embrace inefficient and highly polluting large-scale projects as the foundation of the economy. These included fertilizer and other chemical facilities, metallurgical plants, canals, and irrigation systems. Many of Khrushchev's policies were inconsistent, if not far-fetched, especially in agriculture. He maintained his support for Trofim Lysenko, with his neo-Lamarckian view of the environment. His "Virgin Lands" and corn-planting campaigns (see **Box 1**) were unmitigated environmental disasters that resulted in the plowing up and exhaustion of soils followed by rampant erosion.

In the area of environmental concerns, therefore, while scientists, citizens, and decision makers made some strides in reversing several of the most damaging of Stalinist policies, all in all, in the Khrushchev era, the record is a spotty one. Beyond agriculture, the expansion of the military industrial complex (with the nuclear enterprise producing vast quantities of hazardous waste) and of mining, metallurgy, and energy production facilities contributed to the degradation. The failure to develop comprehensive nature protection legislation meant there were few brakes on development. Such unique jewels of nature as Lake Baikal in south central Siberia suffered the consequences.

In this atmosphere, individuals from arts and letters, from science and engineering, from government administration, from within the party apparatus, and ordinary citizens began to consider a variety of environmental topics – and to approach them in a variety of ways. These were Soviet "environmentalists." Within each group of individuals, not surprisingly, a diversity of viewpoints developed. Many engineers, for example, continued to believe that the geographic and meteorological tricks played on the Soviet Union in terms of rainfall, climate, and so on could be overcome with healthy doses of modern engineering practice, including making changes in the landscape to improve agricultural performance. They believed that some technological solutions existed to problems of weak performance, for example, through irrigation, heavier use of fertilizers, and so on. Other engineers used the greater openness of the Khrushchev era to advance projects they believed had been ignored in the Stalin era or had been underfunded in the past. Few engineers openly or aggressively criticized the environmental costs of their labor.

Scientists tended to be more circumspect about the ways in which the central-planning system had contributed directly and indirectly to environmental degradation. Still others, often those connected with such organizations as VOOP (the All-Russian Society of Nature Protection) and MOIP (the Moscow Society of Naturalists) or the Academy and university research programs, worried openly about the destruction of habitat and biodiversity carried out in the name of socialism. In general, many individuals viewed the state as much as part of the solution as a part of the problem. They believed that if the authorities understood fully the environmental costs of a specific program, they only need bring them to the attention of the right bureaucrat and a solution might be found. Like engineers, they believed that science was apolitical, and they might speak up about the scientific foundations of a policy dispute. Public protest over the pollution of Lake Baikal reflected this diversity of views.

Yet, the Soviet state remained intrusive and all-powerful regarding the fate of the environment. The economic and ideological desiderata that prevailed about the pursuit of Communism meant that plans to increase industrial, agricultural, and electrical energy production remained central to the Soviet development model. They were the key to the creation of a proletarian garden of plenty. Granted, the decentralization of political and economic power that occurred under Khrushchev enabled local and regional interests about potential environmental costs of new projects to be considered more systematically. But all too frequently these interests were ignored or downplayed if local "environmental concerns" ran headlong into national, ministerial, or other interests. And if under Khrushchev greater emphasis was placed on the quality of life – consumer goods, public health, and housing – then also the fate of the environment was often trumped by national interests – precisely in the huge agricultural and other programs he favored.

Questions of environmental problems, of what came to be known as sustainable development, of biodiversity, and so on, played out against this background. In the third program of the Communist Party, passed at the twenty-second congress in 1961, officials promised "to accelerate the social and economic development of the country" by reorienting the economy toward "the intensification of production" and focusing on "qualitative factors of economic growth." Science and technology were keys here, for they exerted "strong influence on all aspects of present-day production, on the entire system of social relations, on man and his environment."[1] Ultimately, however, Khrushchev's emphasis on economic reforms and economic growth stultified environmental progress to a great degree.

Khrushchev Era Economic Reforms: Impact on Environmental Policies

Khrushchev introduced an administrative reform of 1957 to force the pace of modernization. He ordered the transformation of some thirty highly centralized, Moscow-based national ministries connected with various sectors of the economy into 105 new bodies called *sovnarkhozy* (regional economic councils) that were intended to integrate the economy across the previous ministries on a regional basis, not on the basis of bureaucratically distinct economic sectors. Khrushchev's main objective was

---

[1] Third Program of the Communist Party of the USSR (1961), http://www.xs4all.nl/~eurodos/docu/cpsu-texts/cpsu86-21.htm#SocialistProductionRelations.

rationalization. He gave numerous examples of waste and duplication arising from the operation in parallel of the great centralized ministries, each building its own empire. The sovnarkhoz reform was an unprecedented attempt to decentralize economic decision making, and had it worked it might have contributed to improvement in environmental circumstances.[2]

In spite of the continuing emphasis on heavy industry, the government began to pay attention to the production of consumer goods. The goal was to raise wages and the standards of living and reward citizens with the fruits of communist labor. But production of these fruits always lagged behind demand, their quality was poor, and when Khrushchev introduced price reforms of 1962 to make prices more closely reflect scarcity values, this led to mass protests; in Novocherkassk, Russia, Khrushchev ordered police to use live weapons to put down a local demonstration, and dozens of workers were killed.[3]

Rapid economic growth put great pressure on the environment and on natural resources. From 1954 to 1964, energy production increased fivefold, oil production two and a half times, and cement production over three times. Steel production doubled.[4] However, industry lagged decades behind the United States in mechanization, automation, and production of many goods and services. Extensive, not intensive development characterized the Soviet economy; high rates of growth of output depended on large increases in the human and material inputs and less so on productivity of labor or other efficiencies. The average 8 percent annual growth rate of industrial output came almost entirely from annual growth in labor and capital inputs and little from increases in productivity.[5]

Yet, great achievements in science and technology indicated that the modernization of the economy and improvements in environmental

[2] N. S. Khrushchev, *O Dal'neshem Sovershenstvovanii Organizatsii Upravleniia Promyshlennost'iu i Stroitel'stvom: Doklad i Zakliuchitel'noe Slovo na VII sessii Verkhovnogo Soveta SSSR Chetvertogo Sozyva 7 i 10 Maia 1957 g.* (Moscow: Gospolitizdat, 1957). For analysis of the announcement of the sovnarkhoz reform, see J. Miller, "The Decentralization of Industry," *Soviet Studies*, vol. 9, no. 1 (July 1957), pp. 65–83. For an early evaluation of the forms, see Alec Nove, "The Industrial Planning System: Reforms in Prospect," *Soviet Studies*, vol. 14, no. 1 (July 1962), pp. 1–15.
[3] Samuel Baron, *Bloody Saturday in the Soviet Union: Novocherkassk 1962* (Stanford: Stanford University Press, 2001).
[4] Robert Jensen, Theodore Shabad, and Arthur Wright, eds., *Soviet Natural Resources in the World Economy* (Chicago: University of Chicago Press, 1983). See also Naum Jasny, "Prospects of the Soviet Iron and Steel Industry," *Soviet Studies*, vol. 14, no. 3 (January 1963), pp. 275–295.
[5] Robert Jensen et al., *Soviet Natural Resources*.

protection should not have been distant dreams. If under Stalin new factories had been built and manned by unskilled laborers and peasants that needed experienced foreign engineers to operate them, then by the late 1950s, the Soviet Union had trained hundreds of thousands of highly qualified engineers and scientists and millions of workers with high school and sometimes university degrees. Moscow State University alone turned out 3,000 physicists and 3,000 chemists every year, and the entire Soviet Union graduated 50,000 qualified engineers annually, a level ten times greater than in the United Kingdom. The Soviet Union pioneered the design for the tokamak fusion reactor (1951), brought a small nuclear power reactor on line with a capacity of 5,000 kilowatts (1954), orbited the world's first satellite (Sputnik, 1957), and put Yuri Gargarin into space on April 12, 1961. These feats led many citizens to believe that the country had moved closer to achieving Communism. These qualitative "world's firsts" and quantitative increases contributed to improvements in production processes and the quality of life and, given the general atmosphere of de-Stalinization, provided impetus for the development of environmental thinking.

These same individuals had some reason for hope for that the environment would become a focus of attention because, in the international sphere, Khrushchev abandoned Stalinist inevitability of war between two hostile systems (capitalist and socialist) for "peaceful coexistence." He rejected economic and cultural autarky. Peaceful coexistence meant the resumption of international contacts at all levels of society. This meant participation of Soviet citizens in youth festivals, various exhibitions, and academic exchanges, all of which might promote awareness of environmental thinking. At the same time, the Cold War did not abate. Khrushchev supported the expansion of the military industrial complex, including the chemical and nuclear weapons complexes, at great cost to the environment. He sanctioned the violent suppression of the Hungarian revolution in 1956. And he bore responsibility in 1962 for provoking the Cuban missile crisis and nearly sending the world into nuclear war. For domestic and foreign policy failures, Khrushchev was deposed in October 1964, and his successors referred to his failures as "harebrained schemes."

How did domestic reforms contribute to rebirth in environmental thinking that had been forced underground in the late 1920s? During the thaw, Khrushchev reined in the power of the KGB and considerably extended the range of permissible discussion. He accelerated the rehabilitation of gulag prisoners, living and dead. The entire gulag construction apparatus was dismantled and enveloped by the appropriate economic ministries. Khrushchev allowed the publication of Alexander Solzhenitsyn's *One Day in the Life of Ivan Denisovich* (1962). Still, the Soviet press scurrilously criticized political prisoners and prevented the publication of *Doctor Zhivago*, for which Boris Pasternak received the Nobel Prize in 1957. Censorship extended in particular to environmental issues. According to a State Censorship Committee ruling issued in July of 1957, the government prohibited the reporting on TV or in the print media of any information about forest fires, industrial accidents, military accidents, infant mortality, or radioactive pollution.[6] Soviet censors banned the publication of *The Proceedings of a Conference on Threatened and Endangered Plants and Animals and Unique Geological Sites* (March 1957). The censors ruled that each paper could be published individually, but that as a collection they presented too gloomy a picture.[7] In 1961, the authorities issued a special circular prohibiting publication of any materials about Soviet whaling; for years the Soviet Union had been poaching whales and falsifying data reported to the International Whaling Commission.[8] (See **Box 1** in Chapter 4.) Similar restrictions were imposed on information about fishing other valuable species.

So effective was this prohibition that nothing was published about an explosion of a nuclear waste dump in 1957 in Kyshtym near the city of Cheliabinsk, even though the authorities were forced to evacuate the local population for a radius of at least 200 kilometers.[9] Although Igor Kurchatov and Andrei Sakharov, leaders of the Soviet nuclear program, published articles in Soviet journals critical of atmospheric testing of nuclear weapons, no articles were published on the fallout from atomic explosions conducted in Kazakhstan's steppe, where more than 300 tests were conducted from 1949 to 1962 without any effort to protect public health. A television show, *Kak Eto Bylo* (The Way It Was), that aired in the early 2000s reported on battalions of Soviet soldiers who, like their

---

[6] V. E. Boreiko, *Belye Piatna Istorii Prirodookhrany*, 2nd ed. (Kiev: Kievskii Ekolog-Kul'turnyi Tsentr, "Logos," 2003), p. 289.

[7] Boreiko, *Belye piatna*, p. 289.

[8] Alexey V. Yablokov, "Validity of Whaling Data," *Nature*, vol. 367, no. 108 (January 13, 1994).

[9] Although analysts at Oak Ridge National Laboratory and the Central Intelligence Agency in the United States produced a report on the event, they kept it secret lest it generate fear among U.S. citizens that produced an antinuclear power backlash. See Zhores Medvedev, *Nuclear Disaster in the Urals* (New York: Norton, 1979) for a reconstruction of the accident that works back from a series of individually published articles about pathways of radionuclides in flora and fauna.

American counterparts in Nevada, were ordered to ground zero minutes after a nuclear explosion to evaluate the danger to them of exposure to radioactivity; they had no special equipment. Similarly, throughout the period of atmospheric testing, radioactive debris rained down on unsuspecting citizens from the Far North to Siberia.

Technocratic Euphoria and Indifference to the Environment

Soviet socialism was a scientific, rationalist, and technocratic doctrine. The textbook *Foundations of Marxism-Leninism* (1961) unsophisticatedly suggested that scientific developments would enable a Communist society to solve an impressive range of problems, from the prolongation of life to 150 years to placing at the service of man all the forces of nature, including rendering completely harmless natural calamities and enabling climate control. However, there was a remarkable difference between the Khrushchev and Stalin eras regarding attitudes toward academic science. Under Stalin, science had been one more arena for ideological fighting that demonstrated the primacy of socialism over capitalism. Soviet philosophers rejected a number of cutting-edge fields as "reactionary pseudo-science" and an ideological weapon of imperialist reaction. They drew this conclusion based on the argument that scientific theories were either idealistic or materialistic. In Khrushchev era, science was much more utilitarian and much less an arena for ideological battle; it was mainly a means for the construction of Communism. Former "bourgeois" or idealistic theories, among them cybernetics and genetics, were recognized as mainstream science. By the mid-1950s, cybernetics was portrayed as an innocent victim of political oppression and "rehabilitated," along with political prisoners of the Stalinist regime. Cybernetics was canonized in the Third Party Program as a "science of communism."[10] In this ideological atmosphere, ecology, and other fields of biological science crucial to environmental thought could also be seen as value neutral and objective, not as arenas for philosophical battle.

Yet for all of his efforts to improve the quality of life in the Soviet Union, Khrushchev lacked environmental instincts that would have contributed to greater "environmentalism." This was clear from his Virgin

---

[10] M. Sandle, *A Short History of Soviet Socialism* (London: UCL Press Limited, 1999); David Mindell and Slava Gerovitch, "Cybernetics and Information Theory in the United States, France and the Soviet Union," in Mark Walker, ed., *Science and Ideology: A Comparative History* (London: Routledge, 2003), pp. 66–95.

Lands and corn campaigns and from his persistence in promoting the construction of the Lake Baikal pulp and paper combines despite extensive protests from the Soviet intelligentsia. He advanced fantastic figures for increasing agricultural and industrial production, in large part because they corresponded to American economic parameters, not to the Soviet situation. For him, "peaceful coexistence" meant surpassing the West in the production of a variety of products and spheres that would indicate the superiority of socialism. The Soviet Union succeeded only in outproducing the West in concrete and pig iron – each with its own environmental impacts. Like leaders before him, Khrushchev rarely took into account climatic and other environmental constraints on his designs. His naive belief in the potential of science and engineering to overcome any problem overshadowed any concern about the possible impact of technical progress on the environment. Many specialists reflected this same technocratic optimism in various projects – both within and outside the nation's borders – to transform Siberia into a garden, to build a huge hydropower station at the Straits of Gibraltar to provide electricity for all of Europe and to employ peaceful nuclear explosions to blast canals.

Khrushchev was, for all of his reformism, a product of the Stalin era. Like Stalin, he had a hostile attitude toward nature preserves, which he viewed as "wasted," unproductive land. He saw the people working in the reserves similarly as do-nothings. His view of the relationship between humans and nature stemmed from heated discussions at the beginning of the 1930s between an older generation of naturalists and militant young Communists. The former belonged to the pre-revolutionary, liberally minded professorate, while the latter came from students at the Communist Academy or other "red institutes," many of whom had not finished their secondary education. They impatiently called for rapid change in society and nature. As part of the Stalin-era culture, environmental thinking was to be cleared of such bourgeois ideas as gradual change, private property, or the assertion that nature was more powerful than humans. The belief that nature was somehow an enemy of socialism – or at the very least should be transformed rapidly into some kind of machine – persisted under Khrushchev.

In some respects, these positions operated in bourgeois democracies as well – and continue to do so into the twenty-first century. In the United States, for example, environmentalists worry about the irreversible impact of economic development on forests, wetlands, flood plains, deltas, and other ecosystems, whereas such individuals as "property rights" advocates believe that individual rights to own property and job formation

One of Khrushchev's most environmentally-devastating and costly programs was opening "Virgin Lands" to agriculture. The haphazard effort led to erosion and despoliation.

should trump environmental considerations. Still others condemn federal efforts to expand wilderness areas, regulate snowmobiles in national parks, and other usurpations of these rights.[11]

The general indifference of the Soviet people to the environment was also characteristic of the Khrushchev era, especially in the 1950s. First of all, public involvement in the policy process had been virtually eliminated under Stalin. To make matters worse, the Soviet press published little information about air, land, or water pollution that was not highly sanitized. Reporting victories of economic development and goals of plan fulfillment was of greater concern than portraying the deteriorating imbalance between man, woman, and their environment. Very few articles strictly concerning environmental problems of the Soviet Union appeared in the Soviet press until the Gorbachev era.

---

[11] For a discussion of some of these issues, see Norman Henderson, "Wilderness and the Nature Conservation Ideal: Britain, Canada, and the United States," *Ambio*, vol. 21, no. 6 (September 1992), pp. 394–399; Lynn E. Dwyer, Dennis D. Murphy, and Paul R. Ehrlich, "Property Rights Case Law and the Challenge to the Endangered Species Act," *Conservation Biology*, vol. 9, no. 4 (August 1995), pp. 725–741; and Paul Lorah and Rob Southwick, "Environmental Protection, Population Change, and Economic Development in the Rural Western United States," *Population and Environment*, vol. 24, no. 3 (January 2003), pp. 255–272.

Khrushchev's Agricultural Programs

Human activities to grow crops, produce fruit, and raise livestock, of course, have significant impact on land and water. Techniques including the pushing back of forests, slash-and-burn methods of cultivation, the creation of ponds and other impoundments and irrigation systems, the application of manure and other natural and chemical fertilizers, and the use of various tools to plow and harvest have been employed for centuries. In the twentieth century, agriculture became highly mechanical and industrial, from the tractors, combines, and harvesters used to work the land and manage the crops; to the crops and animals that have been hybridized – and now are genetically engineered; this has increased output significantly. Increased urbanization accompanied these industrial changes, with fewer workers required on massive farm operations that are capable of producing enough for domestic and international markets.

In the Soviet Union, these same developments acquired particularly Soviet characteristics given state ownership of the land, the formation of collectivized agriculture, Cold War competition with the United States, and the poverty of the countryside because of the primacy of heavy industry and the city. Periodic – and often economically irrational – campaigns to improve the situation and the resulting overuse of harsh chemical and mechanical methods to overcome shortfalls had little positive economic impact and extensive negative environmental impact, as the environmental history of the Khrushchev era indicates. But Khrushchev and the Soviet leadership were wedded to centralized, technology-based agriculture. They believed that technology could overcome problems of organization or lack of initiative among workers. Indeed, a highly publicized chemicalization program unfolded in the Brezhnev era and led precisely to a rapid increase in the production of chemical fertilizers and biocides, but their overuse at two to five times the level in the United States, and resulting destruction of farmlands.

Khrushchev's agricultural policies had an immediate and devastating impact on the steppe. In 1954, as part of the succession struggle (See **Box 1.**), Khrushchev announced a plan to plow up in short order more than 40 million hectares of the virgin steppe in Kazakhstan and western Siberia as well as long fallow lands elsewhere.[12] The Soviet Union faced

---

[12] W. A. Douglas Jackson, "The Virgin and Idle Lands of Western Siberia and Northern Kazakhstan: A Geographical Appraisal," *Geographical Review*, vol. 46, no. 1 (January 1956), pp. 1–19.

> **Box 1. Political Fighting Inside the Communist Party Leadership and the Virgin Lands campaign**
>
> Although harvests in the year after Stalin died (1954) were relatively good, members of the Politburo engaged in heated discussions about the state of Soviet agriculture and how to improve its performance. Connected to these discussions was an unfolding battle among the successors of Stalin to become the new Soviet leader. At the nineteenth congress of the Communist Party in October 1952, the chairman of the Council of Ministers, Georgy Malenkov, a contender to succeed Stalin, announced that the grain problem had been already solved in the Soviet Union. Nikita Khrushchev, the chief architect of Soviet agricultural policy since 1950 and first secretary of the party since 1953, rejected Malenkov's position in a January 22, 1954, speech. Khrushchev said that grain production in 1952 was only 92 million tons instead of the officially announced 130 million tons. The latter figure was said to be based on estimates of so-called "biological yield" (the maximum possible yield of the standing crop in the field at time of maximum ripeness) and not on the actual harvest. For the first time, a top official had openly revealed that optimistic views of Soviet agriculture were based on distorted agricultural statistics. According to Khrushchev, there was no evidence that Soviet agriculture performed better than in the prewar years. He noted that in 1940 the state procured 35.6 million tons of grain, whereas in 1953 the delivery was only 29.8 million tons. Khrushchev believed that this desperate shortage of grain could not be alleviated quickly by either more intensive or extensive production in the traditional grain areas of the country, as his opponents in the Party elite evidently argued. Proposing the Virgin Lands campaign to increase agricultural output on a new basis gave Khrushchev political advantage over Malenkov.[13]

endemic agricultural shortfalls. Khrushchev believed the current desperate grain shortages could not be alleviated by more intensive or extensive production in the traditional grain areas of the country. Because the purchase of grain on foreign markets was embarrassingly out of the question given the nation's autarkic economy and pronouncements of superiority,

---

[13] Nikita Khrushchev, "Stroitel'stvo Kommunizma i Razvitie Sel'skogo Khoziaiastva," in *Rechi i Dokumenty*, I (Moscow: Gospolitizdat, 1962), pp. 85–100.

any additional grain for human and livestock consumption could only come from exploiting these idle or Virgin Lands. On March 2, 1954, the Communist Party leadership adopted a decree titled "On further increasing grain production in the country and the reclamation of virgin and waste land."[14] The campaign for the cultivation of the Virgin Lands was unprecedented in scale and rate. In the first three years (1954–57), 40 million hectares of new lands had been plowed under, or 23 percent of the total sown area of the Soviet Union.

Officials had high hopes for the Virgin Lands campaign. Accepting Khrushchev's expectation of a minimum average yield of ten centners per hectare from the Virgin Lands for 1956, a gain of 30 million tons would make it possible to solve the grain problem, enable the nation "to catch up with America," as Khrushchev insisted, and to demonstrate that the glorious Communist future was in sight. Excess grain from the Virgin Lands would go to building up a reserve, make possible trade for other goods with the newly socialist states of Eastern Europe, and allow the Soviet Union to export grain.[15] Khrushchev also believed the Virgin Lands were perfect to reorganize collective farms on an industrial basis through the creation of a new type of large-scale agricultural city, the *agrogorod*, that epitomized the glorious tie between industry and agriculture under socialism. Khrushchev had first proposed his dream of agricultural cities years earlier, and the Central Committee had rejected the proposal. Now, as general secretary, he faced no obstacles to his dream – except for nature itself. Nature proved to be quite an obstacle.

The terms "Virgin Lands" and "wastelands" mentioned in the 1954 Central Committee resolution were misleading. The steppe lands in Kazakhstan and western Siberia that were to be plowed under were not virgin, but had been used as pasture or hay fields for centuries. Nor were long fallow lands wastelands. Political expediency, not agronomy, defined the lands, because Khrushchev intended to increase grain production. The slogan "Fallow land is lost land; erosion is a fiction!" found broad play on the front pages of Soviet newspapers. Soviet economists ran convoluted calculations to determine how much grain any region lost when it failed

---

[14] Khrushchev, *O Dal'neishem Uvelichenii Proizvodstva Zerna v Strane i ob Osvoenii Tselinnykh i Zalezhnykh Zemel'* (Moscow: Gosizdatpolit, 1954).

[15] Khrushchev, *Itogi Razvitiia Sel'skogo Khoziaistva za Poslednie Piat' Let* (Moscow: Gosizdatpolit, 1958). On the Virgin Lands campaign, see Nikolai Dronin and Edward Bellinger, *Climate Dependence and ood Problems in Russia, 1900–1990* (Budapest: CEU Press, 2005), pp. 171–218.

to sow fallow land. The fallow lands had been sown mostly with forage, including Khrushchev's new favorite, corn (for silage), that was to be cultivated throughout the country regardless unsuitable climate condition.[16] Finally, these Virgin Lands lacked manpower, equipment, and infrastructure. Grain and corn cultures would have to be created from scratch at great cost while keeping peasants who had been ordered to move to the new regions happy in these difficult circumstances. Everything would have to be perfect to have a good harvest, and, in addition to poorly-thought-out plans and inadequate capital and labor inputs, climate prevented a good harvest.

The Virgin Lands became a region of multiethic, -religious, and -cultural contact as Russians, Chechens, Ingush, Germans, Koreans, Kazakhs, and others mixed there, and hundreds of thousands of people volunteered or were recruited to tame the land. They were to join technology – tractors and other agricultural machines – and were encouraged through a propaganda campaign that brought them "willingly" to the great project. According to the press, the Virgin Lands campaign "received a tremendously positive response." The campaign symbolized de-Stalinization and the hopes of Communism reborn. Yet as many settlers inundated the region, many left – hundreds of thousands of them – because the local authorities were unprepared for the great influx of people and materiel – the trainloads of tractors, tents, and supplies arriving at stations that were never transported to the new state farms; with food supplies that were also inadequate; with rapid turnover of bosses; and with local people having no idea what was happening or why. The result, at least initially, was great confusion, crime, and some ethnic conflict. It was an unsettled and unsettling settlement.[17]

Severe climatic conditions characterized the Virgin Lands, even in comparison even with the older grain-producing areas of the Soviet Union where the soil and climate were marginal by European standards. The geographic location of Kazakhstan, in the center of a huge continent with the axis of extreme barometric pressure that traversed its northern regions, led to dry and windy extremes. This area had a typical continental

---

[16] Radio Liberty Research Institute, *Mistakes in Exploitation of the Virgin Lands*, September 6, 1967.
[17] Michaela Pohl, "The 'Planet of 100 Languages': Ethnic Relations and Soviet Identity in the Virgin Lands," in Nicholas Breyfogle, Abby Schrader, and Willard Sunderland, eds., *Peopling the Russian Periphery: Borderland Colonization in Eurasian History* (London: Routledge, 2007), pp. 238–261.

climate, cold in winter and hot in summer. Winter began in November and ended in March. In the winter, temperatures often reached −40 degrees Celsius, and swirling snow often limited visibility to six feet. Summer provided little relief. Winds often blew at 100 kilometers per hour. The parched topsoil was siphoned off by ominous clouds of rolling dust. Precipitation was the crucial factor for farming, yet annual rainfall ranged from an arid 200 millimeters to a barely sufficient 300 millimeters. In the United States, farm experts hesitated to cultivate nonirrigated land unless rainfall was greater than 400 millimeters.[18] One attempt to plow the Virgin Lands in the period from 1928 to 1933 during collectivization had failed; yields were extremely low – three to four centners per hectare. The Virgin Lands produced a fair crop only in 1956 because of fortuitously good weather. A return to normal conditions meant a steady and spectacular decline, especially in the period from 1960 to 1965 because of poor weather. In 1960, 1963, and 1965, crop failures hit many regions in the country, not only in the Virgin Lands, but in previously fallow lands because of wind erosion.[19] The Soviet press reported the growing threat of wind erosion that accompanied droughts in the virgin and idle land as early as 1955, when about 4,500 hectares were lost. In 1957, an additional 7,000 hectares of wheat and millet were ruined. Furthermore, some lands, press reports cautioned, would lose fertility if cultivated for five years in succession.[20]

The turning point for Soviet agriculture was 1963, which forever dashed hopes for a grain surplus when a bad harvest in the Virgin Lands coincided with a poor harvest in the normally rich agricultural lands of the Soviet Union. In Kazakhstan, many districts failed to harvest enough to meet the following year's seed-grain requirements. In some districts, little rain fell until September 25.[21] The average yield dropped to 4.4 centners.

---

[18] A. S. Uteshev, *Atmosfernye Zasukhi v Kazakhstane* (Alma-Ata: 1963), and *Financial Times*, August 23, 1978.

[19] T. F. Iakubov, "Problema Zashchity Pochv ot Vetrovoi Erozii v Raionakh Osvoeniia Tselinnykh i Zalezhnykh Zemel' Severnogo Kazakhstana," *Pochvovedenie*, no. 10 (1956), pp. 37–39, and *Izvestiia*, January 4, 1958. See also Jerzy Karcz, "The New Soviet Agricultural Programme," *Soviet Studies*, vol. 17, no. 2 (October 1965), pp. 129–161; and Alec Nove, "Soviet Agriculture under Brezhnev," *Slavic Review*, vol. 29, no. 3 (September 1970), pp. 379–410.

[20] Roger W. Opdahl, "Soviet Agriculture Since 1953," *Political Science Quarterly*, vol. 75, no. 1 (March 1960), p. 60.

[21] E. D. Milovanov, ed., *Problemy Ekonomiki Selskogo Khoziaistva na Iugo-Vostoke i v Zapadnom Kazakhstane. Sbornik Nauchnykh Rabot* (Saratov: Saratovskii Selskokhoziaistvennyi Institut, 1967), p. 159.

Wind erosion damaged about 3 million hectares.[22] Western Siberia also suffered the wrath of dust storms, where grain production was only 42 percent of average levels. The drought spread to other key regions: the Central Black Earth, Middle Volga, and Ukraine. The acute 1963 grain shortage forced the Soviet Union to purchase more than 10 million tons of grain from capitalist countries, at great embarrassment to Nikita Khrushchev.[23] One popular political joke of that time asked for the name of Khrushchev's hairstyle (Khrushchev was bald); the answer was "the harvest of 1963." According to another joke, "Khrushchev hokey-pokey" was the sowing of the Virgin Lands, but harvesting in Oklahoma. In 1963, the Soviet Union resorted to panicked slaughtering of livestock, and other food problems were reported. The year 1965 was even worse, as wind erosion destroyed 5 million hectares during a drought.[24] Severe erosion also hit the left bank of the Volga, the North Caucasus, and certain southern provinces of Ukraine. These were similar to those events in the United States in the 1930s, where the rush to bust the sod of the Great Plains states with tractors to plant grain combined with drought and winds to contribute to the Dust Bowl.[25]

From the points of view of agriculture and environment, the Virgin Lands campaign was a fiasco. Grain production in Kazakhstan dropped by 23 percent over the period from 1960 to 1965. Machinery and equipment were inadequate to the task, fell apart, and could not be repaired because there were no spare parts; by 1961, 60,000 harvesters lay idle. Finally, the program led to misuse of land. Spring wheat was sown year after year in spite of warnings by specialists not to do so. The result was weed infestation, soil erosion, and reduced natural fertility.[26] The authorities had to withdraw more 7 million hectares of the newly cultivated land (about 30 percent of the total) from agriculture. In February 1964, Khrushchev admitted the failure and promised that the wheat-growing Virgin Lands plagued by repeated droughts would be turned

---

[22] Radio Liberty Research Institute, *More Data on Dust Storms in the Virgin Lands*, June 24, 1963.
[23] Martin McCauley, *Khrushchev and the Development of Soviet Agriculture: The Virgin Land Programme 1953–1964* (London: Macmillan, 1976); and Naum Jasny, *Khrushchev's Crop Policy* (Glasgow: G. Ouram, 1965).
[24] Milovanov, *Problemy Ekonomiki*, p. 159.
[25] Donald Worster, *Dust Bowl: The Southern Plains in the 1930s* (New York: Oxford University Press, 1979), p. 277.
[26] Alec Nove, "Soviet Agriculture Marks Time," *Foreign Affairs*, vol. 40, no. 4 (July, 1962), p. 579.

back to grazing.²⁷ This was unlikely to happen because of the high social and political costs of such reversal. Hundreds of thousands of people, among them Russians, Kazakhs, Germans, and others, had already settled in the area. The authorities built new infrastructure at a cost of hundreds of millions of rubles. Granted the turbulence and dislocation of the Virgin Lands campaign, many of these people were excited initially by the opportunity to build a new region, create new lives, and define themselves anew.²⁸ In any event, seven months later, in October 1964, Leonid Brezhnev and others removed Khrushchev from office, in part because of this failure of agricultural policy.

After Khrushchev's resignation, the new Soviet leadership undertook several steps to combat soil erosion, first of all in Kazakhstan and western Siberia. The authorities relied on the experience of Western farmers, especially those from Canada. In the western steppe districts of Canada (primarily in Alberta and Saskatchewan), the average yield was much higher than that in the Soviet Union. The Canadian farmers practiced fallow land crop rotation, leaving between 20 percent and 40 percent of their spring grain lands fallow each year. Soviet soil specialists therefore also recommended that up to one-third remain fallow.²⁹ Second, they introduced a new method of soil cultivation. Under Khrushchev, many state farms had plowed up not only fertile tracts of land but also extensive areas, abutting them merely because they yielded readily to the plow. They plowed them too deep, as if they were forest steppe or forest zones, with extensive erosion the result. Adopting Canadian approaches lessened erosion and maintained snow cover into the spring, which protected the soil. These steps somewhat ameliorated the problem of wind erosion in the Virgin Lands and allowed the Soviet Union to use them as a bumper crop region.³⁰ Still, the damage was extensive, and the lands still had not recovered at the beginning of the twenty-first century.

---

[27] Nikita Khrushchev, *Uskorennoe Razvitie Khimicheskoi Promyshlennosti – Vazhneishee Uslovie Podema Selskokhoziaistvennogo Proizvodstva i Rosta Blagosostoianiia Naroda: Postanovlenie Plenuma TsK KPSS po Dokladu Tovarishch N. C. Khrushcheva, Priniatoe Edinoglasno 13 Dekabria 1963 goda* (Moscow: Izdatpolit, 1963); and "Khrushchev Reported Planning To Give Up Virgin-Land Farms; Khrushchev Sees Virgin Farms' End," *New York Times*, February 23, 1964.

[28] Pohl, "The 'Planet of 100 Languages.'"

[29] Milovanov, *Problemy Ekonomiki*, p. 159.

[30] For comparative Canadian-Soviet agriculture, see Radio Free Europe, "The Virgin Lands: Soviet and Canadian, A Comparison of Two Farming Systems," May 1964, in Open Society Archives, Budapest, Hungary.

The Corn Campaign: Grasslands and Maize Malaise

Khrushchev promoted a second "harebrained scheme" – as his successors called it – in agriculture that had significant environmental costs: a campaign to plant corn. Khrushchev knew the United States was rich with corn and assumed the Soviet Union could become a corn-producing country as well. But he was not as knowledgeable about agriculture as he thought he was. The corn campaign, like that of the Virgin Lands, was doomed to failure. Because of radical expansion of the arable area under cereal grains between 1954 and 1975, the Soviet Union faced instead a shortage of fodder at the same time its leaders sought to expand the livestock inventory. As noted, in May 1957, Khrushchev quite unexpectedly set targets to catch up with the United States in meat and milk production per capita within a record time. In the postwar decade, many regions of the world were on the edge of hunger. The Soviet Union performed comparatively well by the late 1950s, with the famines of the two postwar years fading into memory. Although the average world meat production per capita had reached only twenty-one kilograms per capita, the Soviet Union produced about forty kilograms per capita. However, comparison with the U.S. meat sector made Soviet officials desperate: despite livestock inventory that was 22 percent smaller than those of their Soviet counterparts', American farmers produced 50 percent more meat than Soviet farmers did. According to Khrushchev, the low productivity of Soviet herds was mainly due to shortages of fodder.[31] In slogans and speeches, the declaration "Catch the Americans!" meant that the Soviet Union would increase meat and diary production two- to threefold within three years. This target made the problem of shortages of feed much more acute.[32]

To solve the problem of fodder, Khrushchev decided on the radical expansion of the corn crop area – an eightfold expansion that took place in two stages. The first took place in 1957, when 18 million hectares were sown with corn. In 1961–1962, a new wave of corn planting resulted in an additional 8 million hectares in the Russian Republic.[33] The expansion

---

[31] Lester Brown et al., *Vital Signs 2002*, Worldwatch Institute (New York: W. W. Norton & Company, 2002); and A. Popluiko, *Soviet and American Agriculture Systems*, Radio Liberty Research, December 4, 1962.
[32] R. W. D., "The Khrushchev Livestock Plan," *Soviet Studies*, vol. 7, no. 1 (July, 1955), p. 117.
[33] *Narodnoe khoziaistvo RSFSR v 1965* (Moscow: Gostatizdat, 1966).

of corn was at the expense of grasslands, fallow lands, pastures, and meadows; the clover and oat crops were replaced by maize and peas; and meadows were plowed up regardless of local natural features.

The personal dedication of Khrushchev to corn cultivation became the subject of numerous jokes. One of them concerned Khrushchev trying to plant maize on the surface of the moon. Another concerned his threat to the Soviet national soccer team to plant all soccer fields with corn if they did not win an international match. Khrushchev certainly exaggerated his expectations for the corn harvest. He reported a high yield of maize at 27.5 to 30.0 metric tons per hectare, even though the United States, with a much better climate, averaged 20 to 22 tons per hectare. In fact, the Soviet yield was only 10 to 12 tons per hectare in 1965. In total, the production of silage increased fourfold, but the increase did not compensate for losses of hay production. The amount of feed available per head of livestock in 1965 reached only 70 percent of the norm, in part because of the growth of the livestock inventory and in part because of the failure of corn. Corn simply was not suitable for the climate conditions prevailing in the central and northern regions of the country. In forest zones of European Russia, yields were half of official expectations. The negative environmental impacts of the corn crop included erosion, reduction in area of grasslands and destruction of biodiversity, and lack of interest of farmers in any improvement of pastures and meadows.

In the case of hay fields and pastures, forty million hectares of forest and forest steppe zones had become overgrown with bushes and trees.[34] Instead of melioration of these areas, the farmers were ordered to plow them for corn, with the result that hay production, more important for milk cow breeding, dropped. In Vologda province, which had always specialized in milk, butter, and meat, livestock indices stagnated or fell. Khrushchev's agricultural policy and his struggle against the grassland system were responsible.[35]

To the south, in the Central Black Earth region, the expansion of areas under corn cultivation meant further negligence of pastures and grasslands, with hay production falling by 50 percent. The region had always suffered from shortages of hay and grass, and Khrushchev's programs exacerbated the situation. A wasteful attitude toward meadows, especially toward alluvial lands that were good for hay cultivation, accelerated

---

[34] M. Mondich, *Fear of Famine and Panic-Buying in the USSR*, Radio Liberty Research Institute, October 17, 1963.
[35] Radio Liberty Research Institute, *State of Agriculture in Vologda Oblast*, May 16, 1963.

the formation of swamps and overgrowth that were no longer suitable for mechanized harvesting. Thousands of hectares of upland fallow lands were lost because of the disastrous effects of soil erosion. This left meadows and pastures on the sides of hills, on river banks, and around ponds exposed. The exposed areas should have been planted with grass, but the authorities did not provide resources to do so. Specialists in fact decided to recommend no action to avoid incurring the wrath of Khrushchev. "The timing was bad. We would have been harshly criticized," one specialist said.[36] Instead, they imported hay from other regions at great expense.

Shortages of hay reserves put greater pressure on pastures throughout the country. In 1961–1963, thousands of farmers from Stavropol, Dagestan, Astrakhan, Rostov provinces, and even the Caucasus republics brought their flocks of sheep to forage on the remaining pasture in the Kalmyk province. The mild, relatively snow-free winter and the bountiful pastures of Kalmykia permitted livestock to be provided with grass almost year round. But the pastures became overgrazed and eventually exhausted. Windblown sand became a problem. Again, the authorities did nothing to improve the situation.[37]

The Assault on Forests

In 1956, Leonid Leonov published the novel *The Forest* (*Les* in Russian) in which the two protagonists, both forestry specialists, debated the best method to meet planned targets in the Soviet industry. One of the specialists advocated tried but untrue methods that wasted resources, whereas the other, progressive forester pushed a scientifically based plan in which the harvest would never exceed the annual growth of the forest. The Khrushchev reforms in fact permitted a debate over how to manage forests, with the latter approach being adopted in many regions. Yet the wood, pulp, and paper industries remained plagued by rapacious harvest, insufficient infrastructure of narrow-gauge railroad and roads to permit more rational approaches, and such wasteful practices as the spring floating to move lumber downstream. In some areas, poaching was an endemic problem. Ultimately, planners realized that huge losses of lumber during harvest, transport, and processing were inevitable in their system, so they included those losses in five-year and annual plans. Still, conservation became more widely practiced in the Khrushchev era.

---

[36] *Komsomol'skaia Pravda*, June 23, 1964.
[37] *Pravda*, December 21, 1964.

Large-scale industrial exploitation of the forest resources of the Soviet Union began at the end of World War II. As a result, paradoxically, of undermechanization; the use of fellers, haulers, and other equipment that used tank treads instead of low-pressure tires; excessive cutting close to roads and railroads; and poorly trained, underpaid, and undermotivated loggers, Soviet logging enterprises destroyed the forest wherever they harvested. They clear cut; they counted on losing 30, 40, even 50 percent of the harvest to rot on the forest floor or sink to river bottoms during the spring float. And the industry grew and grew. In 1946, the industry felled 214 million cubic meters of lumber; by 1960, it had grown by 70 percent to 347 million cubic meters; and by 1964 it had reached 385 million cubic meters.[38] Postwar reconstruction; the expansion of housing; and a huge increase in newspaper, magazine, and book publications led to the high rate of growth, although the average annual rate of growth slowed in the 1960s.[39] Generally speaking, the pressure on Russian forests reached its maximum during the Khrushchev era.

Theoretically, forest resources were inexhaustible. Practically, the pressure on these resources was a function of their accessibility to roads, equipment, and workers. With the exception of the Russian Republic, most Soviet republics had limited forest resources that suffered from extensive logging operations and forced them to import wood and wood products from Russia. As for the Russian Republic, given its extensive resources, it should have been possible to harvest "scientifically," taking mature, harvestable timber. Yet because of an absence of modern equipment, including forestry roads, two-thirds of logging in the Russian Republic was concentrated in the European and Ural regions. Most Siberian forests were inaccessible to lumbering enterprises because of the lack of roads and rails and the high capital and labor costs. The permafrost forests in Siberia were of low quality. The high proportion of mountain forests in the Soviet Union in the Carpathians, Caucasus, Altai, and Cikhote-Alin' Mountains offered some promise. Yet the industry could not use them in a sustainable fashion because of the absence of the special machines needed to avoid soil erosion and ensure favorable conditions for reforestation.

---

[38] Brenton Barr, "Soviet Timber: Regional Supply and Demand, 1970–1990" *Arctic*, vol. 32, no. 4 (December 1979), pp. 308–328.

[39] *Komsomol'skaia Pravda*, October 14, 1970; *Narodnoe Khoziaistvo SSSR v 1972* (Moscow: Gosstatizdat, 1973).

The presence of regulations that specified types of forests did little to protect the forests from overcutting. Specialists who were concerned about this problem had limited influence in improving the situation. Lumber enterprises regularly went beyond the quotas, leading to a high rate of depletion of forest resources in their proximity in many regions. Government planners themselves violated the regulations by adopting higher and higher quotas. At the beginning of 1955 the government adopted a decision to increase felling over quotas by 50 percent to 100 percent for forests in the second group of densely populated regions for the period from 1955 to 1960.[40] As a result, lumbering activities were eight times more intensive in the more accessible, but less thickly wooded forests of central regions of European Russia than in the country as a whole. Coniferous and oak forests were significantly degraded,[41] and enterprise personnel ignored deciduous species or left them to rot. This resulted in a worsening of the species and age composition of the forests.

Other regions suffered the same degradation of forests because of the insatiable appetite of the central government for construction, pulp, and paper products; and the failure of the government to abide by its own laws for conservation of resources. Once again, the short-term plan meant more than long-term resource management practices. From 1955 through 1960 in Ukraine, the authorities sanctioned the harvest to exceed the quota by up to two times. When, in 1960, the Russian Republic failed to meet the plan to supply Ukraine with lumber, the central forestry authorities in Moscow recommended an increase in felling targets throughout Ukraine, targets that could be met only in the ecologically sensitive Carpathian Mountains.[42] The Carpathian Mountains were one of the most humid regions of the Soviet Union, where annual precipitation reached 1,600 millimeters and occasionally the monthly average of precipitation fell in one day. The results of systematic overcutting in this region were erosion, windfalls, and destruction of river banks. Things got so bad that the May 7, 1963, front page of *Pravda Ukrainy*,[43] the leading republican Communist Party newspaper, was devoted to an appeal for the conservation of Ukraine's natural resources, in the

---

[40] A. I. Bovin, V. P. Tsepliaev, and D. T. Kovalin, *Lesnoe Khoziaistvo SSSR, 1917–57*, (Moskva: Goslesbumizdat, 1958), p. 274.
[41] *Komsomol'skaia Pravda*, July 27, 1965.
[42] *Literaturnaia Gazeta*, August 30, 1966.
[43] *Pravda Ukrainy* (front page), May 7, 1963.

first place its forests and bodies of water. Prominent Ukrainian scientists and men of arts and letters (but no women!) signed the appeal. They protested that the Carpathian forests were the victim of systematic despoliation.[44] But because protest, let alone public protest, had a limited impact in postwar Soviet politics, large-scale felling in the Carpathians Mountains ceased only in the late 1970s. Unfortunately, it resumed again in independent Ukraine in the 1990s when state control of the forests collapsed.

A similar situation occurred in Abkhazia, a northern region of Georgia in the Caucasus with a forested area of 512,000 hectares. Between 1955 and 1965, annual cutting grew twelvefold, from 300 to 3,700 thousand cubic meters. The lumber enterprises foolishly attacked forests near such famous resorts as Lake Riza and Pitsunda simply because roads enabled felling.[45] By the end of the 1960s, the forested area in Abkhazia had decreased by 25 percent, but felling was banned only in late 1970 after repeated protests of Georgian scientists.

No region was immune from the loggers' axe and chainsaw. In the Altai Mountains in Siberia, where 97 percent of forests of the Altai province were located, lumber enterprises assaulted vulnerable cedar forests. Before 1937, felling of cedar was forbidden by law. After World War II, large-scale felling commenced, even though the Altai cedar forests were crucial to the health of the water regime of the entire Irtysh River basin. They covered the mountain slopes along the watershed, preventing erosion and flash floods, and maintained the water level of regional rivers. In the early 1960s, Soviet journalists began to publish reports about the irreparable destruction of the cedar forests. They reported that the Baikal region was at risk because of lumbering practices in the mountains that destroyed root systems, accelerated erosion, and prevented reforestation.[46] Reforestation lagged because of difficult Siberian climatic conditions and the failure of forestry officials to allocate labor resources to replanting.

Khrushchev's sovnarkhoz reforms also had a negative impact on the forestry industry.[47] This reorganization prevented the development of national, long-term scientific management of forest resources and the construction of modern pulp and paper mills because local concerns

---

[44] Radio Liberty Research Institute, *Annihilation of Natural Resources in the Ukraine*, May 17, 1963.
[45] *Literaturnaia Gazeta*, June 23, 1966.
[46] *Komsomol'skaia Pravda*, July 27, 1965.
[47] *Pravda*, December 18, 1965.

and interests, including the fulfillment of plans to which bonuses were tied, predominated over the more universal concerns of contemporary forestry conservation practices. That is, although the ministries had been blamed for empire building and bureaucratic inefficiency, the regional economic councils favored regional or local interests. The Soviet press referred to the poor state of cedar forests in the Altai as an example of mismanagement because of the decision of regional bureaucrats to permit lumbering enterprises to engage in intensive felling in the most valuable plantations. A local party leader in the Altai remarked that "handing over the forests to the lumbering concerns (enterprises) proved unjustified under our conditions. It merely led to an even greater mess in the forests."[48]

The sovnarkhozy also had no effective control over so-called self-packers. "Self-packers" were organizations that had gained the authority to conduct their own forestry operations. They provided their own personnel and equipment to engage in logging for their own short-term needs. These organizations included railway construction brigades. They cut only the best trees, decimating forests without any regard for their conservation. Areas exposed to this kind of logging had the greatest incidence of forests fires. Self-packing developed with increasing frequency from 1959 onward, after which millions of hectares of forests were clear cut each year, with only some 40,000 undergoing selective harvest, which permitted the gradual replanting of cut areas. Many Soviet foresters and officials believed that only a central authority interested in the long-term health of the forests might end this practice.[49]

We cannot blame the sovnarkhoz system for all of the troubles that plagued the Soviet forest industry in the Khrushchev era. After all, the central government sanctioned felling that far exceeded quotas for many of the forests, including such environmentally crucial ones as those in the Carpathian and Altai Mountains. Many of the wasteful and inefficient practices had existed long before the appearance of the sovnarkhoz system and persisted long after its abandonment. The reckless timber floats along European and Siberian rivers had nothing to do with this Khrushchev reform. Only one problem of poor forest management can be entirely attributed to the sovnarkhozy system. This problem resulted from the combining of timber industrial units with forestry enterprises. The result was organizations interested solely in harvesting

[48] *Nash Sovremennik*, no. 6 (November/December 1963).
[49] *Komsomol'skaia Pravda*, July 27, 1965.

and production, while reforestation and scientific management practices lagged.

The highly damaging practice of the spring timber float exacerbated problems in the forest. The float was the primary means to transport timber from sites of felling to lumbering bases. In the 1950s, about 75 percent of all timber was floated down rivers. Hundreds of thousands of cubic meters of pine and aspen sank to the bottom of the Volga. Everywhere, huge losses of timber accompanied the float, which in turn triggered increased felling and more losses to the river bottom. These losses were not only acknowledged, but also accepted and included in plans.[50] The float contributed to the worsening of the species composition of the forests. Such species as oak, ash, and larch quickly sank. Loggers therefore preferred to cut cedar and pine.[51] Throughout the nation, the loggers overcut coniferous species.

The rivers suffered as well. Many rivers in European Russia and Siberia developed "wooden" beds and banks. In Arkhangelsk province, wood accumulated to the depth of several dozen meters in the Pinega River not far from its mouth. Six meters of sunken timber covered the bed of the Kama River in the Ural foothills. In several places by the early 1960s, sunken timber shortened the navigation period to a few weeks only when water levels were very high during the spring thaw; spawning grounds were decimated.[52] In the Lake Baikal basin, the local timber industry began floating logs into the lake in the 1930s and increased the float significantly after World War II to meet the demand of newly built pulp mills. Between 1958 and 1968, an estimated 1.5 million cubic meters of timber sank to the bottom of Lake Baikal and in nearby rivers,[53] once again destroying important feeding and spawning areas.

In March 1964, leading government, sovnarkhozy, and party officials of the Russian Republic met to discuss the tremendous losses that accompanied the spring float. They passed a resolution finding the sovnarkhozy responsible for the excessive losses of timber during the float. The enterprises under their jurisdiction left roughly 50 percent of the lumber to rot in the forest to accompany that which sank in rivers. Enterprises failed to develop alternative transportation over land and did not dry lumber

---

[50] L. Bek, *New Forestry Decree*, Radio Liberty Research Institute, March 12, 1964.
[51] *Pravda*, December 1, 1974.
[52] *Sovetskaia Rossiia*, June 27, 1961.
[53] *Financial Times*, May 11, 1978.

sufficiently in case they used the float. In 1964, timber float lost 3 million cubic meters of the 385 million cubic meters felled that year.[54] The float was banned in 1970 but still practiced on many rivers through the end of 1980s.

Destructive lumbering practices extended to poaching. In the late 1950s, the Soviet authorities recognized that illegal trading and felling of timber was going on in collective farm forests. As before, in the Khrushchev era, these farms received inadequate financial, technical, and material support from the government, whereas the purchase prices for their agricultural products established by the central authorities were too low to generate any profit. Under these circumstances, collective farm forests became an important source of income. Strangely for the socialist Soviet Union, market pressures led collective farm boards to expand logging operations at the expense of farming activities. In some places, able-bodied male collective farmers gave up farmwork for logging to the detriment of livestock, crop yields, and the forests themselves.[55] From the official point of view, the relative poverty of the collective farms and farmers hardly excused lumbering activities. The authorities accused the farmers personally involved in wood trade of being "speculators" – that is, they tarred them with the same term that was applied liberally to any and all perceived criminals during the Stalin period.

The forests under collective farm administration amounted to nearly one-sixth of the Soviet Union's forest resources, almost all of which was accessible to horse and motor transport. Accordingly, the farms' forests provided a large share of the lumber used in private and cooperative construction projects.[56] In theory, the authorities permitted lumbering so long as it did not exceed annual quotas. Yet, many collective farms cut three or more times the annual norm, and in some areas six times more. Illegal commercial lumbering expanded in spite of the ecological damage and official pronouncements that this activity was "speculation." Direct-trade contracts between collective and state farms with inadequate forest resources and those rich in forest resources developed. Wood trade operations were conducted privately and quietly through cash. In 1962,

[54] L. Bek, *New Forestry Decree*; and A. F. Kiselev, and E. M. Shagin, eds., *Khrestomatiia po Noveishei Istorii Rossii, 1917–2004* (Moscow: Drofa, 2005).
[55] Radio Liberty Research Institute, *Russ Woodlands May be Taken From Collective Farms*, July 8, 1961.
[56] Radio Liberty Research Institute, *Russ Woodlands*.

prosecutors reported that criminal cases involving trade with a total return in 1.5 million rubles ($1.65 million) were under investigation.[57] They were obliged to reforest but did little. These practices raised havoc with climatic conditions, had an inevitable impact on flora and fauna and their habitat, accelerated erosion, and contributed to local "desertification" and drying up of rivers. As the Soviet press put it, "irresponsible destruction" of collective farm forests was endemic.[58]

After Khrushchev was removed in October 1964, the sovnarkhozy were replaced by the former ministries. In 1965, officials adopted new administrative measures in the spheres of forest management and the timber industry that included the separation of the forestry sector in charge of forest management from the lumbering industry in charge of felling. A new Forestry Committee of the Soviet Union gained authority for felling and conservation in many of the nation's forests, and a Ministry of Forestry of the Russian Republic was established "for combating a disorder in forests exploitation."[59] Of course, rapacious use of forest resources persisted in the Brezhnev era because officials and forestry experts saw only short-term goals. They promoted similar environmental disasters in the area of energy production.

The Environmental Cost of Energy Production: The Case of Hydroelectricity

After World War II, the Soviet Union accelerated development of its energy production sector as part of Cold War competition with the United States. From its first days under Lenin, the Soviet Union had sought to increase electricity production manyfold; its leaders viewed electrical energy production as a panacea for all sorts of other problems. Yet through the Khrushchev era, the Soviet Union lagged considerably behind the United States. In 1957, the Soviet Union produced 209.5 billion kilowatts per hour of electricity versus more than 700 kilowatts per hour in the United States.[60] As a result of Stalin's discriminatory policy against agriculture, per capita consumption of electricity in the countryside was considerably less than that in the cities. Although 60 percent of collective

---

[57] *Sel'skaia Zhizn'*, November 27, 1962.
[58] *Sovetskaia Rossiia*, June 29, 1961; and *Lesnaia Promyshlennost'*, May 25, 1961.
[59] *Pravda*, December 18, 1965. See also *Izvestiia*, August 8, 1967.
[60] W. J. Jordens, "Khrushchev Urges More Steam Power; Soviet to Stress Power by Stream," *New York Times*, August 11, 1958.

farms and 96 percent of state farms had electricity, only 33 percent of collective farms and 18 percent of state farms were connected to a central grid. The rest were supplied by tiny hydropower stations and small generators with an average capacity of less than 35 kilowatts that could barely meet the demands of the farms; some villages had but a handful of lightbulbs.[61] Only 15,700 kilometers (roughly one-eighth) of the Soviet railways had electrified lines.[62]

Demand for municipal and household needs grew rapidly. The development of electric transport (trolley buses, trams, and trains) and water supply and the expansion of such consumer goods as radios, record players, televisions, refrigerators, washing machines, and other appliances led to significant shortfalls in production capacity by the late 1950s. Conventional (fossil) fuels accounted for 80 percent of the electric power output in the Soviet Union. Soviet leaders therefore turned to hydroelectricity and nuclear power to augment the supply. They viewed these sources of power as less expensive and a sign of modernity. Furthermore, as a part of the Stalinist plan for the transformation of nature, hydroelectric power stations had assumed special importance as part of a system of stepped reservoirs, not only to generate electricity or to irrigate arid but fertile soil, but also because of their ideological importance in competition with the United States. These projects were at once bold efforts to transform vast ecosystems. Yet many of them were far-fetched projects with great environmental and social costs for farmland, forests, fisheries, and people that were impossible to slow.

In terms of scale, cost, and irreversible environmental impact, many of the projects resemble those of the American West or contemporary China; these were not unique to the Soviet Union.[63] They included a series of stepped reservoirs and stations on the Volga and Don Rivers, associated irrigation systems, and canals linking various waterways. During the Khrushchev era, the earth-moving Ministries of Power, Electrification, Water Melioration, and others turned their attention to the Siberian rivers. Engineers rarely conducted environmental impact statements in preparation for these projects. They considered the projects in and of

---

[61] *Pravda*, September 14, 1962.
[62] *Zheleznodorozhnyi Transport*, no. 3 (1962).
[63] Mark Reissner, *Cadillac Desert: the American West and its Disappearing Water* (New York: Viking 1986); Paul Josephson, *Industrialized Nature* (Washington: Island Press, 2002); and Judith Shapiro, *Mao's War Against Nature* (Cambridge: Cambridge University Press, 2001).

themselves of unquestioned benefit, and they dismissed talk of such costs as ecosystem damage as of local or minor importance. The water reclamation projects destroyed entire riverine ecosystems. But the revolutionary symbolism of dams, hydroelectricity, and canals as glorious achievements of socialism – no less than Sputnik or nuclear power – was more important than these concerns.

A number of projects begun under Stalin were completed only in the Khrushchev era: the Kuibyshev Hydroelectric Power Station (GES, in its Russian acronym) on the Volga, the largest in the world at the time of its completion at 2.3 million kilowatts in 1957; the Stalingrad (later Volgograd) GES (October 1956), with an irrigation system for the Caspian semiarid region; the Main Turkmen Canal (1,100 kilometers from the Amu-Darya River to the Caspian Sea through the Kara Kum Desert), with an extensive network of irrigation channels; the Kakhovsk GES on the Dniepr River (February 1957); the South Ukraine and the North Crimean Canals (with a total length of 550 kilometers); and the Volga-Don Canal (101 kilometers) near the Tsimlianskoe Reservoir (1952). This meant that eight major dams with a total capacity of 10 million kilowatts were under construction. In keeping with the Soviet approach of serving the political, economic, industrial, and urban elite, the new stations would supply Moscow, Stalingrad, Kuibyshev (Samara), and a few other smaller cities but leave the countryside impoverished. Planners ultimately hoped to extend power lines to the railroads and the farms. They also intended to irrigate 28.3 million hectares of land to promote a technological revolution in agriculture, with yields to increase up to two times. Planners forecasted building another 44,000 (!) ponds, reservoirs, and other impoundments.[64]

When Khrushchev took power, the Soviet Union had seven major dams in operation with a total generating capacity of roughly 2 million kilowatts (total electrical capacity was approximately 48.5 million kilowatts). On the one hand, Khrushchev happily endorsed Stalin's "hero projects" because of the central place of electrification in the construction of Communism, and he desired to be identified with hero projects of his own. On the other hand, he worried about their cost and long construction time. Stalin's engineers had greatly underestimated the cost of "transformation," perhaps to avoid his wrath. Whatever the case, the Soviet economy lacked the capital, labor, and investment resources to

---

[64] A. A. Sokolov, *Gidrograifiia SSSR* (Moscow: Gidrometeoizdat, 1954).

build them all simultaneously. Cost and other problems delayed the dedication of the Kuibyshev GES until the end of 1957, although an entire army of prisoners had been involved at the site. To complete the project, the authorities had to abandon the Main Turkmen Canal for a later date, shift thousands of workers to Kuibyshev, and postpone the completion of the Stalingrad GES from 1956 to 1960. Projections for the economic value of the Volga-Don Canal also turned out to be highly inflated, and it was never used for transport to the extent that planners had envisaged.[65] The figure of 28.3 million hectares of irrigated land to be achieved through the various hero projects was totally unrealistic; in 1965, the total area of irrigated land in the Soviet Union was less than 10 million hectares.

Khrushchev recognized many of these problems in a speech he offered at the dedication of the Kuibyshev power giant at the end of 1957. His speech shed light on the major problem in the economy: how to secure a rapid return for each ruble of investment, how to choose among the many valuable projects, and whether large-scale projects were simply too large scale. Khrushchev made clear that the Kuibyshev station would be the last of the great Volga power stations. He suggested that the future emphasis would be on smaller thermal power stations based on cheap coal, natural gas, or oil that were faster to build and cost less. Khrushchev dismissed the argument of hydroelectricity being less expensive, which only one year earlier had been considered axiomatic. He stressed the advantages of thermal power stations, repeating the position of most experts on the subject that was clear from even a cursory reading of newspapers and magazines. As an example, he announced that several gas-based thermal power stations would be built in place of a planned hydroproject near Saratov.[66]

Yet, Khrushchev and other opponents of these big projects (as opposed to his own favorite big – agricultural – projects) found it difficult to overcome the institutional momentum they and the ministries behind them had acquired. A number of major bureaucracies had come into existence whose raison d'être was large-scale geological engineering projects, and they were powerful lobbying groups. Through the 1950s, engineering organizations that had spun off from the secret police and

---

[65] H. Schwartz, "Soviet Dam Work Far off Schedule; Delay in Kuibyshev Project Illustrates Time Factor Noted by Khrushchev," *New York Times*, August 11, 1958.
[66] *Christian Science Monitor*, August 22, 1958; and "Man in a Hurry," *Time*, August 25, 1958.

their gulag system were responsible for virtually all of the hydraulic engineering in the Soviet Union. The managers of the major organization, the Main Hydrological Construction Agency, were major generals in the NKVD (Narodnyi Kommissariat Vnutrennykh Del).[67] Many of its workers had been prisoners. The major engineering design institute, the Zhuk Institute, was named after the major general responsible for supervising the White Sea–Baltic Canal. Other ministries tried to avoid conflict with the agency by pursuing their own projects. This enabled the ministries to move ahead virtually unopposed with some of their projects throughout Russia, Siberia, and Central Asia into the mid-1960s, which had the effect of slowing down the development of the thermal power sector.[68]

Khrushchev opposed hydroelectricity because of its high capital costs and the lengthy construction times, not because of environmental considerations. For example, engineers discussed the possibility of utilizing the vast hydroelectric potential of Tadzhikistan in the 1930s, but plans to build a huge station arose at Nurek, a small township a few miles from the capital, Dushanbe. Construction began in the late 1950s based on optimistic estimates of capital and generation costs for the 5,000-megawatt station provided by the Central Asian Branch of Gidroenergoproekt, and operation by 1965.[69] Although announced with great fanfare in November of 1960, the Nurek GES took twenty years to build. By comparison, the 5.2-million-kilowatt Churchill Falls hydroelectric power station in Canada took only ten years to build.[70] Similarly, workers commenced construction on the Inguri GES in Georgia in 1961 with a target completion date of 1966. But engineers overlooked very complicated geological structures and seismic activity in the area that demanded extensive additional work to guarantee the station's safety; the station was fully operational only in 1980. Furthermore, central authorities built the Inguri station, one of the highest in the world, against the wishes of local

[67] B. Komarov, *The Destruction of Nature in the Soviet Union* (White Plains, NY: M.E. Sharpe, 1980), p. 150.
[68] *Komsomol'skaia Pravda*, June 17, 1989.
[69] David Dyker, "Industrial Location in the Tadzhik Republic," *Soviet Studies*, vol. 21, no. 4 (April 1970), pp. 494–495.
[70] Lilia Malik et al., "Development of Dams in the Russian Federation and Other NIS Countries," *WCD Briefing Paper* (Capetown, South Africa: World Commission on Dams, n. y), and Stephanie Joyce, "Is It Worth a Dam?" *Environmental Health Perspectives*, vol. 105, no. 10 (October, 1997), pp. 1050–1055.

residents.⁷¹ Several Soviet economists estimated that the real period to cover the full expenditures for construction of a hydroelectric station was twenty years, although it was profitable only if completed in around eight years.⁷² In the United States, the major organizations involved in the construction of hydroelectric, irrigation, and flood control projects, the Army Corps of Engineers and the Department of Interior Bureau of Reclamation, have similarly faced sharp criticism for underestimating the true costs of projects, running roughshod over the environment, and endangering various species.⁷³ In the case of the Tennessee-Tombigbee Waterway, the Corps came into direct conflict locally and nationally with environmentalists.⁷⁴

In fact, Khrushchev's determination to limit the construction of huge hydroelectric power stations could not withstand the flood of projects from huge engineering organizations in the 1950s and 1960s. The most ambitious Soviet hydro-project, the construction of the Bratsk GES on the Angara River (at 4.5 million kilowatts, twice the capacity of the biggest dam in the United States, the Grand Coulee), revealed other problems with the Soviet system. Unlike the Grand Coulee, the Bratsk station was completed in a record-short time of eight years when it came into service in 1963. Yet the authorities had failed to locate the station near energy demand; the large production capacity simply went unused, waiting for the day when smelting and other large industrial complexes were built. If the industrial facilities had been under construction simultaneously, the planners would have required significantly more labor and capital resources than those at hand, slowing the construction of all of them. Instead, the ecogeography of a huge river basin was irrevocably altered without any economic reason.⁷⁵

Furthermore, Soviet designers failed to include a series of real costs other than capital costs in their projections for their projects: site preparation, costs to resettle the ousted inhabitants ("oustees"), and the value

---

[71] Yuri Badenkov, "Sustainable Development of the Mountain Regions of the USSR. The Realities, the Role of Science, and Research Orientations," *Mountain Research and Development*, vol. 10, no. 2, Transformation of Mountain Environments (TOME). Part One (May 1990), pp. 129–139.
[72] *Komsomol'skaia Pravda*, June 17, 1989.
[73] Mark Reissner, *Cadillac Desert*.
[74] Jeffrey Stine, "Environmental Politics in the American South: The Fight over the Tennessee-Tombigbee Waterway," *Environmental History Review*, vol. 15, no. 1 (Spring 1991), pp. 1–24.
[75] *Komsomol'skaia Pravda*, June 17, 1989.

of agricultural, timber, and other production lost in the submerged lands. They certainly never calculated the environmental impact on the water regime of the river basin in question. Ultimately, Khrushchev opposed hydroelectricity projects because the Soviet Union did not have enough investment capital to undertake all of them while simultaneously pursuing expensive military, agricultural, and other industrial projects, especially because there seemed to be no prospect for a quick return on the investment in hydro. (Russia in the twenty-first century continues to struggle with the legacy of Soviet hydroelectricity. A disastrous accident involving an explosion and flood of the machine hall of the Saino-Shushenskaia GES on the Enisei River, rated at 6,400 megawatts of electrical power, one of the largest in the world, in 2009 killed 75 people and led to the shutdown of the station, with repairs not to be completed until 2014[76].)

In 1961, Soviet specialists and journalists engaged for the first time in a serious public discussion of the relative merits of hydroelectric and thermoelectric power stations, especially those constructed on flood plains. The Soviet technical journal *Nauchno-tekhnicheskie Obshchestva SSSR* (*NTO SSSR*, Scientific-Technical Societies of the Soviet Union) published a series of articles criticizing the essential "gigantomania" of hero projects. The editors asked the rhetorical question: Why had huge hydroelectric power stations been built without a complete consideration of their economic benefits and their social and environmental costs? The publication of such editorials as these likely reflected a policy of stimulating discussions among engineers and scientists about the ways to promote the development of their particular technical specialties. The scientific-technical societies were NGOs under direct control during the Stalin era, but after Khrushchev they were encouraged to engage in technical discussions and mass activities – once again, with the understanding that Soviet Communism was based on objective science.

In the particular sphere of energy production, the discussion concerned a choice between hydroelectric power stations and thermal power plants. Economic considerations as always came first, but environmental issues concerning dam construction inevitably found a place. One scientist provided an analysis of these real and heretofore ignored costs of hydro-project costs. According to his calculations, by 1975, the total

[76] "Results of Probe Into Russian Hydropower Plan Disaster Announced," http://en.rian.ru/russia/20091003/156338526.html, October 3, 2009.

agricultural area inundated by reservoirs would be 30 million hectares, including some 12 million hectares of cultivated and 15 million hectares of wooded land. The author referred to the Virgin Lands campaign, pointing out that thirty billion rubles had been invested since the launching of that program to open up 12 million hectares for cultivation. The land lost to inundation behind hydroelectric power stations provided far more reliable harvests than the Virgin Lands because it was often in fertile flood plains and other agricultural areas. But, the scientist continued, the economic losses due to the inundation of the surrounding agricultural and forest areas had been always underestimated by the authorities in order to keep the electrification effort going full speed ahead.[77]

In another issue of *NTO SSSR*, a specialist calculated that the economic losses to the agriculture and fishing industries connected with the construction of another Volga River power station far exceeded a sum sufficient to build a thermal power station with a capacity of 3 million kilowatts – more than the Volga station itself. The specialist expressed strong opposition to the construction of a proposed Kiev GES, noting that it would cost twenty times more than a thermal power station of the same capacity. He then estimated the total cost of the Dneprodzerzhinsk and Saratovsk GES at 464 million rubles, with which it would have been possible to build thermal power stations with total capacity of 8.4 million kW,[78] and without the extensive loss of land.

Scientists from other specializations – and with other economic interests – weighed in on the debates. The head of the Fisheries Research Institute of the Ukrainian Academy of Sciences published an article in 1960 titled "Treasures under Water." He addressed the environmental impact of the construction of reservoirs on the Dnieper and Volga rivers, noting that tens of thousands of hectares of fertile farmland had disappeared under excessively large reservoirs. He claimed that if the area of the Kakhovsk reservoir were reduced by 70,000 hectares, it would permit a large area of agricultural land to remain in service without any significant loss of water resources.[79] Any further loss of agricultural lands in the European Soviet Union due to construction of a series of stepped hydroelectric power stations would likely force the government to sooner

---

[77] *Nauchno-tekhnicheskie Obshchestva SSSR*, November 11, 1961.
[78] Ibid.
[79] *Izvestiia*, May 11, 1960.

or later refrain from construction on flood plains in favor of locating them in mountainous and sparsely populated regions.

Many engineers referred to relatively low capital costs of hydroelectricity in Siberian power stations compared to those located in the European part of the country because of the relatively few settlements and limited agricultural land. Yet they overlooked the fact that millions of cubic meters of timber would be lost in the zone of any planned Siberian reservoir, and tens of thousands of hectares of land had to be cleared of bushes, trees, and debris at great cost. Before this happened, a new, huge lumbering enterprise had to be created in short order at the construction site. The engineers had to include housing and roads, stores, electrical substations, schools, hospitals, and so on in their design costs. This social capital always lagged behind. And to recoup capital costs, they sought to generate electrical energy as soon as possible. This often led them to start filling the reservoir before lumbering activities ended and before housing was complete. Their solution? They designed floating lumber mills and employed frogmen and loggers, and workers and their families lived in barracks and tents. According to some estimates, the cost of clearing the Bratsk reservoir of forests equaled the amount spent on the construction of the station itself. Felling and removal of timber lasted longer than construction of the station, and large areas of forests were submerged under water when the GES went on line. In the case of the Bratsk GES, the amount of lumber inundated by the waters backing up behind the dam equaled the annual harvest of all lumber enterprises in the province.[80] The scale of preparatory work needed to build any new reservoir and hydroelectricity power station had to be considered on the basis of more realistic calculations.[81]

Another environmental cost associated with hydroelectricity was the loss of fisheries. This cost affected all bodies of water, from the great Azov and Caspian Seas to ponds and rivers. For example, the Don River played a central role in the stability of the Azov fisheries. In the 1930s, the catch of valuable fish species ranged from 400,000 to 600,000 centners. During the 1960s, the figure fell to 40,000 centners. This sharp decline was connected to the construction of the Tsimlianskoe Reservoir in 1953. Flooding of the alluvial plain ceased in eleven of the next fifteen years, including six years in a row, because of the low flow of the Don. In 1963

---

[80] O. V. Afanasov, *Angarlag i Ozerlag pri Stroitel'stve GES Angarskogo Kaskada v kontse 1950 – nachale 1960-x gg* (Irkutsk: Ottisk, 2008), pp. 104–109.

[81] *Izvestiia*, December 8, 1973.

and 1968, the flood plain was covered by water in spring for one to two months, with a good effect on the spawning of Azov fishes. Higher salinity of the Azov Sea resulted from diminished river inflow; this increased the flow of saltier water from the Black Sea into the Azov Sea, and this obviously had a deleterious impact on Azov inland fisheries. Agricultural and industrial pollution also negatively affected the health of the fisheries.[82]

Construction of GES had a similarly deleterious impact on the water balance of the Caspian Sea. At one time, rivers contributed roughly 390 cubic kilometers of the annual inflow of water into the Caspian. The construction of a cascade of thirteen major dams on the Volga reduced the quantity and quality of flow into the Caspian, whose level began to decline sharply in the late 1950s. This reduced the size of the spawning areas in the Volga estuary. In the 1960s, the catch in the Volga basin and the northern Caspian dropped from 5 million to fewer than 1 million centners of fish.[83]

The Soviet Union produced about 95 percent of the world's black caviar, almost all of which came from Caspian Sea sturgeon that spawned in the fresh water of the Volga. The sturgeon population of the Caspian Sea dropped precipitously when their spawning grounds were destroyed by hydroelectricity stations, reservoirs, and irrigation projects, and because of extensive pollution. In the early 1960s, Soviet scientists opposed a plan for a new power station on the lower Volga River because of their calculations that it would reduce spawning grounds by another 96 percent – from roughly 400 hectares to a miniscule fifteen hectares. Individuals from a variety of backgrounds reacted in horror to this information. In 1961, the newspaper of the Union of Soviet Writers, *Literaturnaia Gazeta*, carried an article by eight authors – most of them prominent biologists – who attacked the project for its devastation of sturgeon spawning grounds. In this case, the environmentalists won the battle. The Volga remains free flowing, but only from Volgograd to the Caspian Sea. Officials from the State Committee for Fisheries acknowledged that the Central Committee of the Communist Party and the Council of Ministers had concurred in the decision not to build the station.[84]

---

[82] N. V. Melnikov, ed., *Regional'nye Problemy Razvitiia Proizvoditel'nykh Sil SSSR* (Moscow: Nauka, 1972).
[83] Melnikov, *Regional'nye problemy*.
[84] *New York Herald Tribune*, April 28, 1965.

Soviet planners, hydrologists, and others with vested interests in geo- and hydroengineering responded to these problems of water shortages in the Volga and Don basins by drafting a hubristic project to divert the flow of northern rivers to the south, one of many increasingly aggressive projects to redraw nature through more geoengineering; they believed there were always technological solutions to problems of technological origin. The project envisioned the diversion of the Pechora and Vychegda Rivers into the Kama and the Volga Rivers, both for irrigation (2 million hectares of arid land in the Volga region) and to compensate for the loss of water to the Caspian Sea. In the late 1950s, the Zhuk Hydrological Design Institute designed the project. The plans included a new reservoir (Pechorskoe) that would be constructed at the expense of an equivalent area of land under cultivation. The area included 1.4 million hectares of forests, roughly 5 percent of all the farmland in the Komi Autonomous Republic and Perm province, and 8,000 farmsteads. Moreover, forests in the upper reaches of the rivers would disappear, the Pechora would lose its ability to support transport, fish would vanish and, of course, climate would be adversely affected.[85]

Other engineers recognized that the withdrawal of a large amount of water for irrigation purposes in the Central Asian republics for cotton culture would result in shrinking of the Aral Sea. Hence the idea of the diversion of Siberian rivers to Central Asia, first proposed at the end of the nineteenth century, was rekindled. In the Khrushchev era, the era of sputnik and nuclear power, engineers discussed the diversion of Siberian rivers as a technological solution to a problem of technological, economic, and political origins.[86] We discuss the projects for the diversion of northern rivers that became a source of a great controversy for environmentalists and others in the next two chapters.

The Struggle to Protect Nature Reserves Renewed

An analogue of what today we call NGOs, in this case with interests in nature protection, existed in the Soviet Union, but they played only a formal role because they were under close political supervision or had been co-opted by the authorities. The largest was the All-Union Society for the Conservation of Nature (VOOP), which came into existence in

---

[85] *Nash Sovremennik*, no. 3 (1963).
[86] A. A. Sokolov, *Gidrografiia SSSR* (Moscow: Nauka, 1954).

1924. It was the equivalent of the United States' Sierra Club, but it lacked autonomy and authority. In the late 1960s, it had nearly 19 million members in the Russian Republic, 6 million in Ukraine, and smaller numbers in other republics. In theory, members of VOOP should have been consulted during the drafting of any plans having an environmental impact. Its members might have advocated changes in policies and plans, but this did not happen. VOOP leaders had little to say in public about such important issues as the pollution of Lake Baikal. Occasionally, they spoke up about pollution in sensitive conservation areas.

To the extent that there were any environmental lobbyists in the Khrushchev era, they could be found among biologists of the Academy of Sciences in a series of temporary working groups and commissions headed by civic-minded scientists-turned-activists from the older generation. The groups were interdisciplinary teams capable not only of communicating with the central authorities but also of drafting environmental laws. One of the most important of these groups was the Commission on Nature Reserves that was established in 1955. The commission was one of the key actors in the postwar debates over resource use and nature protection. The commission united activists from such associations as VOOP, MGO (the Moscow Branch of the All-Union Geographical Society), and MOIP (the Moscow Society for the Admirers of Nature). Because the VOOP leadership had been co-opted by party bureaucrats, other groups and associations were extremely important in the generation of public discourse on environmental issues.[87]

MOIP was one of the few public organizations that maintained a modicum of autonomy from the Party-state machine in both the Stalin and Khrushchev eras.[88] Its charter principles were democratic, and its members did not have to be professional scientists, but merely actively involved in the scientific life of the society. MOIP was simultaneously an important theoretical center for biological sciences through the efforts of its members to advance modern understandings of ecology during a time when genetics remained repressed under Lysenko and the leading force in nature protection activism of the Soviet Union through its unbreakable support of zapovedniks.

Conflict raged between Khrushchev and Soviet biologists over many issues, especially Lysenkoism and the place of zapovedniks in Soviet

[87] Weiner, *A Little Corner of Freedom*, pp. 186–190.
[88] Ibid., pp. 211–216.

society. Khrushchev still supported the odious biologist Trofim Lysenko, who was personally responsible for the repression of many Soviet geneticists. Geneticists were able to function largely only in physics institutes under the protection of nuclear physicists who required research on radiation genetics. By 1956, several hundred Soviet scientists had signed a letter criticizing Lysenko that they sent to Khrushchev, unfortunately without the desired result of Lysenko's ouster.[89] Similarly, the biologists disagreed sharply with Khrushchev over zapovedniks. They wished to protect and expand them, but Khrushchev saw them as a waste of land and money because they contributed nothing directly to the economy. Khrushchev had a deservedly bad reputation in this regard. In 1951, he had been involved in the work of a commission that in a Stalinist fashion closed 88 nature reserves of a total of only 128. Khrushchev's rise to power made matters worse. In 1954, officials at such nature reserves as Askania-Nova in Ukraine complained that illegal grazing, agricultural, and lumbering activities had encroached on the reserves.[90] There is no evidence that Khrushchev personally sanctioned this encroachment, but his negative attitude toward the reserves was shared by many local officials who likely tolerated this illegal activity. Indeed, Khrushchev's first action as leader toward the reserves was to permit hunting in some of them. In 1957, three major reserves – the Belovezhsky (Belarus), Crimean, and Azovo-Sivashsky (Ukraine) – gained the new status of "hunting natural reserves." Needless to say, hunting reserves were accessible only to the Soviet elite, and hunting killed animals.[91]

Soviet environmentalists had long been concerned about the future of the nature reserves. They concentrated efforts on restoration of several old reserves and establishment of a small number of new ones so as not to upset the authorities. In 1954, the members of MOIP, VOOP, MGO, and other scientists organized a conference on nature preserves, hoping to reassert their rightful position to influence policy by advancing ecological science now that Stalin was dead. The conference also represented open protest of the scientific intelligentsia against the Virgin Lands campaign. The scientists were successful to a certain extent. They successfully promoted an increase in the network of nature reserves during the thaw.

[89] Zhores Medvedev, "Two Decades of Dissidence," *New Scientist*, vol. 72 (November 4, 1976), pp. 264–265.
[90] V. I. Ponomareva, ed., *Ekologiia i Vlast', 1917–1990* (Moscow: Mezhdunarodnyi Fond "Demokratiia," 1999), p. 428.
[91] Boreiko, *Belye piatna*, p. 289.

If in 1953 there were only thirty-eight nature reserves, by 1960 the number was eighty-five,[92] and many of those closed in 1951 were reopened (the Lazovsky, Lapland, Central Lesnoi, Zhigulevsky, Bashkir, Altai, and others).

Another victory of the Commission on Nature Reserves of the Academy of Sciences was the passage of the first republican law on nature protection, "On the Protection of Nature in the Estonian SSR," adopted by the Supreme Soviet Republic of Estonia in 1957. The Baltic republics were in the vanguard of the nature and cultural monument protection movement. An active group of environmentalists had gathered at Tartu University, where they pushed an agenda to expand nature preserves and promote environmental protection legislation. The Estonian law paved the way for similar laws to be passed in other republics simply by indicating that it could be done. The Russian Republic adopted such a law in 1960. Khrushchev considered the laws to be excessive. He responded to the environmentalists' initiatives at a plenary session of the Central Committee in 1961, where he criticized nature reserves in a typically expansive manner:

> Too many nature reserves have appeared throughout the country. As you probably have done, recently I have seen a documentary film about a reserve located in the Altai mountain region. This film is well made. And it shows us a very healthy man, certainly a scientist – if this is a reserve, then all people who are in it are scientists [laughter in the auditorium] – [who] is lying on a rock and [watching] a squirrel crunching nuts through binoculars. Then the man turns his binoculars to look at a bear walking. For what purpose does this reserve exist? It seems to me that it exists only for the scientists who live there. These people are simply "grazing" there, and they do this better than the bear or the squirrel! But if all these biologists left the reserve nothing would happen there – the squirrel would crunch her nuts as she did it before. It makes no difference for [the] squirrel whether the scientist is in the place or not. The difference is only for the scientist because when he is watching the squirrel he earns money, by the way, very good money.... We have some reserves which are worthy in terms of scientific and state interests. But large numbers of existing reserves have no any excuse. What can happen in forests if they are not protected as nature reserves? Nothing! Nature, of course, needed to be cared for and preserved but not by organizing of a large number of reserves....[93]

---

[92] A. M. Krasnitski, *Problemy Zapovednogo Dela* (Moscow: Lesnaia Promyshlennost', 1983), p. 190.

[93] N. S. Khrushchev at *Plenum Tsentral'nogo Komiteta KPSS, 10–18-ogo Ianvaria 1961 g., Stenograficheskii Otchet* (Moscow: Gospolitizdat, 1961), pp. 601–603.

Other members of the Central Committee met the speech of Khrushchev with jolly approval. Sanctions against some reserves followed, including the liquidation of sixteen of them, among them the large Altai reserve that had irritated Khrushchev so much. The areas of eight other reserves were radically cut.[94] These actions, evidently, violated the laws on nature protection adopted in 1957–1960. Clearly, several of the reserves were closed in favor of the interests of the timber industry.[95] As for the Commission on Nature Reserves, Khrushchev first ordered it transferred from the Academy of Sciences to Gosplan, the state planning administration, with its innate economic interests, and then, in October 1963, he shut it down.

Legal scholars criticized several of the statutes as being too weak.[96] Several Soviet jurists – and environmentalists – found the laws too declarative and pointed out that they did not cover the full variety of complex relationships in many spheres of nature conservation. The laws defined neither the content of offenses nor the liabilities, sanctions, and responsibilities for breaking laws. Yet, their significance was clear in giving nature reserves legal status and protecting them from economic activity forever. Taking into account the opposition of the Soviet leader to the laws, their passage indicated a new power among the environmental lobbyists, even though that power was hardly noticeable to an outsider.

Big Projects, the Environment and Nature Under Khrushchev

Because of the absence of public involvement, lack of interest of political leaders, and continued emphasis on heavy industry, problems with the environment grew worse and worse under Khrushchev, especially as agricultural, water melioration, and power industry projects gained momentum. Soviet cities continued to suffer from high levels of industrial emissions, despite the promulgation of regulations that restricted the location of industrial enterprises to specific regions. Still, many of the enterprises dated to the first Five-Year Plans and the war years; they lacked any pollution-control equipment. They were built by poorly motivated workers, kulaks, and prisoners, who had no interest in issues of

---

[94] Boreiko, *Belye Piatna*, p. 289.
[95] Komarov, *The Destruction of Nature*, p. 150.
[96] Boreiko, *Belye Piatna*, p. 289.

environmental safety when their own lives were at risk. The industrial centers built during the postwar reconstruction period and those of the 1950s in Siberia and 1960s were also heavy polluters.[97] Meanwhile, a number of researches by Soviet hygienists had revealed harmful effects of air pollution in industrial cities on the health of the population, especially children (see **Box 2**). Rivers and lakes fared no better than air and land. A 1960 census showed that only 40 percent of the cities and suburbs in the Russian Republic had sewage treatment facilities, whereas in the Soviet Union as a whole, only 35 percent of the urban housing units were connected to any kind of sewer system. Only 30 cities from a total of 1,763 (or 2 percent) had biological treatment facilities.[98]

The senseless destruction of land and waterways gained significant public attention even under censorship. On April 22, 1960, the Council of Ministers of Soviet Union adopted a resolution, "On Measures to Put in Order the Utilization and Strengthen the Conservation of Water Resources of USSR," that forbade putting into operation any industrial enterprise until waste treatment facilities were installed. The resolution named all major rivers of the country as severely polluted. It ordered an increase in fines for the discharge of untreated industrial effluents and blamed the sovnarkhozy for not taking into account the transboundary character of waterways. The resolution also ordered the introduction into programs in several technical higher educational institutes of special courses on methods of cleaning of industrial effluents. However, most industrial enterprises continued to function without treatment of their discharges, and many new ones came into operation without installed purifying systems.[99]

From 1962 to 1965, the Council of Ministers of RSFSR continued to issue resolutions that referred to the discharge of untreated industrial effluents into virtually all of the main rivers of the Russian Republic: the Kama, Moscow, Neva, Ob, Irtysh, Volga, and Don Rivers, and into the lakes and reservoirs in European Russia. According to the official figures, in 1967, about seventy cubic kilometers of waste water was discharged into these bodies of water virtually untreated, whereas only fifteen cubic

---

[97] Komarov, *The Destruction of Nature*, p. 150.
[98] Marshall Goldman, *The Spoils of Progress: Environmental Pollution in the USSR* (Cambridge: MIT, 1972); and Kiselev and Shagin, *Khrestomatiia po Otechestvennoi Istorii*.
[99] Goldman, *The Spoils of Progress*; and Kiselev and Shagin, *Khrestomatiia*.

> **Box 2. Soviet Research on the Impact of Industrial Pollution on Public Health in the late 1950s**
>
> The Institute of General and Community Sanitation of the Academy of Medical Science of the Soviet Union (established in 1944) conducted a study in 1958 on air pollution from thermal power stations, combined with mass clinical and x-ray examinations of children, twice over a period of three-and-a-half years that showed that exposure to various aerosols could have a detrimental effect on children's health. In 1959, they observed that industrial emissions from metallurgical plants unfavorably affected the health of the population (not surprisingly) within a three-kilometer radius of the plants. Research to determine the effect on public health of phosphorite dust established that there was a correlation, also revealed by x-ray examination, between the extent of the dust burden in the atmosphere and respiratory disease among the population, even in adults who did not experience excessive exposure but lived constantly in a dust-laden atmosphere.
>
> Mottled dental enamel was found in children exposed to industrial emissions containing fluorine compounds, and signs of hypotension, leucopenia, and thrombocytopenia (i.e., signs of the specific effect of hydrocarbons) were observed in children exposed to pollution from the petrochemical industry. Soviet hygienists studied the effect of asbestos dust on children's health, finding that control subjects from districts with clean air put on weight faster and were less prone to acute catarrh of the upper respiratory tract than were children living in districts exposed to emissions from asbestos dressing plants.
>
> In a study of the effect of emissions from a superphosphate plant on the health of people living in the vicinity, scientists discovered that exposure to concentrations of sulfur dioxide, fluorine compounds, sulfuric acid mist, and nitric oxide increased morbidity among children and adults by a factor of 1.6 to 5.7 over that for people living in districts with clean air. This increased morbidity was manifested in children by higher incidents of catarrh of the upper respiratory tract, enlargement of the lymph glands, changes in the bone structure (scoliosis, residual rheumatism), and tuberculosis. The report concluded: "Wherever this type of investigation was carried out, results indicated

> the possibility of penetration into the body from the polluted atmosphere of lead, mercury, phenol, fluorine, sulfur dioxide, chloroprene, and other compounds of industrial emissions, and high morbidity indices were found among children in areas with a polluted atmosphere than in control areas."[100]

kilometers (or only 21 percent) had been treated.[101] Pollution was the heaviest in the European Soviet Union, the area embracing 80 percent of the nation's industrial production and a home to 70 percent of the population. This was also the region where hydropower stations destroyed thousands of hectares of alluvial meadows and spawning grounds for sturgeon all the way to the river's Caspian Sea delta, and where irresponsible lumbering in the Carpathian Mountains and the Caucasus led to erosion, blowdowns, and erosion of river banks.

The negligence of the Khrushchev regime toward the environment must be understood in a broader context; in fact, throughout the world, environmental concerns and environmental movements developed essentially after the Khrushchev era – at the end of 1960s and the beginning of the 1970s. The movement followed the mobilization of public interest in the wake of the publication of Rachel Carson's *Silent Spring* (1962),[102] the first book in biology to reach such a wide audience. It prompted U.S. officials to take immediate action against water and air pollution. In the European community, the public also responded with vigor. Unlike in the West, public pressure was virtually nonexistent in the Soviet Union. Nor was *Silent Spring* or any similar book published. The authorities classified copies of *Silent Spring* or put them in special, closed library repositories.

In addition to the absence of a broad-based public component to environmentalism, another difference between the Soviet Union and the West was the lack of concern of the Soviet leadership. This meant that neither preventive action nor preliminary assessment of the impact of Soviet economic development practices on the environment had ever been considered in a systematic fashion, nor did they commence in the Khrushchev

---

[100] N. F. Izmerov, *Control of Air Pollution in the USSR* (Geneva: World Health Organization, 1972).
[101] Melnikov, *Regional'nye Problemy*.
[102] Rachel Carson, *Silent Spring* (New York: Fawcett Crest, 1962).

era, even as engineers continued to launch typically environmentally risky and expensive projects. These projects frequently had significant direct and indirect losses, including environmental, social, and political costs. The condition of the forests deteriorated. Huge energy complexes resulted in the destruction of valuable agricultural lands and fisheries. Wind erosion and droughts that followed the Virgin Lands campaign resulted in the food crisis, which was only partly relieved by the first massive grain import from the capitalist countries.

Khrushchev ran into a series of obstacles in his efforts to reform the Soviet political and economic system, all of which exacerbated environmental problems. The major obstacle was the power of conservative party officials with entrenched interests opposing any change that weakened their influence and perquisites. In addition, Khrushchev faced the constraints of trying to compete with the United States in the international arena during the Cold War. This required him to pursue policies that favored big industry, big agriculture, and big forestry. Highly polluting metallurgical combines expanded production. Ministries concerned with power generation also embraced highly ambitious projects to increase capacity and power production. Pollution and land use problems grew more severe.

Environmental problems changed the country in one very important way. During the "thaw," the public slowly became involved in various efforts to promote reforms. In the late 1950s, public opposition to Khrushchev's initiatives was hardly noticeable. But by the early 1960s, an explosion of critical articles appeared in the Soviet media, and many of them concerned the wide spectrum of environmental problems. The media had investigated the scale of erosion in Kazakhstan, the uselessness of cultivating maize in forest zones of Russia, the rapacious felling of the forests in the Carpathian Mountains, the threat to Lake Baikal, the decline of fisheries in the Azov Sea, and many other issues. These publications indicated that serious opposition to Khrushchev's policies had grown among all strata of Soviet society, including writers, local administrators, party officials, agricultural and forestry experts. Some of them had joined environmentalists, which meant the expansion of their activity beyond wildlife conservation and nature preserves to a critical analysis of the environmental impact of economic development programs. Two additional factors must be mentioned. First, this public opposition was possible because of relatively free media during the Khrushchev's era. After his removal from office, Party control of the media became much stronger and included the release of environmental information. Data pertaining to environmental issues were frequently classified. Second, although environmental concern grew among the Soviet intelligentsia, including engineers, it remained rare among ordinary people.

The Virgin Lands campaign, and postwar Soviet agriculture generally, was driven by a harsh effort to apply chemical fertilizers that in fact ruined the soils, not produced a great grain harvest as this poster promised.

Leonid Brezhnev, who presided over what came to be known as the "time of stagnation" (1964–1983), pursued policies toward the environment that led to the promulgation of new laws and statutes to ensure sustainable growth, but that were powerless to halt the increasing pollution and land-use problems that characterized Soviet development programs. He constantly referred to "hard work" and the need to perfect existing mechanisms, not pursue reform.

4

Developed Socialism, Environmental Degradation, and the Time of Economic "Stagnation," 1964–1985

Soon after Leonid Brezhnev and his allies in the party deposed Nikita Khrushchev in 1964, they claimed that the country had entered the stage of "developed socialism." Khrushchev had embarrassed them by promising in 1961 to achieve Communism by 1980, clearly a difficult goal given the poverty in the countryside, the shortages of consumer goods in the cities, the growing costs of waging the Cold War, and increasing awareness of extensive environmental problems. In their claim of "developed socialism," they sought to convey the message that socialist society had transformed into something qualitatively more advanced than in the Stalin and Khrushchev eras and rivaled the capitalist West. Developed socialism became a frame of reference throughout Brezhnev's days in power against both the nation's own and Western achievements in a variety of areas. Economic growth, progress in culture and science, and advances in the areas of environmental protection and rational use of resources – all of these things indicated such achievements. Yet, environmental problems grew worse in the Brezhnev era, the pronouncements of the Soviet leaders notwithstanding. Erosion, deforestation, and pollution accelerated. The priority of economic development left the land disfigured, the water poisoned, the air polluted. Whether agriculture and its excessive use of chemical biocides, forestry and its indiscriminate clear cutting and waste, or industry and its mortal contamination, the Soviet system may have been "developed," but it was also increasingly polluted "socialism." The citizen – the ostensible beneficiary of the leadership's enlightened rule – lived in an increasingly dangerous environment.

Granted, the Soviet Union had become a major world power. It was free from political and economic instabilities. Its scientists rivaled those in the Western democracies in number if not always productivity; with one-third of the world's engineers and one-quarter of its physicists, the Soviet Union was a pioneer in atomic energy, space, and other fields. Its dancers, pianists, and writers captivated international audiences. In the environmental sphere, ecologists worked with jurists to pass progressive legislation. Scientific and national societies of environmentalists grew in membership. Yet, at the same time, the nation experienced growing environmental degradation, rising infant mortality, an epidemic of alcoholism, and stagnating quality of life that leaders could not ignore, even as they sought to obscure public awareness of these problems.

Under Brezhnev, the government began a concerted effort to promulgate legislation to protect the environment, establish protected areas, and regulate industry and agriculture through laws that limited pollutants and fined lawbreakers. The domestic and international determinants of these efforts – for example, the need to live up to the rhetoric of "developed socialism" and to keep up with legislation being passed in dozens of Western nations – contributed to the effort, as did a burgeoning environmental protection movement with activists both within and outside official government channels. The lack of true openness regarding the extent of environmental degradation and the insistence of the leaders in emphasizing tried-and-true, but environmentally costly, development programs doomed those efforts to failure. These included a continued emphasis on heavy industry and the so-called Brezhnev Food Program, which contributed to the poisoning of agricultural lands. At the end of Brezhnev's tenure, the problems of air, land, and water pollution had only grown worse.

The Brezhnev administration turned to the promulgation of laws and statutes, joined international regimes, and otherwise pursued the reality – and façade – of legalistic language of environmental protection for three reasons. First, the Soviet Union was a far different country in the 1970s than it had been in the 1930s or the 1950s. This was a modern industrial and urban society, a society of well-read and well-educated citizens whose expectations for daily life and long-term hopes reflected neither the fear of Stalinist coercion nor the uncertainties of Khrushchevean reforms. Second and similarly, "developed socialism" was not only a slogan; it also reflected policies based on science, not on whims, formulated with the input of specialists, not only of party officials who exhorted citizens to engage endlessly in campaigns to increase production. Production would increase, but on the basis of careful, scientifically founded, and efficient utilization of capital, labor, and resource inputs.

Last, in the era of détente, ideological competition between the United States and Soviet Union extended to the sphere of environmental protection, and officials were determined to use laws and statutes – including the new Brezhnev constitution – to demonstrate the superiority of the Soviet system.

The Legacy of Heavy Industry

By the 1970s, Soviet leaders, economic planners, and industrial managers had succeeded in their long-term goals of achieving military and economic parity with the West. According to the official statistics, Soviet industrial output had reached 80 percent of that of the United States. The Soviet Union was the global leader in production of coal, iron ore, tractors, cement, coke, cotton, and wool, and it outstripped the United States in rates of annual growth for the production of oil, cement, cast iron, steel, mineral fertilizers, and train locomotives.[1] Gigantic new industrial combines and infrastructural projects – the Urals-Kuznetsk Combine, the Angara-Enisei Industrial Belt, West Siberian oil and gas in the Tiumen region, power stations, roads, bridges, power lines, a new trans-Siberian railroad, the Baikal-Amur Magistral, known by its acronym as BAM, and massive hydroelectric power stations-spread from the European Soviet Union into Siberia, the Far East, and Central Asia. During the Stalin era, industrialization was financed by artificially low purchase prices in the agricultural sector to extract capital from the sector and by limited investment in housing and light industry. Now, the discovery of oil and natural gas fields in western Siberia provided a new source of financing; wasteful and haphazard mining, drilling, and transport practices accompanied the opening of each new field, leaving despoiled fragile taiga and tundra behind.

Socialist legalities notwithstanding, the scale of these projects, together with typical Soviet operating practices, ensured extensive and spreading environmental damage. The Kansk-Achinsk mining operation in Siberia left behind plundered landscapes, torn geological substrates, and disrupted subterranean water systems as machines and workers scraped off surface deposits with heavy machinery. Abandoned mining operations covered more than 5.5 million acres of the Soviet Union by the mid-1970s. Or, consider the tireless pursuit of electricity: the Bratsk Reservoir

---

[1] V. A. Golikov, ed., *Sovetskii Soiuz: Politiko-Ekonomicheskii Spravochnik* (Moscow: Politizdat, 1975).

covered 2,000 square miles of good bottom land; by the mid-1970s, in total, Soviet reservoirs covered more than 20 million acres. Because the natural flow of rivers was disrupted, nutrient matter important to downstream marine life had precipitated, adjacent groundwater levels were disrupted, downstream of stations ice-free periods increased, local warming occurred, and winter fog often resulted.[2] This was not good for people or fish.

The largest project, potentially the most disruptive of them all, was the Siberian river diversion project that will occupy our attention in several places in this chapter. The authorities planned the redistribution of the flow of water from Siberian rivers – up to 10 percent of the Ob River alone – through massive transfer canals into Central Asia to feed burgeoning agribusinesses. The environmentally devastating project, symptomatic of other projects that acquired huge inputs of manpower, capital, and funding, moved ahead in the absence of sufficient public discussion.

The rich endowment of fossil fuels and other energy resources strengthened the economic position of the Soviet Union during the world energy crises of the 1970s. Soviet oil and gas output both met domestic demand and generated hard currency through exports. Yet, the exploitation of these resources sustained the Soviet economy when there were indications that the economy had, in fact, stagnated and environmental problems were growing, including in the fossil fuel industry. Furthermore, an increase in oil and gas production was not sustainable. Once easily accessible deposits were exhausted by wasteful methods of extraction, for example the injecting of water into wells, the cost of production nearly doubled in the 1970s.[3] The colossal distances fuels had to be transported, often in extreme conditions, compounded problems.

Industrial growth slowed from 8.4 percent per annum in the second half of the 1960s to 3.5 percent in 1981–1985, whereas the agricultural sector slowed from 4.3 percent to 1.4 percent.[4] When ambitious industrial development plans could no longer be met with available labor, capital, and natural resource inputs, planners paradoxically adopted more wasteful resource use practices and encouraged managers to ignore pollution laws, both of which ignored short- and long-term natural resource costs.

---

[2] Norman Precoda, "Winds of Change Blow in Siberia... as Viewed from Within," *Environmental Review*, vol. 3, no. 1 (autumn 1978), pp. 2–19.

[3] G. M. Lappo, V. I. Kozlov, and N. I. Mikhailov, eds., *Sovetskii Soiuz – Obshchii Obzor, Rossiiskaia Federatsiia*. Series "Strany i Narody" (Moscow: Mysl, 1983).

[4] Gosudarstvennyi Komitet RSFSR po Statistike, *Narodnoe Khoziaistvo RSFSR za 70 Let: Statisticheskii Ezhegodnik* (Moskva: Finansy i Statistika, 1987).

Two trends contributed to this wasteful and dangerous approach. By the 1960s, population growth started to decline, leading to a shortfall in labor inputs. By 1970, fully 90 percent of the working-age population was already employed. This worried Soviet economic planners, because much of the population growth took place in Central Asia, the locus of the empire's Moslem population, rather than in the industrialized – and Slavic – regions of Russia and Ukraine. Second, natural resources like timber, fisheries, productive farmland, and freshwater that were located near population centers, and the infrastructure to tap those resources, had been depleted or destroyed. All of this led policy makers and ministerial leaders to pursue rapid development of resources in the Far North, Far East, and Siberia.

Under Brezhnev, state planners also increased the economic power of Gosplan (the State Planning Agency) and ministerial offices located in Moscow. But the highly centralized system proved incapable of managing a modern economy, determining scarcity values, efficiently allocating resources, or encouraging innovation. Officials in these bureaucracies undervalued resource costs and ignored environmental and quality-of-life issues – as they had since the 1930s. The central ministries, constantly driven to maximize the output of their own sectors, showed little regard for the interests of the locales where their enterprises were situated, the people who lived and worked there, or the environment. They wanted production, and they were willing to pay for natural resources no matter the cost – and they often secured them through bribes.

The environmental implications of these developments were profound. Soviet leaders increasingly depended on the export of natural resources and raw materials to generate investment income. Yet severe environmental impacts accompanied the haphazard and breakneck extraction and processing of resources, especially in remote areas with fragile ecosystems where workers left mine tailings, polluted streams, oil spills, and rusted equipment behind; their bosses encouraged this behavior because the punishment for missing targets was greater than that for polluting. Increasingly sophisticated industrial and agricultural activities generated an ever-more toxic mix of waste and runoff. The military-industrial complex, although reaching the goal of achieving parity with the United States in the fields of nuclear, chemical, and biological weapons, was a huge offender, as was its counterpart, the U.S. military-industrial complex. It left a legacy of thousands of square kilometers of land and water that are so radioactive or filled with heavy metals that they may never be habitable, and radionuclides continue to migrate into populated regions.

Soviet Environmental Policy from the 1960s to 1980s

The other side of this picture of extensive environmental degradation was a modern legal system to prevent and mitigate the costs. Soviet leaders established a nationwide environmental policy and the laws to support policy in the 1970s, perhaps in response to similar trends in the West, for example, the National Environmental Protection Act (1969) in the United States.[5] Even if the laws and policies were rarely followed, they reflected rising concern and awareness at least in some circles of environmental problems in the Soviet Union. As noted in the previous chapter, the Supreme Soviet of the Estonian SSR passed a law "On the Protection of Nature in the Estonian SSR" in 1957, followed by a similar law in the Russian Republic in 1960, and framework laws on land (1968), water (1970), and nature protection legislation in several other union republics soon thereafter. Granted, a number of legal specialists and environmentalists criticized these laws for failing to assign any legal responsibility for committing an environmental crime. Nor did the first laws prevent Khrushchev's government from dissolving a large number of nature preserves.[6]

Still, as an aspect of developed socialism, Party ideologues declared a healthy environment to be an important objective, along with those of providing each family with its own apartment, equalization of rural and urban living standards, and other milestones on the way to Communism. Reference to environmental issues began to appear in the major Party pronouncements. At the twenty-fourth Party Congress in 1971, General Secretary Leonid Brezhnev indicated that "while taking measures to speed up scientific and technological progress, it is imperative to do everything in order that this progress should be combined with a proprietary attitude towards natural resources, that it should not serve as a source of dangerous air and water pollution and land degradation."[7]

A number of other environmental laws and decrees followed – in the Soviet Union and abroad – that indicated growing recognition of the need

---

[5] For commentary from the Environmental Protection Agency's (EPA's) first administrator on the nature of the agency's responsibilities, see William Ruckelshaus, "From Careless Indifference to Remedial Action," *Journal of Water Pollution Control Federation*, vol. 44, no. 4 (April 1972), pp. 523–530. For early commentary on the U.S. environmental movement leading to the formation of the EPA, see David Sills, "The Environmental Movement and Its Critics," *Human Ecology*, vol. 3, no. 1 (January 1975), pp. 1–41.

[6] V. V. Dezhkin, *V Mire Zapovednoi Prirody* (Moscow: Sovetskaia Rossiia, 1989).

[7] V. Larin, R. Mnatsakanian, R. I. Chestin, and E. Shvarts, *Okhrana Prirody Rossii: ot Gorbacheva do Putina* (Moskva: KMK, 2003), p. 416.

to protect biospheres among leaders, scientists, and citizens in many countries. The United Nations (UN) Educational, Scientific and Cultural Organization (UNESCO) biosphere project (1974) confirmed this recognition, which was accompanied by a sense that environmental issues are international, not only of local or national concern. In the Soviet Union, nature conservation and environmental issues appeared in planning documents and budgets. For the first time, Gosplan included expenditures for environmental protection in the state budget for the ninth Five-Year Plan (1971–1975). Finally, in September 1972, likely in response to a UN meeting earlier that year,[8] at a joint session of the Central Committee and Council of Ministers, Party officials discussed environmental issues at length. This led to a joint resolution in December, "On Intensifying the Conservation of Nature and Improving the Utilization of Natural Resources." In October 1973, the country's most influential law journal, *Sovetskoe Gosudarstvo i Pravo*, published an article that indirectly criticized the 1972 resolution because it did not establish a ministry for environmental affairs.[9] Similar resolutions in each of the Soviet republics followed that identified key environmental and resource-use issues and established specific responsibilities of individual ministries for pollution control and other concerns. Furthermore, the Soviet Union began to contribute to several international environmental organizations and programs.

According to a leading Soviet biologist, nature conservation had assumed central importance in the Soviet Union owing to the geographical diversity of the nation. Through a series of legal and economic planning documents, the government had turned to preventative approaches: maintenance of ecological balance regionally, conservation of renewable resources, protection of the gene pool, conserving ecosystems and geosystems, and provision of recreational opportunities. The "Main Guidelines" for economic development of the country from 1981 to 1985 set forth measures in particular to protect agricultural lands from erosion, salinization, desiccation, and waterlogging; to improve water quality control in the rivers flowing into the Black, Azov, Caspian, and Baltic Seas, as well as into the Arctic and Baikal basins; to expand efforts on forest protection and afforestation; to pursue air and water pollution control; and to increase the numbers of preserves and parks.[10]

---

[8] Larin et al., *Okhrana prirody Rossii*, p. 416.
[9] *Sovetskoe Gosudarstvo i Pravo*, no. 10 (1973).
[10] Yu. A. Isakov, "The Protection of Nature in the U.S.S.R.: Scientific and Organizational principles," *Geoforum*, vol. 15, no. 1 (1984), pp. 89–94.

In the West, public concerns that followed the publication of Rachel Carson's *Silent Spring* (1962) triggered a search among citizens and officials for environmental policies that led to the establishment of departments and bureaucracies to carry out the policies and the passage of a variety of clean air and water laws in the United States and Europe. In the European Union, an opinion poll conducted in 1973 indicated that environmental pollution was among the top-ten public concerns, ahead of such traditional concerns as poverty, inflation, and unemployment.[11] In the Soviet Union, the public had a limited role. Granted, articles on national environmental problems, for example, the pollution of Lake Baikal, had been published in the Soviet press since the late 1950s. Yet the popular Soviet weekly magazine *Ogonek* published only two articles on environmental problems between 1960 and 1979.[12] Outside official channels, there were no pressure groups or lobbyists to push the leadership to consider environmental issues. Still, Soviet leaders had apparently realized the economic and social consequences of accelerating degradation and resource scarcity (overfishing, deforestation, and so on). Environmental quality was also an aspect of health and quality of life in general. Finally, external factors stimulated policy. In environmental protection, as in many other areas, the Soviet Union strove to catch up with Western nations and their progressive legislation.

The environment had become an issue of global significance – and competition between socialism and capitalism. The UN sponsored a conference in 1972 in Stockholm on the Human Environment. The conference organizers intended its discussions to be ideologically neutral. But they could not succeed toward this end for two reasons. First, any discussion of economic development and environmental degradation is necessarily political, for it involves choices of values, targets, and goals. Second, Cold War politics were ever present. The preliminary program indicated that all industrialized countries (whether socialist or capitalist) should assume responsibility for the protection of the environment and assist developing countries in solving environmental problems. But official participation in UN conferences was limited to states that were already members of the UN or its agencies, such as West Germany, but excluded East Germany, which was not. The Soviet Union threatened to boycott the conference on this basis. Western nations considered this a

---

[11] C. C. Park, ed., *Environmental Policies. An International Review* (London: Croom Helm, 1986), p. 315.
[12] V. I. Sokolov, *Pressa i Ekologiia* (Moscow: Geoss, 2000).

pretext for pressing the recognition of the so-called German Democratic Republic.[13]

On the eve of the conference, the Soviet Union and its allies withdrew from participation, and during the conference the Soviet media reported about it in an exclusively negative tone. The Soviet press reported that the conferees failed to meet any goals for the meeting. This reporting ignored the fact that the Soviet Union bore some responsibility for the success or failure of the conference. It had been on the organizing committee and participated actively in the preparation for the Stockholm meeting. Still, soon after the conference, the UN environmental program offered membership to the Soviet Union on its governing body as a vice president. The Soviet Union started to participate actively in international environmental organizations and initiated highly visible if largely unsuccessful environmental policies at home. These policies were partly based on familiarity with the western approaches to environmental management that its experts had gained in the run-up to Stockholm.

Soviet representatives began regular participation in various UNESCO conferences, as well as collaboration with the UN Economic Commission for Europe in the field of ecology, and with the International Union for Conservation of Nature and Natural Resources. Despite these institutional ties, Soviet environmental policy at the international level was geared to achieve diplomatic success rather than solve domestic or such international environmental problems as transborder pollution. The Soviet Union subsequently signed the Final Act of the Helsinki Conference on Security and Cooperation in Europe in 1975. The accords granted each nation the right to noninterference in domestic human rights concerns, a position that the Brezhnev administration used to mistreat dissidents. The accords also mentioned environmental issues as a human right and included measures to save rivers, lakes, and seas from pollution. Later, the Soviet Union passed several decrees targeting pollution problems in the Black, Azov, and Baltic Seas.

On the international scene, Soviet representatives had been vocally most active in the cause of wildlife conservation. In 1964, the Soviet Union and Norway agreed to suspend all hunting of the Greenland seal for five years. In 1966, the Soviet Union announced that it had banned the catching and killing of dolphins, and between 1967 and 1969, it joined with the United States in two pacts for the conservation of crabs and

---

[13] *The Economist*, January 15, 1972.

certain fish in the Pacific and the Atlantic.[14] These were all very good-sounding initiatives, but they represented areas of marginal economic concern for the Soviet Union. The Soviet record on whaling stood in stark contrast – illegal, secret, and rapacious harvest of whales was the rule.[15] (See **Box 1**.)

The slaughter of whales was hidden under very public efforts by the Soviet Union to identify and protect endangered species. The Convention on International Trade in Endangered Species of Wild Fauna and Flora (CITES) drew attention to increasing threats of extinctions – and extinctions. The United States signed in 1973 and the Soviet Union in 1974 after some preparatory work. In 1972 and 1973, scientists in the Ministry of Agriculture, the Komarov Botanical Institute, major botanical gardens, and other institutions prepared for the convention by examining ecosystems and then recommending various species for listing. Building on the momentum of CITES, the scientific community took important steps to protect species. The Hydrometeorological Service gained expanded responsibilities for monitoring and establishing standards. In 1975, under the editorship of A. L. Takhtaian, the government published *Red Book – Native Plant Species to be Protected in the USSR* and listed 600 species. Preparation for the publication indicated that many species were already extinct and others threatened, including several of great economic significance. The committee that produced the book recommended fully protecting a species within a reserve, creating permanent or temporary preserves as needed, restricting collection of rare and threatened species, and organizing qualified botanists to evaluate populations of rare and threatened species.[16]

---

[14] *New York Times*, December 24, 1964; *International Herald Tribune*, March 14, 1966; and Keith Bush, "Environmental Disruption: the Soviet Response, Supplement to the Research Bulletin," *Radio Liberty Research Institute*, February 9, 1972.

[15] See Y. V. Ivashchenko, P. J. Clapham, and R. L. Brownell, Jr., *Scientific Reports of Soviet Whaling Expeditions in the North Pacific, 1955–1978*, trans. Y. V. Ivashchenko (Washington, DC: NOAA, U.S. Department of Commerce, July 2007) for publication of internal government reports from 1955, 1956, 1966, 1967, 1969, and 1970 that indicated the extent of Soviet illegal whaling. The decline of whale populations was "in part an inevitable consequence of the Soviet system of industrial planning. The government set annual targets for quantities of whale products...." The following year's plan was always racheted up. Crews were forced to kill more and more whales to obtain bonuses. The reports reveal declines not only in populations, but also in average size and age of animals, disappearance from critical feeding grounds, and so on.

[16] Thomas Elias, "Rare and Endangered Species of Plants – the Soviet Side," *Science*, vol. 219 (January 7, 1973): pp. 19–23.

> **Box 1. Soviet Whale Poaching**
>
> Whales were a valuable source of meat and by-products for the Soviet Union. The Soviet whaling fleet was modern, impressively equipped, and represented a substantial postwar investment. Whales have now been so overhunted that many species face extinction. The tragedy was that the signing of the whaling convention in 1946 was followed by two decades of barbaric slaughter of these great sea mammals. Even the modest controls of the commission were flouted by two of its member nations, with tragic results. The two countries, the Soviet Union and Japan, accounted for 84 percent of the 42,000 whales killed in 1970, not to mention some 200,000 dolphins and porpoises. The Soviet Union did everything to avoid implementation of more strict international sanctions to whale poaching. In 1961, a secret government circular prohibiting publication of any information about Soviet whaling was disseminated. After the breakup of the Soviet Union, a special commission of the Russian parliament under the prominent environmentalist Alexei Yablokov was organized to investigate the true scale of whaling in the Soviet Union from 1947 to 1972. In 1995, the Commission concluded that the Soviet Union had largely falsified the official data reported to the International Whaling Commission (IWC). The Soviet Union had reported that 141,390 whales had been killed from 1947 to 1972, but the real figure was 193,048. In the case of the humpback whale, the corresponding figures were 2,705 and 33,249 whales. The Soviet Union joined the IWC's moratorium on commercial whaling in 1986 only after intense international pressure.[17]

In several cases, the authorities had species protected – and brought them back from the brink of extinction – because of their economic value to the state, especially for export markets through which the government earned hard currency. For example, by pursuing a lucrative sable fur market, the Soviets were able to save the sable.[18] Through industrial management, other species of economic significance also increased

---

[17] *New York Times*, November 30, 1971; Center for Ecological Politics, *Materialy po Promyslovoi Deiatel'nosti Sovetskikh Antarkticheskikh Kitoboinykh Flotilii, 1947–1972* (Moscow: Tsentr Ekologicheskoi Politiki Rossii, 1995).
[18] "Sable or Lynx the Prize of Soviet Hunters," *Los Angeles Times*, August 30, 1985.

in number. By the late 1970s, there were some 900,000 wild reindeer of six subspecies (synonymous with the caribou, R. *tarandus*) and 2.5 million domesticated reindeer. The numbers increased rapidly in the postwar years, especially in the Far North, where competition from domestic reindeer was limited. These northern herds were administered and managed through systematic aerial counting and annual harvests. A large number of them were in the Taimyr – about 460,000 animals in 1978, up from 110,000 in 1959 and 252,000 in 1967. "Harvest" meant shooting from helicopters with dog teams, sled reindeer, or over snow vehicles becoming less and less important.[19] Unfortunately, a large number of other species – lynx, sea otters, various bats, voles and other rodents, Siberian tigers, and dozens of other species – remained in crisis.

After the 1972 Stockholm meetings that the Soviet Union boycotted, the United States and Soviet governments nonetheless commenced unprecedented bilateral cooperation on a number of scientific issues including environmental matters. At a May 1972 summit in Moscow, Leonid Brezhnev and Richard Nixon signed agreements to establish the technological, commercial, and diplomatic framework for the policy of détente, including the creation of a Joint Committee on Environment Protection.[20] Some forty cooperative scientific projects were established. Joint teams of scientists discovered seismic activity beneath the Nurek Reservoir in the Tadzhik Soviet Socialist Republic, behind the world's highest earthen hydroelectric dam, unfortunately too late to prevent its construction. Experts met to discuss stratospheric pollutants, organized highly successful clean air exhibits, and exchanged delegations to study pipeline construction in the permafrost areas of Alaska and Siberia. Both sides signed a convention in November 1976 on the conservation of migratory birds and their environment. The scientists met annually under these bilateral agreements until the Soviet invasion of Afghanistan in 1979 that led the United States to immediately withdraw from Joint Commission activities. The relationship was revived in 1984, and two years later officials from both sides agreed to broaden cooperation on environmental

---

[19] David Klein and Vladimir Kuzyakin, "Distrubtion and Status of Wild Reindeer in the Soviet Union," *Journal of Wildlife Management*, vol. 46, no. 3 (July 1982), pp. 728–733. In the United States, shooting of wolves from helicopters in the name of protecting cattle or for sport resumed under the administration of George W. Bush – without any scientific justification and clearly for sport, not to protect cattle.

[20] P. Bryan, "US-USSR Cooperation on Environmental Protection: Limits and Potential," *Radio Liberty Research Institute*, August 16, 1973.

issues over the following five years.²¹ The Soviet Union also signed bilateral agreements on environmental protection with other Western nations including Great Britain, Sweden, and France.²²

The Soviet Union pursued cooperation in other arenas of environmental policy even as it pursued an aggressive military policy that mirrored that of the United States. Although acid rain had never been a central issue in the socialist bloc, surprisingly given the contribution of such coal-hungry nations as Poland to the growing problem of acid rain, in the late 1970s, the Soviet Union supported international environmental conventions on stratospheric ozone depletion and global climate change.²³ It went so far as to initiate a pan-European meeting on cooperation in environmental protection in Geneva in 1979, underwriting the costs of preparatory meetings.²⁴ Many Western observers believed this course of action was meant to deflect serious concern among Western countries that the Soviets and their allies flouted the human right sections of the Helsinki accords. Sensing a means to defuse this concern, Brezhnev put forth proposals for three pan-European conventions on transportation, energy, and the environment. The West boycotted the Soviet proposals until the Scandinavians, who suffered directly from Polish coal, urged it to convene a conference on acid rain.²⁵

By the mid-1970s, the Soviet Union, realizing a shortage of financial and other resources to compete with the United States in the Cold War, sought ways to curtail the arms race. According to Western estimates, the Soviet Union's military budget was approximately $200 billion, whereas the United States spent roughly $300 billion a year, and the Soviet Union likely spent one-third of its much smaller economy on the military.²⁶ Maintaining parity in the arm race simply put a greater burden on the Soviet economy than on the U.S. economy. The Soviet Union used every occasion, including environmental forums, to discuss disarmament issues. For example, in 1980, Soviet representatives proposed a UN General Assembly session to discuss the "Historical Responsibility of States for

---

²¹ P. Shabecoff, "U.S. and Soviet to Study Ways to Save the Ozone," *New York Times*, December 20, 1986. For a critical history of the value of the U.S.-Soviet scientific exchanges, see Yakov Rabkin, *Science Between the Superpowers* (New York: Priority Press, 1988).
²² *Trud*, May 12, 1976; and *Pravda*, December 8, 1980.
²³ F. Szegi-Tóth, "A globális környezeti kockázatok kommunikációja: a savas eső a magyar sajtóban," *Jel-kép*, no. 3 (1996), pp. 81–108.
²⁴ *TASS*, February 12, 1979.
²⁵ *International Herald Tribune*, November 16, 1979.
²⁶ V. Goldansky, "Logika Razuma Vmesto Logiki Egoizma," *Izvestiia*, February 10, 1989.

the Preservation of Nature for Present and Future Generations."[27] The proposal linked environmental protection problems to disarmament, an assertion that delegates from several countries questioned because the Soviet Union, along with the United States, was the major contributor to the arms race.[28] Still, in October 1980, a Soviet resolution passed with no votes cast against it, although Western and many third world countries abstained. Soviets officials next linked arms control endeavors to environmental ones at a 1982 UNEP gathering in Nairobi, a 1984 Munich conference on reduction of sulfur dioxide emissions, and a session of the coordinating council for the UNESCO program "Man and the Biosphere" that same year. Western countries generally rejected these initiatives as either politically disingenuous or inappropriate given the scope of the forum.[29]

Domestic Determinants of Environmental Policy

During the Brezhnev era, environmental policy and actions amounted to a series of bold resolutions and governmental orders to improve the situation, but little concrete action. Discussion of environmental problems at home widened after Leonid Brezhnev signed the Helsinki Accords on human rights in 1975.[30] Over the next decade, the government passed a series of laws and resolutions intended to demonstrate that the Soviet Union was at the forefront of the world's environmental protection activities. These resolutions accomplished more in word than in deed. Managers found it better to pollute and pay miserly fines than to miss production targets; rapacious harvest of natural resources continued. Enterprises in the military and nonmilitary sectors alike disposed of hazardous waste in a haphazard fashion with the complicity of government officials, subjecting water, air, land, and the people and animals living within them to dangerous levels of harmful chemicals.

At the twenty-fifth Party Congress in 1975, within the "main directions of the development of the economy of the Soviet Union for 1976–1980," the delegates mentioned nature protection as a policy objective. As Stalin

---

[27] A. Gromyko, "Sberech' Prirodu Zemli dlia Nyneshnish ii Buduschish Pokolenii," *Pravda*, August 17, 1980, p. 4.
[28] "A Soviet-Initiated Resolution," *Radio Liberty Research Institute*, October 31, 1980.
[29] "Governments Worry about Environment," *Radio Liberty Research Institute*, May 19, 1982; "Munich Ecological Meeting Seen a Success," *Radio Liberty Research Institute*, June 28, 1984; *Izvestiia*, December 11, 1984.
[30] Sokolov, *Pressa i Ekologiia*.

had in 1936 with the promulgation of the so-called Stalin constitution, under Brezhnev, party functionaries approved a new constitution that touted the advantages of the Soviet system. In addition to human and civil rights that found a prominent place in the Brezhnev constitution but were ignored in practice, the constitution included two articles on environmental rights – both of which the government also ignored. Article 18 stated, "In the interests of the present and future generations, the necessary steps are taken in the USSR to protect and make scientific, rational use of the land and its mineral and water resources, and the plant and animal kingdom, to preserve the purity of air and water, and ensure production of natural wealth, and improve the human environment." Article 67 stated that "citizens of the USSR are obliged to protect nature and conserve its riches."[31]

In December 1978, the government passed yet another resolution, "On Additional Measures to Intensify the Conservation of Nature and Improve the Utilization of Natural Resources." The fact that the resolution had nearly the same title as the one issued six years earlier suggests that officials and bureaucracies had accomplished little in the intervening years. The resolution obliged all ministries whose enterprises polluted immediately to establish environmental offices in their headquarters. The resolution further obliged the State Committee for the Protection of Nature, a precursor to the State Environmental Protection Agency set up under Mikhail Gorbachev, to coordinate the efforts of the ministries and engage in a press campaign to instill a "communist attitude" toward nature in the population, especially among young people. In 1980, two new environmental laws attempted to put teeth into previous regulations by coordinating the numerous requirements adopted by various ministries and agencies. One law, "On the Protection of the Atmosphere," set air quality standards for the country as a whole, as opposed to industrial regions alone. The law also stipulated setting up water and air pollution permits and specified the creation of an "environmental cadastre" of all enterprises in the country. A second law, "On the Protection of Animals," followed.

The Soviet Union had introduced a system of fines for pollution in the Stalin era. But it remained cheaper to pay fines than risk punishment

---

[31] Constitution of the USSR Adopted at the Seventh (Special) Session of the Supreme Soviet of the USSR Ninth Convocation On October 7, 1977, http://www.departments.bucknell.edu/russian/const/1977toc.html.

for not meeting production targets.[32] In the 1980s, however, Soviet legislation introduced more substantial fines for environmental pollution, even if officials had a poor record in collecting them. In 1987, for example, Uzbek authorities fined a biochemical plant 1 million rubles ($1.6 million) and threatened it with closure if it continued pumping waste into the main irrigation canal of the city of Tashkent. Earlier fines had been significantly lower, at 18,000 rubles ($28,000) and 57,000 rubles ($91,200), and had had no effect.[33] In 1988, a gigantic metallurgical plant in the industrial city of Cherepovetsk (375 kilometers north of Moscow in Vologda province) was fined 20 million rubles ($33 million) for dumping waste from coke production into the Rybinskoe reservoir on the Volga River, destroying the already threatened fish habitat and putting fishermen out of work. This second case illustrates a paradox of the Soviet system. Here, the Ministry of Water Resources and Land Reclamation of the Russian Republic, itself well known for destruction of the environment in many regions of the Soviet Union, had levied a fine on an enterprise under the jurisdiction of the Ministry of Ferrous Metallurgy.[34] In a word, any ministry might fine another ministry because each was likely negligent toward the environment and in violation of the law, and it seemed that none could be forced to the pay; fines had no force of law.

Might the Soviet Union establish pollution standards to limit degradation? Like the European democracies, Canada, and the United States, in the mid-1970s, the Soviet Union also began the process of setting effluent standards. Previously, policy makers had accomplished little in this area; Western specialists considered this a major weakness in the Soviet system.[35] Such issues as controls on vehicle exhaust emissions were not even discussed. Still, the first tentative steps occurred in May 1974, when the Ministry of Water Resources, the Ministry of Health, and the Ministry of Fishing adopted new rules for the protection of surface waters. In 1976, the authorities urged the various ministries to draft standards for air, water, and soil pollution within eighteen months.[36] The new rules set an important precedent by categorically prohibiting the opening of any new facilities wherever health experts had not yet established maximum

---

[32] *Planovoe khoziaistvo*, no. 7 (1970).
[33] J. Whitechurch, "New Soviet Anti-Pollution Decree," *Radio Liberty Research Institute*, July 15, 1985.
[34] *Reuters*, February 16, 1988.
[35] Bryan, "US-USSR Cooperation on Environmental Protection."
[36] *Christian Science Monitor*, October 20, 1976.

permissible concentrations (MPCs) for the associated pollutants.[37] A significant problem with all of this activity was that monitoring equipment remained in short supply.[38]

The business of setting standards proceeded at a crawl, and the Soviet Union squandered a crucial opportunity for cooperation with the United States in 1972. According to Boris Komarov,[39] the United States proposed a cooperative venture to set MPCs for almost all known pollutants at that time, but Soviet authorities refused to participate. The U.S. program cost some $2 billion, and within two years it yielded standards for 17,000 pollutants. By comparison, Soviet scientists in these two years set MPCs for only 15 of the 300 most important pollutants. High cost and secrecy prevented cooperation.

The Soviet authorities announced at this time that a substantial percentage of the budget was being allocated to the environment at both national and republic levels.[40] An environmental budget of more than 6 billion rubles (about $7 billion), a sum that was modest but not out of line with the levels in other countries, was included in the ninth Five-Year Plan (1971–1975).[41] They opened hard currency reserves to enable the purchase of pollution control equipment in the West. Although still a miniscule amount, imports of U.S. air-pollution control equipment rose to $4.5 million in 1976 from less than $1 million in 1974. In the tenth Five-Year Plan (1976–1980), the environmental budget of the Soviet Union reached 11 billion rubles or $16 billion. But much of the environmental budget of the Soviet Union was to be spent on land-reclamation projects.[42] It is difficult to evaluate these expenditures because there were also indirect investments in nature conservation. Some conservation measures, such as reforestation or reclamation of agricultural land, were included in ministerial budgets. According to one source, the total direct and indirect expenditures for conservation, including the construction of purification

---

[37] B. Komarov, *Destruction of Nature in the Soviet Union* (White Plains, NY: M.E. Sharpe, 1980), p. 150.
[38] Murray Feshbach and Alfred Friendly, Jr., *Ecocide in the USSR: Health and Nature Under Siege* (New York: Basic Books, 1992), p. 376.
[39] Komarov, *Destruction of Nature*, p. 44.
[40] Bryan, "US-USSR Cooperation on Environmental Protection."
[41] "Environmental Disruption: the Soviet Response," supplement to the research bulletin, *Radio Liberty Research Institute*, February 9, 1972; and Tsentralinoe Statisticheskoe Upravlenie SSSR, *Narodnoe Khoziaistvo SSSR, 1922–1982: Iubileinyi Statisticheskii Ezhegodnik* (Moskva: Finansy i statistika, 1982), p. 623.
[42] *Wall Street Journal*, November 21, 1977.

facilities, land reclamation, and reforestation, was roughly $48 billion, or roughly twice the official military budget.[43]

Yuri Andropov, who served as general secretary of the Communist Party for eighteen months after Brezhnev's death (1982–1984) until his own death from kidney failure, had little time to devote to, or interest in, environmental issues. He considered the lack of discipline among workers to be the fundamental problem facing the Soviet economy and one of the reasons for environmental problems. He likely based this view on an accident on the Dniester River in September 1983 that resulted in an explosion in a fertilizer plant and spread dangerous chemicals far and wide. As is often the practice in industrial societies throughout the globe, the authorities blamed people – the managers – for negligence leading to the accident, not the technology that lacked safety redundancies and filtering equipment (see **Box 2**). In December 1983, Andropov addressed the Party leadership, stressing the need to reduce environmental pollution and follow existing laws.[44] He dropped the very costly and wildly adventurous plan to divert water from Siberian rivers to Central Asia through massive transfer canals, although the next – also short-lived – general secretary, Konstantin Chernenko (1984–1985), revived the river diversion project,[45] based in part on a scathing report of the lead ministry for the project, Minvodkhoz, produced by the KGB (see **Box 3**).

In sum, Soviet environmental policy was inadequate to the task. First, it was not based on a commitment to cleaning up the environment or protecting nature, but rather on the political goal of demonstrating that environmental standards met those in the West. Environmental laws and resolutions were Soviet "Potemkin villages," that is, false fronts to present a good side.[46] Moreover, many environmental policy tools were copied from free market democracies, but they simply could not function in a centrally planned economy, were not intended to function, or were too costly. Surely, the situation would have been worse in the absence of the laws and regulations the Soviets had passed in the Brezhnev era, but that was little consolation to the people, wildlife, or trees.

The major reason for this failure of laws to be enforced may have been the absence of any central environmental protection agency until 1988. Environmental management functions were dispersed across many

---

[43] *Bulletin APN*, October 10, 1980.
[44] Whitechurch, "New Soviet Anti-Pollution Decree," *Radio Liberty Research Institute*, July 15, 1985.
[45] *Economist*, August 11, 1984.
[46] Feshbach and Friendly, *Ecocide in the USSR*, p. 376.

> **Box 2. The Dniester River Disaster of September 1983**
>
> In one of the worst ecological disasters in Soviet history, more than 500 kilometers of the Dniestr River were poisoned in September 1983 following the collapse of an earthen dam holding concentrated salt wastes from a fertilizer plant at Stebnik that permitted 4.5 million cubic meters of potassium sulphate fertilizers to flow into the Dniestr. Because the salt solution (250 grams per liter) was very heavy, it did not mix with the river water but sank to the bottom and migrated downstream. No one at the plant died in the misfortune, but the spill disrupted water supply to millions of people. To supply clean water to the towns and factories along the Dniester, engineers diverted water from other rivers and bored emergency wells. The salt killed virtually all life along a huge stretch of the river; farm animals and birds that drank the water died. More than 1 million cubic meters of the waste accumulated on the bed of the Novo-Dniester dam. This prevented it from reaching the cities of Kishinev (the capital of Moldavia) and Odessa (on the Black Sea).
>
> An investigation found plant managers responsible for the disaster. But blaming managers rather than inadequate technologies and processes was standard Soviet operating procedure and thinking. Waste processing facilities at the plant were totally inadequate. Plant production had increased more than sixfold in fifteen years, but waste control equipment was never updated. The decision to store the waste in an artificial lake held back by a primitive earthen dam was clearly inappropriate. Concentrated sulphates were known to destroy soil structures, so the dam was bound to collapse sooner or later. Federal inspectors had warned the plant managers in 1982 and in May 1983 that the dam was in danger of failure. The managers in this case were negligent. Eight people from the plant were arrested. Their trial was widely covered in the press. Later, they were all exonerated after a determination that the complex geology of the area was the main cause of the accident. Once again, the Soviet Union had an environmental crime with millions of victims but no one held responsible for it.[47]

---

[47] Serge Schmemann, "Russians Report Vast Waste Spill in West Ukraine," *New York Times*, October 29, 1983. See also Theodore Shabad, "Five Jailed for '83 Disaster," ibid., July 17, 1985; and J. Whitechurch, "New Soviet Anti-Pollution Decree."

> **Box 3. The KGB against the Ministry of Water Resources**
>
> Professor Nikita Glazovsky, a prominent Soviet geographer, gave an interview in which he related a remarkable story about informal meetings of scientists who discussed environmental damages associated with Minvodkhoz projects. The discussions took place during the evening hours at the office of Academician A. L. Ianshin, a vice president of the Academy of Sciences. Academician Ianshin had been admitted to the highest echelons of Soviet power and understood the rules of politics. When he suddenly learned that the KGB "was interested" in these evening meetings, Ianshin immediately wrote a letter to the chairman of the KGB, Iurii Andropov, briefly leader of the Soviet Union after the death of Leonid Brezhnev, in which he explained the goals and subjects of the meetings. Soon after, he received a polite response in which Andropov expressed his concerns about Minvodkhoz projects. Moreover, the letter indicated that KGB experts had made their own investigation of Minvodkhoz activities and had come to similar conclusions about the huge environmental costs of the projects. Andropov attached a technical document produced by KGB experts. As Glazovsky put it, the document confirmed "many our conclusions" about Minvodkhoz. It is likely that Andropov's strong opposition to the very idea of the diversion of Siberian rivers was based on that KGB investigation.[48]

ministries, even though the need for a single agency responsible for pollution monitoring and control had already been raised in the Soviet press beginning in the 1960s.[49] The result was regulation by the regulated, conflicts of interest, confusion over responsibilities, and incentives to continue polluting, not to clean up. The primary responsibility for protecting and conserving water resources, for example, belonged to the infamous Ministry of Land Reclamation and Water Management (Minvodkhoz), whose goal was to develop water resources. For some projects, a dozen ministries and republican-level state planning agencies might be involved in a poorly defined process of consultation. The Soviet Ministry of Public Health and its local counterparts, including its Sanitary Epidemiological

---

[48] V. Larin et al. *Okhrana Prirody*.
[49] *Ekonomicheskaia Gazeta*, no. 40, 1969; Bush, "Environmental Disruption: the Soviet Response," and A. Smirnov, "Nashi Iuzhnye Moray," *Pravda*, October 7, 1968, p. 3.

Service, technically enforced regulations, but they were powerless against ministries with industrial and other economic functions.[50]

There was some hope. A forerunner of an environmental protection agency, the Hydrometeorological Agency (hereafter, Gidromet), gained expanded powers for monitoring background air and water pollution levels in the 1960s. Through an extensive network of weather stations, it gathered data on enterprises that violated the pollution control laws, providing information on background air, water (and later, soil) pollution. In 1967, Gidromet issued its first annual *Review of the State of Atmospheric Pollution in Cities and Industrial Centers of the Soviet Union*. Gidromet later issued reviews of water pollution (1978), background environmental contamination (1983), radioactive contamination (1987, in the wake of Chernobyl), and soil contamination (1990). Yet, Gidromet reports were classified so that information could not be openly published or cited. Moreover, the number of copies of these reports ranged typically between 20 to 200 – absurdly low for a country of 250 million people. Only a very small circle of top officials saw these reports, and their concerns were production, not regulation or cleanup. Gidromet had, however, established a reputation – by forcing three hundred highly polluting enterprises to relocate outside the Moscow city limits after 1950, along with several other high-profile actions.[51] Unfortunately, this suggests that Moscow residents – and the central political apparatus – received better pollution control than did residents of other cities.

Another agency that collected and analyzed environmental information was the State Committee on Statistics (Goskomstat). It played a key role in the centrally planned economy by gathering data on all aspects of production and consumption, most of which again were classified. It collected annual information from roughly twenty thousand large enterprises on resource consumption, pollution discharges, and waste practices. Most of its data were obtained by extrapolation rather than measurement and consequently were inaccurate. Military enterprises were not included in Goskomstat's reporting system; military bases, weapons testing sites, and other facilities are a unique and troubling case of pollution throughout the globe in many, many countries. In April 1973, the government announced that the responsibility for monitoring environmental pollution

---

[50] D. R. Kelley, "Environmental Problems in the USSR as a Political Issue," *Radio Liberty Research Institute*, September 19, 1975.

[51] Kelley, "Environmental Problems in the USSR as a Political Issue."

would remain with several other ministries and agencies whose activities were in no way coordinated.

Specialists had for decades called for the creation of a federal environmental protection agency. In 1976, a jurist pleaded for the adoption of a national environmental law. He argued that "the principle of priority of nature protection is difficult to reconcile with the concentration of the functions of nature protection exclusively within the competence of those organs whose basic task is its exploitation."[52] In December 1975, A. Grigorian, the chairman of the board of the Union of Architects of the Armenian Soviet Socialist Republic, called for the creation of a nationwide committee for nature protection that would coordinate all environmental management activities. A subsequent survey of letters to the editor published on February 7, 1976, noted that "many readers of *Pravda*" supported Grigorian's proposal. On August 10, 1977, the Union of Soviet Writers joined the call for the creation of a nationwide system of environmental control similar to those found in other countries in their professional organ, including an amendment to Article 18 of the new Soviet constitution to that effect.[53] We must keep in mind the central role of writers from a variety of genres and political positions in pushing environmental protection.

On March 31, 1978, perhaps in response to the growing discussion, the Presidium of the Supreme Soviet of the Soviet Union issued a resolution that upgraded Gidromet to the State Committee for Hydrometeorology and Environmental Regulation. The main duty of the new body was to monitor the condition of the atmosphere, water, and soils throughout the country. This was no small task at the end of the 1970s, when there were already 1,000 permanent air monitoring stations in 250 Soviet cities, 1,200 freshwater monitoring stations, 1,600 sea monitoring stations, and 2,700 soil monitoring posts.[54] The head of the new state committee, Yuri Izrael (still active in 2009 as a vice chairman of the Intergovernmental Panel and Climate Change; see Chapter 6) was the leading Soviet delegate to all major international environmental meetings.

Unfortunately, Gidromet was handicapped from the start. It lacked the status and clout of the ministries it strove to regulate. Its chairman did not have ministerial rank, nor had he been elected to any leading organs of

---

[52] *Sovetskoe gosudarstvo i pravo*, no. 4 (1976).
[53] V. Mutian and Iu. Shemshuchenko, "Ekologiia i Pravo," *Literaturnaia Gazeta*, August 10, 1977, p. 2; and "Konstitutsiia (Osnovnoi Zakon) SSSR," *Pravda*, October 8, 1977, pp. 3–6.
[54] *TASS*, February 12, 1979.

the Party apparatus, whereas members of the Council of Ministers of the Soviet Union, irrespective of whether they might be ministers or chairman of a state committee, held party membership in the Central Committee of the Communist Party.[55] In 1983, one more committee responsible for environment conservation was organized – the Commission on Conservation and Rational Use of Natural Resources of the Presidium of the Council of Ministers of the Soviet Union. The idea was that the new body would have control over the implementation of government resolutions on environmental conservation by ministries.[56] There is little evidence that it had any greater power to enforce laws than Gidromet.

Hence, in spite of the creation of a number of agencies with environmental functions, real decision-making power remained in the hands of economic ministries. In contrast to free market societies, where independent regulatory agencies exist along with ministries, Soviet ministries were responsible for planning, executing, and regulating all economic activities in specific sectors. The largest of these were Minvodkhoz, the Ministry of Medium Machine Building (Minsredmash, responsible for nuclear facilities), the Ministry of Agriculture, the Ministry of Forestry, the Ministry for Fisheries, and the Ministry for Oil and Gas Industry. These were huge organizations with great bureaucratic momentum. For example, Minvodkhoz employed 1.5 million people and controlled roughly 28 percent of the total investment in agriculture from 1966 to 1986 (with little to show for it),[57] and its officials pushed the infamous river diversion project.

Soviet managers were aware of the legal requirements on them and the damage they did when they ignored them. In 1974, Petr Neporozhnyi, who headed the Ministry of the Energy Industry and Electrification for many decades, explained in a Soviet newspaper that the replacement of thousands of small thermal power stations by several large power plants in Moscow would improve air quality. He added that the search for ways to protect pristine environments was one of the most important tasks of the energy industry.[58] But during the same period, his ministry oversaw the construction of huge energy complexes (highly polluting thermal plants in the Baltic states, the Kansk-Achinsk lignite combine, and many others)

---

[55] C. Duevel, "The Tortuous Road to Appointment of an 'Environmental Tsar' of the USSR," *Radio Liberty Research Institute*, May 26, 1978.
[56] A. Yablokov, "Sud'ba Prirody – Sud'ba Strany," *Komsomol'skaia Pravda*, February 14, 1986, p. 2.
[57] Feshbach and Friendly, *Ecocide*, p. 376.
[58] P. Neporozhnii, "I Elektrichestvo, I Teplo," *Izvestiia*, December 12, 1974, p. 5.

that triggered extensive environmental problems. In a word, like their counterparts elsewhere, Soviet officials always maneuvered within a web of conflicting interests. But unlike their counterparts in other countries, it was better for them to meet production targets and ignore laws pertaining to environmental protection.

What level of spending was needed to sustain the environmental status quo? What were Soviet goals? To reduce present pollution? To redress past damage? In isolated cases, the Soviet Union allocated up to 30 percent of capital investment for environmental protection when building such showcases of industry as a "clean" oil refinery in Perm, an industrial city in the west Ural Mountain foothills.[59] According to official Soviet data, from 1976 to 1980, Soviet capital investment exceeded 600 billion rubles, and about two-thirds of the sum was allocated to industry. Yet, the share of direct investment in environmental protection (that is, directed mainly for construction of purification facilities and waste-free technologies) was less than 2 percent of this total, a figure significantly smaller than in the United States or Japan.[60]

Economic Disincentives to Rational Environmental Policy

Because of the attempt to base all prices according to their labor component (the Marxist labor theory of value), economic planners levied no – or insignificant – charges on such natural resources as water and timber. The fact that they were essentially free – except for the labor to extract them – meant that there was no incentive to protect them from pollution or wasteful use, especially because they were also regarded as inexhaustible. The first public glimmering that this way of viewing things was mistaken appeared in press reports in the late 1960s. Several economists and environmentalists argued that it was irrational for enterprises to receive water free of charge while the state spent billions of rubles on the creation of reservoirs and other impoundments, not to mention constructed public works to provide clean water. They pointed out that as long as water was supplied free of charge, the enterprise director had no incentive – other than a moral one – to economize in its use.[61] The same was true for other natural resources, with predictable

---

[59] Komarov, *Destruction of Nature*, p. 150.
[60] Ibid.
[61] N. Mel'nikov, "Prezhde Chem Reki Povernut' Vspiat," *Literaturnaia Gazeta*, July 12, 1967, p. 11.

consequences. As one Soviet newspaper explained, "At present, mining industry enterprises receive the value of ores free of charge, although society expends substantial funds on their discovery and prospecting. Since the enterprises receive [these resources] from the earth for free, they have no incentives to exploit the deposits to the utmost, since costs increase and profitability declines the further into the mineshaft or oil-stratum. For that reason up to twenty-five percent of existing reserves remain in the deposits."[62]

These resource pricing issues came to the fore in the effort to improve agricultural performance with large-scale programs that eerily reminded some individuals of Khrushchev-era campaigns. In March 1974, the government adopted a fifteen-year plan to develop the non–Black Earth Regions of the Russian Republic. Despite the intensive resource cost per calorie, increased livestock production was made a high priority. In addition to the construction of large livestock complexes, the program promoted land reclamation projects and encouraged the application of more chemical fertilizers. Although it has a relatively short growing season, the zone has the highest average annual rainfall of any agricultural area in the European part of the Soviet Union. Government agricultural officials hoped that the area would become a stable base for grain output and livestock production. But instead of instituting changes in management practices, it amalgamated already giant production units into new ones, creating massive and more inefficient agri-industrial complexes. Herds grew in size but declined in average weight. To the outside observer, this was reminiscent in scale and irrationality of the Khrushchev-era Virgin Lands campaign.

A new policy for developing the Soviet livestock breeding sector was another threat to the environment. Officials at the October 1968 plenary session of the Central Committee of the Communist Party ordered the complete industrialization of livestock breeding within three or four years. Officials simultaneously called for the development of a specialized meat production branch in the livestock sector (the majority of Soviet cattle were of a dual-purpose variety (dairy and meat) in which productivity was generally very low. This required the full mechanization of all operations as well as year-round production of meat, milk, and eggs. The plan envisaged the accelerated construction of 1,170 large state industrial livestock complexes and the construction or expansion of 585 poultry

---

[62] *Ekonomicheskie Nauki*, no. 10 (1971). See also Keith Bush, "Environmental Disruption: the Soviet Response."

enterprises throughout the country. Such large complexes had no provision for treatment for their wastes, and they polluted nearby groundwater and rivers.[63]

In 1982, delegates at a plenary session of the Communist Party adopted a new Food Program with great fanfare that came to be associated with Brezhnev's name. The Brezhnev Food Program sought to improve production significantly through extensive investment in the agricultural sector. The program had symbolic importance to indicate how the Soviet citizen benefited from developed socialism. The citizen's diet was rich in cereals and starches, whereas efforts to increase per capita meat and vegetable consumption had slipped. In addition to combating large crop losses on the way from producer to consumer, the Food Program sought to decrease food imports from Western countries. By some estimates, roughly 90 percent of gold mined annually in the Soviet Union was sold for hard currency to cover grain imports. The Food Program had important environmental consequences because of the usual practices of heavy use of chemical fertilizers and biocides. In their resolution to adopt the program, the delegates asserted that rational use of natural resources was required, and one way to accomplish this was the introduction of a system of charges. Yet the regime failed to act again, constantly postponing any decision about how to determine rational prices for resources.[64] As discussed in greater detail in Chapter 5, the Communist Party finally adopted a more radical approach by implementing fees for the use of natural resources only in 1988, too late to have any impact, for the State Committee on Environmental Protection announced that the system would commence working only in 1991 – and then the Soviet Union collapsed.

When Khrushchev pursued his wild schemes to transform Soviet agriculture, their appeal often lay in the misleadingly low initial investment costs; in Chapter 3 we discussed the huge outlays in the Virgin Lands campaign connected to the overuse of chemicals, destruction of grasslands, erosion, and so on. During the 1970s, Soviet agricultural policy returned to a level of capital investment not seen since the establishment of Machine Tractor Stations during forced collectivization. The authorities poured more than $500 billion into agriculture before the decade was over. This represented an increasing share of scarce resources being funneled into a sector with continually diminishing returns. The primary motive was geopolitical: leaders wished to achieve autarky in staple

[63] *Pravda*, January 18, 1971.
[64] *Novoe Vremia*, October 7, 1988.

foodstuffs over the long term, even at the cost of a medium-term boost in imported goods (e.g., American wheat). It was also ideological: to feed a population increasingly aware of the low quality, variety, and nutritional value of food in their stores. Investment enabled the authorities to supplement "Fruit and Vegetable," "Meat," and other generic stores with several scores of "Ocean" stores in major cities with the promise of seafood, as well as new chicken, egg, and livestock production facilities. But long lines for tasteless, and even spoiled, foods persisted.[65]

Soviet agricultural, irrigation, and land-use policies reflected the dangers of centralized, state-determined production targets and massive nature transformation projects in the absence of public discussions of the costs and benefits of those projects. Neither groups of citizens nor engineers with access to plans and data were able to mount critical discussions of them. The result was the destruction of agricultural lands on a scale rarely seen in history except, perhaps, the great Dust Bowl in the United States. The negative impacts of erosion, poisoning of soils, poorly designed irrigation systems, and hasty reclamation of wetlands persist into the twenty-first century.

Government specialists did not give up. They proposed two ambitious reclamation programs in the early 1980s to create within a decade an immense agricultural region responsible for the production of one-half of the nation's agricultural output, which was supposed to be immune to the caprices of weather. Even more grandiose than the Stalinist plan for the transformation of nature, the plans aimed to add 50 million hectares of land from reclaimed marshland and irrigation of huge tracts of arid land. One project required a new northern rivers diversion project. The other involved diversion of water from the Danube into the estuaries of the Dniestr and Dniepr. The disruption of ecosystems that resulted would have been an unmitigated disaster.[66] Reclamation and irrigation

---

[65] On reforms in Soviet agriculture before and after the Brezhnev Food Program, see Alec Nove, "Soviet Agriculture under Brezhnev," *Slavic Review*, vol. 29, no. 3 (September 1970), pp. 379–410; and Robert Miller, "The Politics of Policy Implementation in the USSR: Soviet Policies on Agricultural Integration under Brezhnev," *Soviet Studies*, vol. 32, no. 2 (April 1980), pp. 171–194.

[66] "V Tsentral'nom Komitete KPSS i Sovete Ministrov SSSR. Postanovlenie 'O Dolgovrremennoi Programme Melioratsii, Povysheniya Effektivnosti Ispol'zovaniya Meliorirovannyh Zemel' v Tseliakh Ustoichivogo Narazchivaniya Prodovol'stvennogo Fonda Strany,'" *Pravda*, October 27, 1984, pp. 1–2; "Dolgovremennaia Programma Melioratsii Zemel,'" *Ekonomicheskaia Gazeta*, no. 44 (1984), p. 1; and Sergei Voronitsyn, "Renewed Polemics Over Siberian Water Diversion Scheme," *Radio Liberty Research*, August 6, 1982.

created their own series of problems, including salination of the soil. The irrigation systems that were built before the Soviet collapse were often hastily constructed, poorly monitored and run, and led to significant water loss and soil poisoning.

Soviet agriculture before, during, and after the Food Program was an exercise in overuse of chemicals and ruination of farmland. Recall that Khrushchev initiated the wide application of chemical fertilizers in the Soviet Union. In the wake of crop failures in 1963, Khrushchev, at the December 1963 Plenary Session of the Central Committee of the Communist Party, proposed a grandiose program for production of mineral fertilizers, herbicides, and other chemicals needed for an increase in agricultural production. Party officials and ideologues altered Lenin's injunction that "communism equals Soviet power plus electrification of the entire country" to read "communism is Soviet power plus electrification of the entire country, plus chemicalization of agriculture." In 1965, the country produced 31.2 million tons of fertilizers, but by 1975 there were already 90 million tons. During this period, mineral fertilizer applications developed at higher rates than in other developed countries, including the United States.[67]

In a typical tale of unclear responsibilities, brute force approaches, and regulation by the regulated, the distribution and instructions for application of pesticides and fertilizers was centralized, but there were no laws to regulate their use. Separate agencies and organizations acted in accordance with standards issued by the central authorities, even if their actions were dangerous. Pesticides like DDT were often sprayed indiscriminately from helicopters or airplanes, polluting surrounding settlements, forests, and meadows. Faulty transportation and storage of fertilizers and pesticides exacerbated these problems. There was also a qualitative difference between pesticides produced in the Soviet Union and those produced in Western countries. The latter were much safer, much more efficient, and required much lower proportions of active ingredient per unit area. Some of the best pesticides required only several hundred grams or less of active ingredient per hectare, whereas in the Soviet Union, farmers applied pesticides at a rate of five, six, or even eight kilograms per hectare, and in

---

[67] Marshall Goldman, *The Soviet Economy: Myth and Reality* (Armonk, NY: M.E. Sharpe: 1968), W. N. Parker, *The Superpowers: The United States and the Soviet Union Compared* (New York: MacMillan, 1972), and *Narodnoe Khoziaistvo SSSR, 1922–1982 gg.* (Moscow: Gosstatizdat, 1982).

extreme cases, well over thirty kilograms were required to achieve the desired pest control effect.[68]

The Brezhnev Food Program led to excessive agricultural specialization in the various republics, with monocultures of various cash crops leaving those republics to face disaster should there be a food crop failure. Georgia, for example, was forced to become ever more reliant on citrus fruit, wines, tobacco, and flowers to the exclusion of all else because it was obliged to meet the needs of the entire Soviet Union. After independence in 1991, Georgia was hard-pressed to achieve self-sufficiency in food staples again, an experience repeated in Armenia, Tajikistan, Kyrgyzstan, Turkmenistan, Azerbaijan, and Uzbekistan.

Roughly half of the land in use in Armenia was exposed to erosion as a result of intensive and irrational agricultural practices. To begin with, barely a quarter of arable land was on level ground, whereas another quarter under cultivation lay in steep terrain. More than a quarter of Armenia's agricultural land was lost to erosion, with damage particularly evident on highland pastures. Erosion also hit Azerbaijan, with the loss of more than two-fifths of its arable lands, pastures, vineyards, and forests, especially to gully erosion processes. Even on the plains, poorly managed cotton irrigation eroded soils and carried away expensive fertilizers. Some regions of Belarus saw up to two-fifths of their total land area eroded, whereas in Ukraine, one-third of agricultural land remains subject to erosion, not to mention the impact of Chernobyl in spreading radionuclides in the soil (on Chernobyl, see Chapter 5).[69]

Even before Khrushchev's departure from power, the spectacular negative environmental consequences of his Virgin Lands campaign in Kazakhstan had become apparent. Seduced by the initial high returns, he disregarded expert advice and insisted on planting monocultures year after year on marginal soils. The press chronicled the disaster as millions of tons of topsoil blew away in huge dust storms. New agricultural technologies later in the 1960s prevented further storms, but water erosion continued to plague fully one-fifth of the nation's arable land, and large swaths of land remained vulnerable to wind erosion as well. Two decades later, the exhausted soil often returned yields that barely covered seeding

---

[68] C. Csaki, *Armenia: The Challenge of Reform in the Agricultural Sector* (Washington, D.C.: World Bank, 1995), p. 198.
[69] Csaki, *Armenia*; State Report. Azerbaijan, 1993; Country Overview. Belarus, 1998; and State of Environment Ukraine.

norms. To this day, Kazakhstan can actively cultivate only about one-fifth of its 25 million hectares of arable land.[70]

Soviet laws eventually reflected growing concern with land degradation. The Supreme Soviet adopted legislation in 1968 that explicitly identified implementation of land protection and rational use policies as a concomitant responsibility of state ownership. A nationwide land cadastre was established in November 1972 for these purposes. But creating an inventory and doing something useful with that information were, as usual, two different things. The authorities developed plans in all regions to reduce soil erosion and allocated annual funds to construct facilities and plant trees to protect soils. By 1990, nearly 36 million hectares were newly irrigated and 15.2 million hectares drained, yet their productivity often did not meet the cost of reclamation. On the eve of the breakup of the Soviet Union, the authorities were planning more programs, more fertilizers, new reclamation projects, user fees, and so on to save what they had already lost.[71]

The Urban Environment

Waste of resources extended from the government to the ministries to the consumer to the cities. Consumers rarely engaged into resource conservation. There was no reason; home heating and water were too cheap. In 1981, Gosplan proposed to install water meters in every apartment, because many residents simply left the taps constantly open. The government also sought to introduce higher prices for "excessive" use of hot water. But citizens were very sensitive to rising prices. Stable prices for staple foods, industrial commodities, and services – many of which had not changed for decades – were regarded as one of the main achievements of socialist society, for only in a capitalist society did increased demand give rise to indiscriminate price rises. The common sentiment was that a socialist economy was supposed to raise production to meet demand. How it would do this with increasingly scarce natural resources remained unclear.

Soviet specialists assumed they had an advantage over the West in domestic home heating and refuse disposal because of lower demand and

---

[70] *State of the Environment of the Republic of Kazakhstan* (Almaty, 1993) (in Russian); *Official Statistics of the Countries of the Commonwealth of Independent States*, (Moscow: Statprogress, 1999).
[71] *Environmental Status Report of the USSR* (Moscow 1990) (in Russian).

lower waste. About one-tenth of all air pollution in the United States and likely a higher share in Western Europe and Japan stem from domestic uses. Soviet specialists believed that centrally supplied hot water for heating and washing enabled energy use savings over Western systems, in which each home had a furnace and water heater. Many of the water heaters were not on-demand models but more inefficient storage tanks. Furthermore, more than five thousand small coal-fired heating plants were eliminated during the 1970s alone in Moscow as boilers switched to natural gas. Moscow and Leningrad reportedly reduced the level of air pollution by more than 80 percent through this switchover.[72] Yet, in fact, Soviet central heating was highly inefficient, with tremendous heat loss visible in swaths of muddy or green land through the middle of deep snow drifts in winter that appeared above poorly insulated ducts and conduits. Many of the ducts were also above ground. Soviet apartment complexes were notoriously unevenly heated, with many residents leaving windows open to get some comfort, and others freezing, no heat control valves on radiators, and no incentive to conserve hot water because it was so inexpensive to the consumer.

As for garbage, waste, and trash disposal, several advantages over the West really existed. One was the significantly smaller amount of material used for packing. Indeed, the consumer carried a mesh or plastic bag at all times against the chance there would be something somewhere to buy and that item would be packed in old newspaper. There were no throwaway cans and bottles. The bulk of the consumer's food purchases were also unwrapped. As late as 1970, a Ukrainian trade official reported that only 6 percent of the flour, 3 percent of the sugar, and 5 percent of the butter in stores were sold in prepackaged form.[73] Officials hoped to raise the share of packaged foodstuffs nationwide to 60 percent to 65 percent by 1975; the belated adoption of self-service stores and Western-style supermarkets obliged foreign trade organizations to order packaging technology from Eastern and Western Europe.[74] Although the Soviet people were spared the highly wasteful consequences of consumerism, they still managed to fill their cities with garbage, and a film of noxious dirt that included heavy metals and particulate rained down all day, every day.

Environmental degradation plagued Soviet cities, especially outside apartments. There were fewer automobiles than in the West, which should

---

[72] *Gazovaia promyshlennost'*, no. 4 (1970), p. 37.
[73] I. Kuleshov and V. Burlai, "Bez Priemki," *Pravda*, March 5, 1970, p. 3.
[74] V. Sukchanov, "Preodolevaia Inertsiiu," *Izvestiia*, November 18, 1971; and Bush, "Environmental Disruption: The Soviet Response."

have indicated cleaner air. But the 1.5 million vehicles as of 1970 (the number of automobiles in the United States in 1913) had heavily polluting engines. Impure leaded gasoline, sustained use of old vehicles, inadequate servicing, and miserable carburetion all contributed to air pollution. With the debut of the Zhiguli automobile, the output of higher-octane fuel was stepped up, while servicing facilities slowly improved. Moreover, a drive to expand export markets for Soviet-made cars obliged manufacturers to conform to existing and projected exhaust emission standards in other countries.[75] Manufacturers introduced such emission control devices as catalytic converters on cars and buses in Almaty, Kazakhstan, beginning in the late 1960s, and successfully tested another emissions prototype on Volga sedans. Yet, few of these devices went into wider production, and few Soviet cars ever employed fuel injection.[76]

In theory, public transportation in Soviet cities had distinct environmental advantages over cities in the West, especially in the United States with its poorly developed and supported municipal transport systems. As one example, in the last thirty years of the twentieth century, the U.S. government provided a miserly $40 billion for its passenger rail system, Amtrak, while pouring nearly $4 trillion in automobile, truck, and airline infrastructure (roads, rails, and so on). In marked contrast to the noxious diesel buses in Europe and North America, the Soviets had continued to develop their electric trolley bus and tram networks, adding several thousand kilometers of new lines in the first half of the 1970s.[77] Still, pollution increased when the Soviets imported Czech-manufactured "Ikarus" articulated buses that quickly broke down under the stress of Soviet passengers and roads and belched diesel smoke proudly.

The main polluters were factories that discharged their wastes directly into the rivers and atmosphere. Native animal and plant life died, the water became unsafe to drink, and the usable supply shrank markedly. In 1967, Soviet sources reported that two-thirds of all factories in Russia discharged effluents without any effort to clean them up, and all rivers in Ukraine had been heavily polluted. Municipalities also dumped untreated waste into the environment. Sewage facilities were primitive or lacked

---

[75] Bush, "Environmental Disruption: The Soviet Response. On Soviet automobility, see Lewis Siegalbaum, *Cars for Comrades* (Ithaca: Cornell University Press, 2008).

[76] *Avtomobil'naia promyshlennost'*, no. 2 (1973), p. 13; Trentral'noe Statisticheskoe Upravlenie pri Sovete Ministrov Kazakhskoi SSR, *Narodnoe Khoziaistvo Kazakhstana; Statisticheskii Sbornik* (Alma-Ata, Izd-vo Kazakhstan, 1968), p. 416; and *TASS*, October 21, 1970.

[77] S. M. Ionov. "Piat' let po-Novomu," *Ekonomicheskaia Gazeta*, no. 4 (1972), p. 17.

the capacity to treat all wastes. Until the mid-1960s, 300,000 to 400,000 cubic meters of raw sewage were flushed daily into the Moscow River in the nation's capital. In 1968, one-third of state-owned urban housing in Moscow was not linked to sewage treatment plants, and nearly the same share lacked running water.[78]

More serious for public health, a tendency toward urban concentration of industries emerged with a vengeance at the end of the 1950s that had serious consequences for the public. Soviet planners managed to situate apartments and factories near one another because of both the lack of zoning laws and their hope of keeping commuting times and costs to a minimum. This led to high levels of exposure to industrial chemicals for workers at work and for workers and their families at home. According to Soviet measures, between 1959 and 1969, the number of towns and cities increased from 4,619 to 5,466, and the urban population by 34.2 million people, with two-thirds of them, 23 million people, concentrated in 208 large towns, or less than 4 percent of all urban settlements. Cities thus lay at the heart of the environmental dilemmas faced by the Soviet Union. This level of concentration of industry and population in many cities ultimately far exceeded that in the notorious industrial centers of England, Germany, and the United States: Manchester; Liverpool, the Ruhr region; Pittsburgh, Pennsylvania; Gary, Indiana; and so on (see **Box 4**). The worst case may well have been the Cheliabinsk region, which was regularly covered by a pall of noxious fumes from coal boilers, heavy metal plants, and petrochemical facilities. The residents of Sverdlovsk (Ekaterinburg) lived through exposure to radiation and biological weapons (anthrax) from military facilities.[79]

The reasons for favoring municipal plans in which industry and housing were proximate were quite simple, even if they meant that the worker, the individual in theory for whom the state existed, suffered the consequences. The construction of large-scale industrial firms in small towns would have required a proportionally larger investment in nonproduction infrastructure: housing, water and electricity, roads, bridges, schools, and other cultural services. In big cities, the municipal government and the

---

[78] Marshall Goldman, "The Convergence of Environmental Disruption," *Science*, vol. 170 (October 2, 1970), 37–42; *Moskovskaia Pravda*, September 1, 1968; and *Vestnik Statistiki*, no. 3 (1970).

[79] Josephson, "Industrial Deserts: Industry, Science and the Destruction of Nature in the Soviet Union," *Slavonic and East European Review*, vol. 85, no. 2 (April 2007), pp. 294–321.

> ### Box 4. Combating Air Pollution in Industrial Dneprodzherzhinsk, Ukraine
>
> Dneprodzerzhinsk, situated in central Ukraine on the banks of a reservoir for one of the major dams on the Dnieper River, arose in the late nineteenth century. A metallurgical center for more than eight decades, officials embarked on the first attempts to clean up the city and its surroundings in the mid-1970s. The main sources of pollution were factories built during the first five-year plans whose production capacity had been augmented later by other chemical, fertilizer, building material, and other facilities. More than thirty plants turned out cast iron, rolled metal, steel, paint, mineral fertilizer, railroad cars, and other items through highly toxic and poorly monitored processes. The largest, filthiest plants worked around the clock. After the Council of Ministers of the Soviet Union adopted antipollution measures for Ukrainian industrial cities in 1976, more than $100 million was spent to install smokestack scrubbers and the like over the next five years. Although this greatly reduced emissions, a thorough cleanup required that up to 300,000 persons be moved away from factories and hazardous waste dumps surrounding them. Because of a new Dniepr Reservoir that was built in the 1980s, the people who had been relocated lived in regions of the city stretched twenty kilometers upstream from the old city. Yet, even with funding from the central government, relocation upstream proved to be too expensive, and many people remained living in highly polluted environments. In 1988, Dneprodzerzhinsk was included in an official list of the sixty-eight most polluted cities in the Soviet Union, with average concentrations of NO_2, NH_3, formaldehyde, benzene, and airborne dust all at levels five or more times their maximum allowable doses. Blood pathologies of children were three times the average Soviet figures. Only with the economic collapse of the region in the 1990s did the environmental situation in the city improve, but only because outdated and inefficient industries shut down, leaving workers unemployed but still exposed to waste dumps. Between 1988 and 1998, industrial emissions declined from 336.5 thousand tons to 122.1 thousand tons.[80]

[80] *Izvestiia*, April 18, 1976, p. 2; and *Environmental Protection and Rational Use of Natural Resources in the USSR*, 1989.

Ministry of Urban Infrastructure financed the costs. In smaller cities and towns, industrial firms had this responsibility. (Note the contrast with market-oriented economies, where industries have often fled the older manufacturing centers when rising urban housing costs exerted indirect upward pressures on wages, or tax policy toward new industrial enterprises favored locations other than traditional urban centers.) The irony is that the very collective responsibilities to find homes for the worker that were imposed on Soviet industrial enterprises contributed to trends detrimental to the environmental commonwealth, because they were loath to invest in housing when productive capacity was a priority.

Finally, in response to the pernicious pattern of putting industry and housing side by side, the expansion of industry was forbidden in the Moscow region and thirty-four other large cities of the Russian Republic in the 1970s. Ministries and state and city authorities nonetheless continued to favor this practice. Urban leaders were well aware of environmental degradation but only had recourse to diagnostic measures, not therapeutic ones. They established observation posts and automatic sensors to monitor air quality in many large cities. By 1980, Moscow spent 100 million rubles annually on enforcement and monitoring, a third of which went to combat industrial air pollution. A general plan for the city envisaged dismantling roughly two hundred polluting enterprises situated within the city limits, but results were slow in coming. As late as 1984, such major enterprises as the "Hammer and Sickle" metallurgical plant still belched pollutants in the center of the city.[81]

One of the rare instances of progress in fighting urban air pollution did not target a city as the beneficiary, but rather as a cultural site. This indicated that the work of activists in local lore organizations and in such national groups as VOOP (the All-Russian Society for the Protection of Nature) had had some impact. Smoke from a chemical combine in Moscow regularly blew southeast toward Iasnaia Poliana, the former family estate of Leo Tolstoy. On several occasions between the 1960s and 1980s, this lethal brew of ammonia, carbides, and sulfuric acid killed off large stands of trees at Tolstoy's home. The government issued decrees forbidding this pollution, but in typical Soviet fashion, the connection

---

[81] M. Mkrtchyan, "Methodological Problems of Allocation of Productive Forces," *Planovoe Khoziaistvo*, December, 1969; A. Shapovalov, "Green Necklace," *Moscow News*, July 24, 1971, p. 3; and V. Zagvozdkin, "Chistota Vozdushnogo Basseina," *Moskovskaia Pravda*, December 8, 1984, p. 2.

between Kremlin wishes and local action was achieved only through ad hoc intervention. The writer Yuri Bondarev drew attention to the problem at the 1985 Soviet Writers' congress, and the conservative newspaper *Sovetskaia Rossiia* subsequently published a scathing attack on the hapless plant managers for ignoring Kremlin orders. (Bondarev, a Russian nationalist, served in the rabid Communist opposition to the Gorbachev reforms.) Within a year, a local Party official announced that, following the conversion of two power plants to natural gas, the closure of a blast furnace, and the addition of scrubbers to several smokestacks, pollution in the vicinity had been cut by half.[82]

In the last decade of the empire, flurries of government resolutions obliged local authorities to combat air pollution. Officials allocated substantial funding to clean up cities, and scientists pursued extensive research programs on urban nature conservation. Despite tougher laws, industrial managers ignored instructions to install air filters and in other cases simply refused to hook up equipment. The authorities repeatedly vowed that severe measures would be taken, in one instance publicly reprimanding five industry officials over air pollution in the western Siberian city of Kemerovo.[83] Usually, however, nothing happened.

Acid Rain, Air Pollution, and the Soviet Union

In many conditions, sulfur and nitrogen oxides convert to sulfuric and nitric acids in the atmosphere and then fall to earth as acid rain, where they harm soil, rivers, and lakes. The first references to this scientific literature date to the 1920s. From the mid-1950s to the early 1970s, sulfur levels in the atmosphere over Europe increased by 50 percent, and in southern Scandinavia they doubled. The primary suspects in the latter instance were the tall chimneys designed to minimize ground-level discharge of sulfur dioxide in Britain that sent pollutants far downwind.[84] Anxiety about acid rain was one of the reasons why the Swedish government called for an international meeting, which eventually became the UN Conference on the Environment in Stockholm in 1972. Yet, the subject of acid rain – then regarded as a problem for just a few industrial

---

[82] S. Kalesnik and A. Shnitnikov, "Vnimanie – Priroda!" *Komsomolskaia Pravda*, April 3, 1968, p. 2; and "Air Pollution at Tolstoy's Estate Reported Cut by Half," *Radio Liberty Research Institute*, no. FF133, February 13, 1986.
[83] *Reuters*, January 21, 1984.
[84] *Times*, "Downwind Acid Drops," March 15, 1983.

countries – received scant attention because the participants found so many other important issues to address.

It was another decade before the first Convention on Transboundary Air Pollution came into effect, partly as a result of subsequent UN negotiations that included Soviet diplomats. The purpose of the convention was to achieve a reduction in the high levels of acid rain experienced downwind from factories, smelters, and power stations. Tortuous negotiations stemmed from the quintessentially international nature of the problem: one country's emissions could easily become another country's problem, and attributing the responsibility proved to be contentious. Some of the signatories, including Great Britain and the United States, were reluctant parties to the treaty until the initial proposals had undergone significant revisions because they were major contributors to transboundary pollution. Although many countries of central and southern Europe were among the offenders, the position of Great Britain and the United States was unusual in that they managed to disperse pollutants to the detriment of others without themselves suffering discharges from elsewhere (although New England suffered the consequences of acid rain because of midwestern U.S. coal). They were disinclined to pay for the modifications that would cut gaseous emissions to levels that the victims of pollution, notably Scandinavia and Canada, would have liked.[85]

The Soviet Union had a vested interest in reducing sulfur dioxide emissions because it led the list of downwind victims in absolute terms. Yet, it had mixed motives at the negotiating table because it was also the biggest polluter in Europe in terms of the tonnage of sulfur dioxide generated by its industries. The construction of major new energy production complexes in the 1970s and early 1980s made it all the more difficult for the Soviets to cut pollution. Hoping to exploit the vast reserves of lignite coal in central Siberia, they built the Kansk-Achinsk complex in south-central Siberia in the mid-1970s. Engineers determined to burn low-efficiency lignite on site in a series of thermal electric power plants that have left the region one of the most polluted in the world. They built the plants near the extensive coal fields because transporting the coal to major industrial centers in the European Soviet Union would have been more costly. Shockingly, high voltage transmissions lines to transmit electricity to consumers followed only a decade later. Similar considerations

---

[85] Ibid.

and approaches led to the creation of the Ekibastuz coal-burning complex in Kazakhstan.[86]

A 1983 Swedish report attempted to apportion the responsibility for acid rain to spur further action. Though total Soviet emissions of sulfur dioxide dwarfed all others at 25 million tons annually, many other European nations also pumped significant quantities into the atmosphere. On a per capita basis, the clear leader was East Germany, generating some 527 pounds of sulfur dioxide per person in 1982, followed by Czechoslovakia at 485 pounds per person. The third-place Soviet Union produced only 211 pounds per person by comparison, but it was followed in short order by Romania, Great Britain, Poland, France, Italy, West Germany, and Sweden. The heaviest polluters were largely those in the Soviet sphere. Several others had accelerated production of emissions: Romania's sulfur dioxide emissions had grown by 235 percent, an order of magnitude greater than those of the remaining offenders – Yugoslavia, Portugal, Czechoslovakia, Hungary, Bulgaria, the Soviet Union, Turkey, and Finland.[87]

In Geneva, in 1984, more than forty nations finally formally acknowledged that sulfur dioxide pollution threatened the natural basis of life, though they were not bound to take any specific actions. Nearly half of the participants nonetheless pledged to make 30 percent reductions within a decade. The Soviet Union and the Eastern European countries were among the would-be reformers, but they largely limited themselves to reducing transborder emissions or failed to adopt specific targets at all and certainly pledged to do nothing at home.[88]

Water, Water Everywhere

If the position of the Soviet Union regarding atmospheric pollution was somewhat ambivalent, there was no doubt that Soviet policy was based on rapacious use of water resources with no intention to cut back, and

---

[86] Radio Liberty Research Institute, "Air Pollution Problems at Siberian Power Plants," no. FF058, October 17, 1984; G. Anohin. "V Voskresen'e Budet Pozdno!," *Sotsialisticheskaia Industriia*, November 12, 1989, p. 4; and *Rabotnitsa*, February 1, 1989. The largest sulfur emitter of all, the Astrakhan Petrochemical Plant at the mouth of the Volga River, which went online in 1987, poured more than 1 million tons of sulfur dioxide into the atmosphere in its first year. Respiratory disease has become endemic.
[87] Radio Liberty Research Institute, "Sweden Report on Acid Rain in Europe," no. FF050, July 13, 1983.
[88] S. Voronitsyn, "USSR Assumes Minimal Obligations at Ecology Conference," *Radio Liberty Research Institute*, RL 276/84, July 17, 1984.

the Soviets had very little interest or success in cutting pollution. The Soviet Union had abundant water resources, but they were unevenly distributed. Siberian rivers had copious flow but flowed northward into the Arctic. European rivers had, by the time of the Brezhnev era, been tapped out toward industrial, municipal, and agricultural ends. No major river remained untouched by engineers or devoid of dams or reservoirs. A serious water shortage ensued, not only in arid regions, but also in the areas where the great rivers – the Don, Dnieper, Dniester, and Volga – flowed. The reason was simple: by the late 1960s, water pollution had become the Soviet Union's biggest resource conservation issue. With inadequate waste treatment facilities, the country used greater and greater quantities of water to dilute effluents. As is well known, the solution to pollution is not dilution, but prevention and cleanup. The levels of pollution most likely exceeded those in the Rhine and Danube. The Volga River was a sewer. With 6 percent of the nation's total river flow, it carried half of the country's industrial wastes into the Caspian Sea. As for Siberia, when plans called for 500 percent growth in industrial output in Siberia in the 1970s, additional water equal to the total annual flow of the Enisei and the Ob Rivers would have been required unless radically more efficient processes were introduced.[89]

Belatedly, the authorities realized the need to clean up industrial pollution. They commenced a cleanup campaign on the Volga and Ural Rivers in March 1972. The Volga was the Soviet Union's most vital waterway, carrying two-thirds of the country's river freight. The Volga basin was home to a quarter of a billion people, and twenty percent of Soviet farmland under cultivation abutted it. One billion rubles (roughly $1.2 billion) was allocated for the first part of the task. This may have been inadequate; by comparison, in 1971, the U.S. government approved a $24 billion bill to clean up the nation's navigable waterways and limit pollution discharge.[90] Subsequently, the Soviet government issued decrees and initiated programs by the dozens that addressed pollution on the Volga River, but remediation was modest. The dozen major reservoirs on the river trap pollutants and have so slowed the natural water cycle that recovery seems impossible. Eutrophication resulted. The sturgeon population downriver at the Caspian Sea plunged as a result. Other terrifying

---

[89] E. Agaev, "Kaspii – Vzglyad Realista," *Literaturnaia Gazeta*, no. 10 (1971), p. 11; and Oldak, "Tochka Zreniya Ekonomista," *Nedelia*, no. 12 (1968), pp. 14–15.
[90] "Steps Towards Pollution Control in the USSR," *Radio Liberty Research Institute*, April 6, 1972.

> **Box 5. Combating Water Pollution in the Georgian Soviet Republic**
>
> In the 1970s, Georgian officials could no longer ignore the problem of water pollution along the Kura River. Dozens of enterprises discharged such untreated wastes as phenols, oil, ammonia, copper, and nitrates in Tbilisi, the capital, Gori, and Rustavi. The Rustavi Metallurgical Factory was one of the largest industrial facilities in the country and a major source of pollution. Few plants had purification systems. Reservoirs annually received millions of cubic meters of toxic effluent. The extensive use of chemical pesticides on tea and citrus farms contributed to contamination of the waterways through runoff. Near Batumi, small rivers carried high concentrations of phenols into the Black Sea near popular resorts. The Vere, Debeda, and Mchavera Rivers in the Caspian Sea basin were polluted with urban sewage, copper mine tailings, and ore residues. The Georgian government passed a resolution calling for an end to the situation in 1973. No action followed. As late as 1988, official reports estimated that more than four-fifths of the 317 million cubic meters of pollution flowing into Georgian waterways went untreated – the second-highest figure in the Soviet Union, where the average was 28.1 percent.[91]

examples of the impact of water pollution include the Dnieper River disaster in Ukraine in 1983 (see **Box 4**) and Kura River pollution in Georgia (see **Box 5**).

If waterway pollution by industry in European Russia had caught the attention of Soviet authorities by the 1970s, they did little to combat surface and groundwater pollution caused by agriculture in the Central Asian republics. The pollution of Central Asian rivers increased with the expansion of cotton irrigation. The low immunity of cotton to disease made the drive toward monoculture risky at best and required massive applications of herbicides, pesticides, mineral fertilizers, and defoliants to ease harvesting. Drainage water from irrigation fields discharged these toxic substances into the main rivers.

Inland bodies of water – lakes and ponds – suffered from the same litany of misuse, pollution, and violation of laws. In 1962, the Soviet government had decided to substitute more fish for meat in the national

---

[91] Environmental Protection, 1989; Review of the State, 1990; and State of the Environment, 1989.

diet, ordering an expansion of the fishing industry. A State Committee for Fisheries was organized under Gosplan. The rapid expansion of the fishing industry was almost entirely due to harvesting from the inland waters. Yet the annual catch in Russian seas and lakes reached its peak before the revolution in 1913 at 614,600 tons. Between 1962 and 1968, the total catch dipped from 426,500 to 270,600 tons. In the leading fishery, the Azov Sea, the fish catch was 158,800 tons in 1936, sinking to a mere 14,500 tons in 1965. Pollution, dams, rapacious fishing, and poaching had already destroyed the industry.

In 1969, the editors of the main literary weekly newspaper, *Literaturnaia Gazeta*, published an interview with several government ministers about the terrible state of fisheries. The minister of trade noted that sturgeon and bream catches were half of previous levels, whereas salmon and whitefish stood at one-fifth the level of previous catches, Caspian herring at one-thirteenth, and kerch herring a mere one-fortieth of historic norms. The industrial catch of fish like lamprey, eel, lake salmon, and sterlet (a European sturgeon) had practically ceased. An insignificant increase in the overall fish catch came at the expense of the small fish, thus undermining future fish resources by interrupting the food change. And, of course, the fisheries in no way met consumer demand.[92] In 1970, the Council of Ministers of the Soviet Union decreed that environmental agencies charged with protecting fish resources could fine industrial enterprises as well as collective and state farms that damaged fisheries. The fines would then be used for restoration of the country's fish stocks.[93] As usual, imposition of and collection of fines, restocking of fisheries, and other measures fell far short of expectations and necessity.

"Hero Projects" of the Brezhnev Era: From Central Asia to Lake Baikal and Siberia

Several major cases stand out as examples of how Soviet words and deeds – calls for environmental protection and ongoing devastation – stood in stark relief. They concern Lake Baikal, the Caspian Sea, the Azov and Aral Seas, and the Siberian River Diversion Project. (See **Box 6. A Twenty-Year Battle to Protect Baikal.**) Each project reflected the disjunction between the rhetoric of "developed socialism," with its reference

---

[92] V. Mikhailov, "Edali Vy Pirog s Sigami?" *Literaturnaia Gazeta*, no. 10, March 5, 1969, p. 10.
[93] "Rekam ne Skudet,'" *Pravda*, January 13, 1970, p. 6.

Box 6. A Twenty-Year Battle to Protect Lake Baikal

Lake Baikal holds one-fifth of the world's freshwater. It is 638 kilometers long, with an average width of 48 kilometers, and has the greatest depth among inland waters, varying from 730 to 1,620 meters. Lake Baikal is renowned for the purity of its water and its 1,200 species of animal life, including 700 endemic species. The *Guinness Book of World Records* gives Lake Baikal a double entry – for depth and capacity (5,520 cubic miles [23,008 cubic kilometers] – more than the five Great Lakes combined). The first environmental problems for the lake emerged with the exploitation of neighboring forests. The local timber industry began floating logs into the lake in the 1930s and increased the float significantly after World War II as modern techniques allowed the industry to be organized on a large-scale basis. Between 1958 and 1968, an estimated 1.5 million cubic meters of timber sank in the lake and nearby rivers. With the construction of pulp and paper mills to produce cellulose cord for jet bomber tires, pollution of the lake accelerated with grave consequences. This led to increased government involvement in managing pollution and in construction of waste treatment facilities according to a series of resolutions and laws.

When it became clear that no one was paying much attention to the laws and resolutions, a public movement to protect the lake developed, and no fewer than four national groups came forward opposing the opening of more cellulose and pulp plants and demanding the treatment of plant discharges. By the end of the 1950s, the dispute over the construction of paper mills on the shores of Lake Baikal had become the locus for testing the limits of openness in the mass media. Many writers added their voices to debates about how to protect the world's largest body of fresh water. Nobel laureate Mikhail Sholokhov broached the subject at the Congress of Soviet Writers in 1956. Despite the fact that Khrushchev himself forbade extensive discussion, Sholokhov raised the specter of the destruction of Lake Baikal again at the twenty-second Party Congress in 1966. Over the next few years, party officials increasingly restricted the spread of environmental information in the media. Still, writers often managed to get around these restrictions, occasionally in highly visible venues. Other writers managed to place essays in *Literaturnaia gazeta*. This might be the

(continued)

> Box 6 (*continued*)
>
> equivalent of Aldo Leopold publishing *Sand County Almanac* in the *Atlantic Monthly* or, better still, Rachel Carson publishing excerpts of *Silent Spring* in 1962 in *The New Yorker*.[94]

to the importance of the environment for the citizen, and the mandates of economic growth at any cost that left public health and environment as secondary concerns. These projects reflected a common and troubling pattern for economic development in industrial societies in the twentieth century: the bureaucracies and other vested interests behind the projects acquired significant momentum, whereas the public was excluded from decision making about how and whether to proceed. When it became clear that a project would experience huge cost overruns or that its environmental and social costs were far greater than its proponents claimed, the project almost always – universally, throughout the world – moved ahead in any event with the argument that all investment to date and jobs would be lost were the project to end.

The unique history of environmental problems surrounding Lake Baikal indicates how different a place the Soviet Union had become under Khrushchev and Brezhnev than it was under Stalin, although it still had the same handicaps on environmental policy as central economic planning and reliance on planners' preferences. But in this case, public concern turned into a national movement to save the lake from industrial pollution. Even though it was ultimately unsuccessful, the movement indicated that the public might be involved actively in environmental activities.[95]

In the late 1950s, industrial development along the shores of Lake Baikal slowed, but several cities and factories had grown up along the lake's tributaries. (A far-fetched plan to used 20,000 tons of TNT to open the outflow of Lake Baikal into the Angara River to permit more water to flow through the downstream Irkutsk Hydropower Station was, however, floated in 1954 by a Leningrad design institute.) In the early 1960s, Gosplan adopted a program to develop the Baikal basin, rich in

---

[94] Boris Komarov, *The Destruction of Nature in the Soviet Union* (White Plains, NY: M.E. Sharpe, 1980), p. 150; and *Financial Times*, May 11, 1978. See also Aldo Leopold, *A Sand County Almanac: Sketches Here and There* (New York: Oxford University Press, 1949); and Rachel Carson, *Silent Spring* (New York: Fawcett Crest, 1962).

[95] On Baikal, see Marshall Goldman, *Spoils of Progress* (Cambridge: MIT Press, 1972); Komarov, *The Destruction of Nature*; and P. S. Serebrennikov, *O Baikale – s Liuboviu i s Boliiu* (Ulan Ude: Buriatskoe Knizhnoe Izdatel'stvo, 2001).

fish, forest, and mineral resources. About fifty factories were built, but only a few treated their wastes before they entered the lake. The major threat to the lake was connected with construction of two huge paper and pulp combines, one of them, the Baikal Pulp and Paper Combine, at the southern shore of the lake. Many experts and officials opposed the project of construction of the paper and pulp combines. The real pressure to build the plant originated in the Ministry of Defense. The ministry wanted a domestic supply of durable cord for bomber tires, then being imported from Canada and Sweden. They claimed that only two sites could provide the huge quantities of clean water needed for the manufacturing process: Lakes Baikal and Ladoga. But Lake Ladoga, near Leningrad, was already surrounded by industry and was the chief sources of drinking water for the city.

The first warnings of pollution came in 1961 when the yearly catch of omul, a whitefish species, fell by half. Protests from prominent Soviet scientists and writers had already appeared in Soviet newspapers in the 1950s that untreated industrial sewage wastes would not only destroy the marine life, but might also affect the water supply of the city of Irkutsk. The solution at that time was to treat effluents of the industrial plants and reuse the water, but the process was of limited effectiveness, and paper industry officials refused to put much investment into waste treatment and control. Little happened – except for more construction and production – until a 1969 government decree that spelled out steps necessary to clear up the lake. The decree – "About Measures on Conservation and Rational Utilization of Natural Ecosystems in the Basin of the Lake Baikal" – instructed the Pulp and Paper Ministry to complete in the same year a complex of purification installations at the plants and a complete water recycling plant at the Selenga Pulp Mill, where fish damage was most extensive. Responsible ministries of the Russian Republic were ordered to prevent the discharge of industrial, municipal, and household sewage into the lake through waste treatment facilities. The executives of the enterprises were held responsible for existing problems, although few were punished or fined.[96]

The health of Lake Baikal grew ever more precarious. A series of proclamations followed to force industry to clean up its act, but little happened. A 1971 decree astonishingly similar to the 1969 one recognized that neither the Selenga nor the Baikal pulp water purification projects had been completed as ordered; the plants were given another year to

[96] A. P. Aleinik, "Kommercheskaia Postanovka Dela," *Izvestiia*, February 8, 1969.

finish the job. The decree prohibited the spring log float, requiring rafting instead, and fish spawning grounds were to be cleared of sunken logs, which consumed oxygen during decay and covered breeding areas. The Academy of Sciences was included in all of these projects to ensure the fullest utilization of science and technology in the undertaking. But it appears that little was done from the point of view of industrial sewage treatment. The decree repeated official concern about the deterioration of Lake Baikal, perhaps in recognition of the action of public groups over the previous decade. Several Soviet publications indicated that pollution of the lake had increased, not decreased, so it is doubtful that the 1971 decree would have any greater success in opposing entrenched power of industry and military.[97]

Lake Baikal suffered decades of industrial and other encroachment. A report of the Soviet Academy of Sciences in 1975–1976 concluded that the combine should be shut down and reequipped for environmentally safe production. The report pointed out that a tiny crustacean – the epishure – that formed the first link in a food chain that supported all the fauna of the lake, including omul, grayling, and seals, had been threatened. Epishure were endemic to Lake Baikal but died even if held in pure Baikal water in laboratory test tubes. Epishure are not only a replaceable food, but also act as a potent biological filter, helping to extract about 250,000 tons of calcium a year from the river water that flows into the lake. They are responsible for the unique purity of the Baikal water and for its saturation with the oxygen even in winter. The epishure were dying even in the supposedly purified effluents from the paper and pulp combine, even when the effluents were diluted 100 times.[98]

Although it was meeting production targets, the Paper and Pulp Ministry could not manage to restore the ecological balance in the area. Granted, as a result of public alarm, the pulp plant on the southern shore was equipped with a costly filtering system that some experts said would make the wastewater fit to drink. The cost was high, some 40 percent of the plant's operating budget. By 1978, after tens of millions of rubles had been expended, about 70 percent of the factories in the Baikal basin had water treatment equipment; most tributary streams had been cleared of sunken logs; and no more timber moved by float. A ban on fishing from 1969 to 1977 restored stocks of fish, and the seal population of Lake

---

[97] Radio Liberty Research Institute, "Steps Towards Pollution Control in the USSR," April 6, 1972.
[98] V. Markin, "U Vechnogo Ozera," *Sovetskaia Rossiia*, August 15, 1981, pp. 1, 4.

Baikal rebounded to 70,000. After 1983, limited commercial fishing was permitted once again on the lake.⁹⁹ Pollution had even begun to interfere with the operations of the combine itself, for it could not produce the special cord needed for tires. But this was not really a problem, because since by 1964 those tires were produced from petrochemicals. In 1981, *Sovetskaia Rossiia* reported that the Baikal Pulp and Paper Combine had been online for fifteen years, although its waste treatment facilities – with an impressive sequence of mechanical, biological, and chemical filters – did not function properly. As Komarov wrote, "So we are destroying Lake Baikal for producing about 160,000 tons of ordinary cord – a small proportion of the nation's needs; 3,000 tons of coarse packaging paper; 100,000 tons of nutrient yeast for feeding pigs; and a little turpentine and oil, used in paints."¹⁰⁰

The Soviet government issued its last resolution on Lake Baikal in May 1987. This unprecedented resolution lashed out at scores of government and party organizations and their leaders for failing to stop the pollution of Lake Baikal. It concluded that ministries and local authorities had fulfilled none of the adopted environmental directives to protect the Lake. The resolution reported that the deputy minister of the pulp and paper industry had been fired for irresponsibility; the forestry minister of the Russian Republic had been severely reprimanded, and the Council of Ministers of the Russian Republic had been instructed to review whether he should keep his job. A list of officials singled out for criticism by the Central Committee filled an entire column of a four-column article in the newspaper.¹⁰¹

The story of the Caspian Sea is no less disturbing. As with other bodies of water, Soviet planners saw the sea solely for its immediate economic potential and treated it as a kind of machine for other industries. Its fisheries and underwater oil fields were its two greatest resources. Eventually, nuclear specialists also got into the act, building a breeder reactor on the Mangyshlak Peninsula, the BN-350 (now closed). Breeder reactors produce plutonium that can be used to fuel other reactors or for nuclear

---

[99] D. Shipler, "Siberian Lake Now a Model of Soviet Pollution Control; Ban on Fishing Is Lifted Huge Crevasses May Open Up," *New York Times*, April 16, 1978, p. 10; Radio Liberty Research, April 17, 1978, Nature Protection, 1978–1979; *Financial Times*, May 11, 1978; *Radio Liberty Research Institute*, "Steps Towards Pollution Control in the USSR," April 6, 1972; Serge Schmemann, "Siberian Fighting to Preserve Lake," *New York Times*, December 27, 1981, p. 15; and *Izvestiia*, November 10, 1983.
[100] Komarov, *The Destruction of Nature*, p. 150.
[101] "Rezolutsiia TSK KPSS o Sostoyanii Ozera Baikal," *Pravda*, May 5, 1987.

weapons. The BN-350 produced electricity and used its tremendous heat to desalinate water for municipal purposes. But primarily industrial effluent has been the cause of the Caspian's degradation. Perhaps the earliest case of large-scale water pollution in the Soviet Union, the Caspian, the world's largest inland lake and once the main source of caviar, suffered from oil pollution caused by Soviet oil refineries and offshore rigs. Oil pollution dated to the late tsarist era and the development of those resources in the nineteenth century. By the late Soviet period, an oil slick covered the Bay of Baku (the capital of Azerbaijan), once full of fish but now regarded by local residents as dead. Plumes from the "Black City" refinery near Baku drifted far out to sea. All the effluents from factories and urban sewage disposal along the 3,000-kilometer Volga River ultimately also flowed into the Caspian Sea. (See **Box 7**. Damming of the Kara-Bogaz Gulf in the Caspian Sea.) The most tragic manifestation of this degradation was the drop in the sturgeon catch from more than 50,000 tons a year in the nineteenth century to fewer than 10,000 tons in 1970. By the early 1970s, black caviar production had dipped to as low as eighty-one tons of black caviar, and in 1973, the Soviet Union actually imported seventy-three tons from Iran for resale.[102] Caspian Sea caviar is now largely embargoed by international agreement.

To counteract the decline, fishery specialists ordered the construction of twelve sturgeon hatcheries near Baku and Astrakhan at the Volga delta. The stations released millions of young sturgeon into the sea each year (with four Iranian stations adding another 5 million hatchlings). Soviet experts experimented with crossbreeding the three main sturgeon types: the beluga, which is the biggest and takes up to twenty years to start producing eggs; the middle-sized osetr, which matures in twelve years; and the sevruga, which is the smallest, but matures in seven to ten years. The task was to obtain a hybrid that matured faster, although any success would not overcome the great losses. Lysenko would have been proud – had the breeding program worked. The sturgeon population stood at an estimated 200 million, but according to Russian specialists, twice that number was needed to support the desired 50,000-ton catch. The annual catch plummeted from 7,300 tons in 1993 to a mere 980 tons in 1998. The receding waters of the Caspian Sea, which fell by three meters between 1940 and 1970, in part because of unabated withdrawal of water for industry and agriculture along the Volga basin, had contributed to

---

[102] M. Parks, "Soviet Union Finds Oil and Caviar Do Not Mix Well," *Baltimore Sun*, August 12, 1974; and R. C. Toth, "Hope for Relishes of the 'Black Pearls' of the Caspian," *International Herald Tribune*, March 16, 1976.

> Box 7. Damming the Karas-Bogaz Gulf in the Caspian Sea
>
> Kara-Bogaz Gulf Bay in the Caspian is rich in salt minerals. In some years, up to 70 percent of the total sodium sulphate produced in the country was extracted from the bay's deposits. Industrial extraction commenced in 1924. In 1970, engineers unveiled a project to eliminate the bay in order to raise the level of the Caspian Sea. They estimated that the Caspian Sea lost significant quantities of water annually because of extensive evaporation from the surface of the shallow Kara-Bogaz-Gol. Engineers offered assurances that the Kara-Bogaz-Gol would remain a unique source of salts and that any damage to the bay would be mediated by the completion of another project to divert part of the flow of several northern rivers into the Caspian Sea. In typical Soviet fashion, economic ministries underwrote one risky project while relying on another equally ambitious, equally questionable project. By 1980, when dam construction across the Kara-Bogaz-Gol was well under way, engineers already recognized that the construction would not reverse the drop in sea level. Only the transfer of larger amounts of water from the northern rivers would reverse the Caspian's declining sea level.
>
> The dam markedly changed the composition and quality of the mineral deposits and hence destroyed the region's ecology. Engineers calculated that the bay would dry out over twenty-five years, enough time for the river diversion project to kick in. Yet, in three years, the bay was destroyed, and the chemical industry worried publicly about the future of the unique salt deposits. In a surreal reversal, engineers then proposed to construct a pipeline to resupply the bay with water from the Caspian Sea. In the 1990s, the dam was dismantled by a newly sovereign Turkmenistan.[103]

the destruction of the sturgeon population. Hydroelectric dams on all the main Caspian tributaries prevented migration upriver even where the depth was sufficient. The few sturgeon elevators and ladders installed at some Volga dams, like those elevators throughout the world, have failed.[104]

---

[103] *Komsomol'skaia Pravda*, February 13, 1971, p 4; and *Izvestiia*, March 16, 1980.

[104] R. C. Toth, "Hope for Relishes of the "Black Pearls" of the Caspian, *International Herald Tribune*, March 16, 1976; Parks, "Soviet Union Finds Oil and Caviar Do Not Mix Well"; *State of the Environment of the Russian Federation* (Moscow, 1993)

The destruction of the Aral Sea has been complete. One wonders if its location in the dry steppes of southern Soviet Asia, among Moslem instead of Slavic people, was the reason for the vanishingly small concern expressed by officials in Moscow about its destruction. The Aral Sea underwent a slow strangulation for two decades as the Soviets diverted the rivers flowing into it for the sake of irrigation for grain, fruits, vegetables, and cotton to serve the privileged people in the center. The depth of the Aral Sea decreased by more than 3 meters between 1960 and 1970 alone. What could have motivated such a vast experiment with nature? Inspired primarily by failed harvests in 1963 and 1965 and by the need to import grain from the West at great cost and embarrassment, Soviet leaders announced grandiose projects for reclaiming and irrigating tens of millions of hectares of swampy and arid land in Central Asia. These plans envisioned diverting the flow of the great Siberian rivers to the dry regions of Kazakhstan and Central Asia, with half of the annual flow of the Amu Darya and Syr Darya Rivers to be drawn off, and up to one-quarter of the Volga's water to be diverted to irrigate the lower Volga region.[105]

Many Soviet and Western specialists protested these colossal gambles with natural water cycles and sensitive ecosystems; Soviet technocrats and Party officials downplayed their concerns. In 1980, academician Evgenii Fedorov, a geophysicist and arctic explorer, who was both chairman of the Soviet Peace Committee and a member of the presidium of the Supreme Soviet, conceded that large-scale irrigation would lead to shrinkage of the Aral Sea, perhaps even to the point that it would become two separate lakes. Yet, from his point of view, the irrigation was more important than the conservation of the Aral Sea.[106] By 1982, cotton monoculture and its irrigation practices virtually eliminated runoff water from reaching the Aral Sea. (Israel, the world champion of efficient water use, produces six to seven times more cotton per unit of water than the Soviet Union.) This left only an average of four cubic kilometers of "free flow" that made it to the sea annually, and in dry years like 1986 there was no flow at all. By the 1980s, the sea had lost three-quarters of its 1960 volume. Toxic salts and dust blew from the exposed seabed into the Amu Darya delta. The contamination of drinking water with toxic

(in Russian); and *State of the Environment of the Russian Federation* (Moscow, 1998) (in Russian).
[105] Keith Bush, *Soviet Agriculture in the 1970s* (New York, Radio Liberty Committee, 1971), p. 49; and "Ob' Pridet k Aralu," *Pravda*, December 27, 1971, p. 1.
[106] E. Fedorov, "Put' k Ravnovesiiu," *Izvestiia*, March 16, 1980, p. 3.

chemicals from pesticides and fertilizers most likely caused high rates of intestinal diseases like hepatitis, throat cancer, and even typhoid. Infant mortality in the Karakalpak Autonomous Republic rose to the levels of those in some African countries.[107]

The Kara-Kum Canal in Turkmenistan, at 1,400 kilometers the longest canal in the world, built between 1954 and 1980, symbolizes the irrational and environmentally unsound practices connected with cotton and with the Brezhnev Food Program. Turkmenistan catches about one-third of the natural annual runoff in the region for human use, with the vast majority going to irrigation. The Kara-Kum Canal becomes the repository for much of this water, starting in the east at the Amu Darya River and traversing the whole of Turkmenistan to the Caspian Sea in the west. Along the way, the canal collects water from numerous small rivers and springs originating in the Copet-Dag Mountains. Roughly a dozen cubic kilometers of water pass through the canal annually, although in many places individuals make unauthorized use of water for irrigation purposes. Since 1980, crop yields in the region declined steadily, even as the area of land under irrigation increased because the canal was poorly designed and built. It was not watertight, as a result of which nearly one-third of the water percolated through sandy soils, leading to salinization and waterlogging over vast areas along the canal.[108]

Standard Soviet water management practices contributed to the destruction of the Sea of Azov, too. These included the construction of a cascade of dams and reservoirs on the sea's tributaries, the Kuban and Don Rivers; the siphoning off of water for municipal, agricultural, and industrial purposes; the siting of factories with inadequate pollution control equipment on its shores and on those tributaries; a sense that engineers would find solutions to any unanticipated problems that arose in their projects; and the belief that, in the absence of public outcry, planners could move ahead with impunity. After all, what was more important, public health and a clean environment or jobs? Or was the Soviet system, contrary to public proclamation, unable to seek both simultaneously? In the 1950s and 1960s, commercial fish disappeared from the Sea of Azov. Freshwater extracted from the Don and Kuban for cities, industry, and farms led to a decline in sea level and permitted saltier water from the

---

[107] S. Klotzli, *The Water and Soil Crisis in Central Asia – a Source for Future Conflicts?* Center for Security Policy and Conflict Research, ENCOP Occasional Paper No. 11 (Zurich: Center for Security Studies and Conflict Research, 1994).

[108] Environmental Policy Review, Turkmenistan, 1994; and State of the Environment, Tajikistan, 1995.

Black Sea to flow in. In 1972, the Soviet government announced plans for a gigantic dam at the northern edge of the Black Sea designed to counteract the processes it had unleashed earlier with the river dams. Three miles in width, the dam would have prevented the salty influx from the Black Sea. But this would not have solved the destruction of Azov fisheries, and the dam was eventually abandoned.[109]

Soviet planners and officials never learned from experience. Once again, competing ministerial interests joined with weak laws and lax enforcement to destroy an important ecosystem. Wastewater-carrying pesticides, nitrogen compounds, heavy metals, oil (up to 100 MPCs [maximum permissible concentrations]), and phenols and zinc (40 to 50 MPCs) created dead zones of eutrophication in the easternmost part of the Sea of Azov and Taganrog Bay. The Soviet government announced its intention to prevent all pollution discharges into the Black Sea and Sea of Azov at the 1976 Conference on Security and Cooperation in Europe. But the results were miserly, sea level continued to drop markedly, and pollution increased.[110]

Things were no better in the Black Sea, which, like the situation in the Sea of Azov, created problems for the tourist industry. Erosion along the coast grew extensive by the late 1960s. The Georgian resorts of Gagra, Kobuleti, and Batumi lost coastline at a rate of one to three meters per year. The town of Poti lost 260 hectares to the sea. Scientists disagreed about how best to protect the shoreline. Because the Georgian Republic did not have the resources to wage an effective struggle against this disaster, it appealed to the government of the Soviet Union. The Council of Ministers responded with a resolution in March 1969 obliging all the relevant regional and federal authorities to tend to the threat but did not offer funding until 1976.[111] And amid the resolutions, plans, and construction of reinforced embankments and seawalls, the very designers and builders of Georgian resorts and hotels had literally undermined their own positions by removing sand from beaches. Every year, construction firms removed as much as 1 million cubic meters of sand from Georgian

---

[109] E. Fedorov, "Put' k Ravnovessiu," *Izvestiia*, March 16, 1980, p. 3.
[110] Komarov, *Destruction of Nature*, p. 150; A. Velichko, "Kakim Stanet Klimat," *Izvestiia*, August 20, 1976, p. 2; M. Kriukov, "Pora Spasat' More," *Pravda*, January 22, 1987, p. 6; *The State of the Environment of the Russian Federation* (Moscow, 1993); and *The State of the Environment of USSR* (Moscow, 1989).
[111] "Na Pomosch' Prirode," *Sovetskaia Rossiia*, March 21, 1969, p. 2; and "Georgian Official Presents Economic Demands at the USSR Supreme Soviet Session," *Radio Liberty Research Institute*, no. RL 410/80, November 3, 1980.

beaches to build sanitaria, hotels, and roads. Restoration of the coastal belt required replacement of the roughly 35 million cubic meters removed for construction purposes at a projected cost of about 300 million rubles. The absence of modern sewers and water treatment facilities along the entire Black Sea coast made this major Soviet vacation spot an environmental disaster, no different from any industrial site.[112]

In the effort to open Siberia resources to development, Brezhnev-era policy makers and economic planners joined with ministries growing ever more hungry for natural and mineral resources. Siberia had long been a focus of the planners and visionaries. Many of the projects to secure access to Siberian forests, ores (chemical, ferrous, and non-ferrous), oil and gas, and to take advantage of some regions of Siberia for agricultural purposes, in particular grain, dated to the early years of Soviet power – the Urals-Kuznetsk Combine, for example, with Kuznetsk basin coal going from the Urals to Magnitogorsk (with a new steel mill built locally at Novokuznetsk). Several steps had already been completed – the construction of the Turksib railway (Chapter 2), pre- and postwar efforts to chart Siberia's "productive forces," and the construction of the first hydroelectric power stations on Siberia rivers in the Khrushchev era. Now, these programs expanded rapidly. They reflected a costly new policy to invest in Siberian industry, agriculture, and infrastructure at the same time that the need to rebuild and refurnish capital in the European part of the nation became critical as it aged; by the mid-1970s, roughly one-third of the nation's capital investment was directed to the east. Yet, the authorities thought they could and must develop Siberia. They established a series of huge "territorial-production complexes" – large-scale integrated projects – on which to focus labor and capital inputs. Siberian development had significant environmental impacts, for example, those connected with BAM.

BAM runs roughly parallel to the Trans-Siberian Railroad, from Ust-Kut on the Lena River north of Lake Baikal, past Severobaikalsk at the northern tip of Baikal, endangering local forests and the Baikal basin, then on to Komsomolsk-on-Amur, where it is connected by existing lines to the Pacific. BAM stimulated the development of coal, steel, oil and natural gas, chemical, hydroelectricity, and forestry industries in Siberia connected with the territorial-production complexes, opening up some 100 million acres of Siberian and Far East taiga to the forestry industry

---

[112] A. Tenson, "Who Is to Be Blamed for the Destruction of the Black Sea Resorts?" *Radio Liberty Research Institute*, no. RL 181/84, May 7, 1984.

alone. BAM crosses a seismically active region; 800 earthquakes were recorded in the Baikal region alone in the last 200 years, with up to 2,000 tremors in the BAM zone annually. The authorities accompanied the decision to revitalize BAM in 1973, a railway begun in the 1930s and shelved because of the war, with great fanfare, calling it the "Road into the 21st Century!" and considering it a symbol of Brezhnev's great rule.

BAM subjected taiga, tundra, and permafrost, lakes and rivers to the planner's pencil and the worker's shovel and bulldozer. Each new arrival to Siberia required roughly twenty thousand rubles of capital investment, with attendant changes to ecosystems. Discharges and spills of dangerous petrochemicals accompanied every spike and rail. Thirty thousand workers built and abandoned unneeded vehicular access roads and located fuel and oil dumps on banks of rivers near spawning grounds where rain and spring thaws washed mud and oil wastes into the soil. They moved more than 400 million cubic yards of dirt and built 2,400 miles of auxiliary roads, 2,237 bridges, 1,525 drains, and 2,200 miles of railroad. The authorities involved Komsomol brigades from across the nation in building BAM. The task of forcing the railroad across fragile taiga, over rivers, and through mountains at huge expense led the authorities to ignore geological, seismic, climatic, and other challenges that accelerated environmental degradation obvious to the young workers. The route of the railway did not permit airborne pollutants to disperse because of the series of mountains and valleys.[113]

Forest Resources and Soviet Management Practices

This deadening litany of environmental disasters may have inured some readers to yet another story of profligate use of natural resources promoted by state officials in the name of ever-increasing and ever-more-inefficient Soviet practices. But the point is that the disjunction between rhetoric and reality – between the glories of developed socialism and the declining quality of life, and between the allegedly scientific basis of policies and the fact that they were beholden to economic interests – was one of the most disturbing aspects of the Brezhnev era for a growing number

---

[113] On BAM generally and its significant environmental costs, see Chris Ward, *Brezhnev's Folly* (Pittsburgh: University of Pittsburgh Press, 2009). See also A. Grigoriev, "Doroga Skvoz Taiga," *Selskaia Gazeta*, September 17, 1975; "Industrialnaia Karta Strany," *Pravda*, April 20, 1976; *Sovetskaia Rossiia*, June 13, 1982. See also Norman Precoda, "Winds of Change Blow in Siberia... as Viewed from Within," *Environmental Review*, vol. 3, no. 1 (autumn 1978), pp. 9–15.

Developed Socialism and Environmental Degradation   237

of individuals. The history of forestry management indicates that there had been some improvement in practices since the 1940s, but that the same problems of ruined forests, waste, and pollution remained.

On the surface, things were fine because the timber felled in the Soviet Union annually amounted to only a fraction of a percent of the total reserves, which suggested that forestry specialists should have mastered sustainable forest resource management practices.[114] As a practical matter, because of uneven distribution of those forests and development pressures that differed considerably from region to region, and also because of inefficient management and harvesting practices that relied on brute force of workers and rudimentary machines, waste and degradation of forests resulted. In some places, workers denuded vast areas of trees; others remained inaccessible; and workers employed selective harvest in no regions. Throughout the Brezhnev era, the Soviet Union logged close to 400 million cubic meters of forest annually. Yet, it suffered chronic paper shortages stemming from profligate use and inefficiency. At least one-third of all timber felled in the Soviet Union was either lost in transport or wasted during production, whereas in market economies, mills received – and used in various products – more than 90 percent of the felled timber.

Most Soviet republics had limited forests and imported wood from the Russian Republic. Less than one-fifth of mature lumber was found in European Russia, whereas Siberia's reserves seemed almost endless by contrast. More than 28 billion cubic meters of pine, birch, cedar, fir, and silver fir were spread across the Siberian taiga, virtually inaccessible because of weather and distance from roads and towns. Lack of infrastructure drove up transportation costs for removal to distant sites of consumption. The gigantic size of the total reserves also masked the poor quality of forests in permafrost regions. More than two-fifths of the forested areas were located on mountainous terrain, whether in the Carpathians, the Caucasus, southern Siberia, or the Far East. To log the forests in a sustainable fashion, avoid erosion, and ensure conditions for reforestation, the forestry industry demanded well-designed machinery and trained forestry specialists, both of which were in quite short supply. Who could see the forest for the trees without them?[115]

---

[114] "Berech' Lesnye Bogatstva," *Trud*, July 5, 1973, p. 2.
[115] S. Khlatin, "Les: Kak On Est' i Kakim On Budet," *Komsomol'skaia Pravda*, October 14, 1970, p. 2; and "Les, Chelovek i Pila," *Komsomol'skaia Pravda*, July 27, 1965, p. 2; *Narodnoe Khoziaistvo SSSR v 1990* (Moscow: Finansovaia Statistika, 1991); N. Anychin, "Mohzno li 'Ugovorit' Ekonomiku?" *Literaturnaia Gazeta*, November

Still, the planning process gave local officials every incentive arbitrarily to revise targets upward. The editors of *Literaturnaia Gazeta* commented that all decision making about timber harvesting appeared to take place in the absence of any notion that it might be wise to correlate annual logging quotas with actual annual mature timber growth.[116] As usual, decrees inevitably followed, and failure to follow decrees then inevitably followed because the plan dictated every decision, and the public was powerless to learn about the devastation of forests or to do anything to prevent it. Regulations ignored such regional differences as precipitation or slope of terrain. In Buriatia near the Mongolian border and Khabarovsk in the Far East, clear cutting triggered floods. In time, reforestation came to be widely practiced, albeit with limited success because of adverse climatic conditions or inadequate labor and capital resources devoted to it. In the 1970s, reforestation programs in Karelia and Perm, Cheliabinsk, and Kemerovo provinces suffered failure rates of 20 to 50 percent, a dismal record that appears to have been typical. The prized cedars of the Altai Mountains, crucial to the delicate watershed, were felled indiscriminately, and reforested cedar saplings rarely took root. The story was repeated throughout the empire.[117]

The technological shortcomings of the Soviet pulp and paper industry further encouraged the overharvesting of conifers in comparison with deciduous varieties. It was standard logging practice to cull only the largest and most valuable timber, leaving "substandard" deciduous trees aside or (more frequently) flattening them heedlessly if they were a hindrance to hauling behind bulldozers whose treads destroyed any saplings.

26, 1975, p. 11; and A. Zhukov, "Potentzial Zelenogo Okeana," *Pravda*, February 11, 1970, p. 3.

[116] A. Ditmar et al., "Ostanovite Drovoseka," *Pravda*, December 22, 1983, p. 3; N. Mironov, "Logike Vopreki," *Pravda*, December 29, 1976, p. 3; L. Mikhailov and V. Shumakov, "Les: Rubit' ili ne Rubit'? Konechno, Rubit', No Kak, Gde i Skol'ko," *Komsomol'skaia Pravda*, September 28, 1973, p. 2; V. Travinskii, "Topor iz-za Ugla," *Literaturnaia Gazeta*, no. 27 (July 4, 1973), p. 11; I. Valentik, "V Les s Toporom i Raschetom," *Sovetskaia Rossiia*, January 17, 1968, p. 2; *Pravda*, October 20, 1971; *Pravda*, December 12, 1983, p. 2; *Literaturnaia Gazeta*, no. 49 (1974). See also A. Popluiko, "Annihilation of Natural Resources in the Ukraine," *Radio Liberty Research Institute*, May 17, 1963.

[117] A. Popluiko, "Annihilation of Natural Resources in the Ukraine"; "Les na Perekrestke Mnenii," *Literaturnaia Gazeta*, November 26, 1975, p. 11; S. Khlatin, "Les, Chelovek i Pila," *Komsomol'skaia Pravda*, July 27, 1965, p. 2; N. Pavlov et al., "Spasti Lesa Kazakhstana," *Sel'skaia Zhizn'*, February 16, 1965, p. 2; A. Frolov, "Les Zhalob ne Pishet," *Izvestiia*, October 1, 1984, p. 2; Iu. Tiushin, "Kedrograd Prosit Pomoschi," *Komsomol'skaia Pravda*, June 19, 1975, p. 2; Iu. Yudin, "Velikany Taigi," *Trud*, November 19, 1976, p. 3.

Abundant beech and aspen that met this fate could have been used for plywood production, particleboard, and various chemical products. But the Soviet wood, paper, and pulp industry lacked the technology to process the deciduous varieties. This contributed to recurrent paper shortages, because higher grades of wood were needed to produce the newsprint that most developed countries manufactured with the kind of wood that was left to litter the logging roads of the Soviet Union.[118]

Soviet planners were not entirely blind to the depletion of forests in European Russia, even if they frequently ignored the ecological consequences to which they had contributed. In the 1960s, they began systematically increasing timber-processing capacity in eastern Siberia. Although these plans required the solution of numerous logistical challenges, many Soviet and Western experts saw good prospects for the Siberian timber industry. The world market for pine proved strong enough that the Soviet Union went to extraordinary lengths to overcome the problems of distance and weather. Nuclear-powered icebreakers were dispatched to lead timber-bearing ships along Arctic shipping lanes from Murmansk to Vladivostok. Siberian timber exports more than doubled in the first half of the 1960s, with deliveries to more than fourteen countries, among them Great Britain, Belgium, Cuba, and France. As Siberian logging operations expanded, even the Japanese were courted assiduously. By 1970, however, eastern Siberia still accounted for only about 30 percent of the total harvest, largely because it cost from 50 to 100 percent more to harvest there than in the European Soviet Union.[119]

The Development of Environmental Thinking in the Brezhnev Era

Although the Communist Party held on to its monopoly of power, it abandoned many of the most insidious aspects of authoritarian rule from the Stalin era and under Brezhnev permitted more open discussion of the problems facing the regime in state-approved forums, including the problems of air and water pollution and destruction of the forest. Party activists – many of whom were specialists themselves – encouraged scientists, engineers, writers, and artists to suggest how developed socialism might achieve the goals of stable economic, cultural, educational, and

---

[118] O. Volkov, "Chem Dal'she v Les," *Literaturnaia Gazeta*, no. 49 (1974), p. 11.
[119] D. Miller, "Soviet Union Exploiting Timber of Eastern Siberia," *Herald Tribune Bureau*, July 10, 1964; and R. Bobrov, "Les: Ekonomika i Ekologiia," *Sovetskaia Rossiia*, July 14, 1981, p. 2.

scientific institutions. Dissidents, by definition, were outsiders and not permitted to participate in the policy process. In general, Soviet intellectuals with complex bonds to the state and Party had confidence that the mass repressions of the Stalin era could not be repeated. Many of them actively identified themselves with the intelligentsia of pre-revolutionary times, whether as fierce opponents of the Soviet order, of whom there were comparatively few, or as muted critics of the regime's faulty adherence to Communist ideals, of whom there were many more. Social activism expanded in the Brezhnev period, especially in the area of environmentalism and also in science, culture, education, and local affairs.

Because Marxism-Leninism remained the only official philosophy of the Brezhnev era, discussion of alternatives to the Soviet model of economic development was restricted. Yet, Marx and Engels wrote little about environmental issues, except by the implication that when workers controlled the means of production, greater attention would be paid to their health and safety. The gap between this Marxian sentiment and the Soviet reality gave intellectuals considerable leeway to develop their own theories. Although their theories had to correspond with official ideology about the need for vigilance against the United States and other enemies during the Cold War and the role of class struggle in determining the outcome, they were able to address a variety of issues without raising the specter of heterodoxy. New theories and approaches found a wider audience, especially after the 1972 UN Conference on the Environment, and triggered extensive discussion in such leading journals as *Kommunist* and *Voprosy filosofii*.

Theories of the convergence of the capitalist and socialist systems contributed to discussion of environmental issues among scholars. One such theory concerned an ongoing scientific-technological revolution and suggested common processes at work in both systems. The environmental impacts of rapid technological developments were interpreted as a global problem that could only be solved by international cooperation. This line of thinking found expression in *Civilization at the Crossroad: Social and Human Implications of Scientific and Technological Revolution*, published in 1967 by Radovan Richta's well-known sociology research group in Czechoslovakia. The authors argued that scientific and technological progress was based on human creative potential and, above all, rationality. Where democratic societies gave free rein to this potential in the anarchy of markets, socialist societies were vulnerable to the bureaucratic irrationalities of authority derived from political loyalties. The ongoing revolution could transcend the weaknesses of both systems, thus making socialist reform a problem of competent management, a position that held immense appeal for intellectuals in Eastern Europe.[120]

In 1968, the physicist Andrei Sakharov (1921–1989), father of the Soviet hydrogen bomb and, since the late 1950s, increasingly a critic of the Soviet Union, published a major treatise on what he claimed was the convergence of political systems and the need for human rights titled *Progress, Coexistence, and Intellectual Freedom*. The book was published underground in the Soviet Union and later in the West. Although environmental issues were a small facet of *Progress*, Sakharov had become attuned to environmental dilemmas because of debates with the American biologist Linus Pauling and others about the long-term effects of low-level radiation. Sakharov had worked hard to ban atmospheric nuclear tests before this time. He wrote,

> We live in a rapidly changing world. Industrial and water management projects, cutting of forests, plowing up of virgin lands, the use of poisonous chemicals – all this is changing the face of the earth, our "habitat." Scientific study of all the interrelationships in nature and the consequences of our interference clearly lag behind the changes. Large amounts of harmful wastes of industry and transport are being dumped into the air and water, including cancer-inducing substances. Will the safe limit be passed everywhere, as has already happened in a number of places? Carbon dioxide from the burning of coal is altering the heat-reflecting qualities of the atmosphere. Sooner or later, this will reach a dangerous level. But we do not know when. Poisonous chemicals used in agriculture are penetrating the body of man and animal directly and in more dangerous compounds, are causing serious damage...[121]

Sakharov also addressed haphazard dumping of wastes, including radioactive wastes. He likened the source of these problems, including that of the "sad fate" of Lake Baikal, to "local, temporary, bureaucratic, and egotistical interest[s]."[122] Like the Richta group, Sakharov therefore called for global cooperation in environmental issues. He pointed to the glaring deficiencies of a Soviet regime unwilling to cede any prerogatives

---

[120] R. Richta, *Civilization at the Crossroads: Social and Human Implications of the Scientific and Technological Revolution* (White Plains, NY: International Arts and Sciences Press, 1969), p. 372. This book quickly become a manifesto of the reform movement in Czechoslovakia and soon appeared in a Russian edition, where its antibureaucratic arguments enjoyed considerable favor among Soviet intellectuals. After the suppression of the "Prague spring" in 1968, however, many of the authors, including Richta, were forced to revise their views.
[121] A. D. Sakharov, *Progress, Coexistence, and Intellectual Freedom* (New York: Norton, 1968), pp. 48–49.
[122] Sakharov, *Progress, Coexistence, and Intellectual Freedom*, pp. 48–49.

in the international realm – although many democratic nations also have demonstrated this unwillingness.[123]

Several intellectuals adopted a pragmatic approach to finding practical solutions to the most significant existing and emerging environmental problems of the Brezhnev era. Taking advantage of short-term economic reforms initiated under Prime Minister Alexei Kosygin, they emphasized the need to forecast the consequences of technological progress for nature and the society. They drew on the growing interest in large-scale mathematical and computer modeling – which would appear to be apolitical – to explain the rationality of their approaches. In 1966, the leading cyberneticians V. M. Glushkov and G. S. Pospelov suggested a complex model for forecasting the dynamics of the economy, scientific and technological progress, natural resources, and other key development factors. They argued that this model should be included in a nationwide automated computer system to manage the centrally planned economy. Central political leaders rejected the model, fearing the diminution of their influence in planning, but other mathematical forecasting tools became widespread and often included environmental factors.[124]

The interest in forecasting was in line with the global modeling boom, for example, the well-known model used in the *Limits to Growth* (1972) Club of Rome report.[125] Soviet scientists dealing with interdisciplinary mathematical modeling maintained professional contacts with their Western colleagues, including the Club of Rome. At the center of these contacts was academician D. M. Gvishiani, the vice president of the State Committee for Science and Technology and Prime Minister Kosygin's son-in-law. Gvishiani was the driving force behind the establishment in 1972 of the International Institute of Applied System Analysis (IIASA) in Laxenburg, Austria, with the aim of providing a platform for interdisciplinary collaboration between Eastern and Western scholars intent on solving global problems, which operates to this day. Despite the declared interest of political leaders in objective forecasting, the results of scientific studies often came into harsh conflict with the political and economic interests

---

[123] Sakharov was not a radical environmentalist by any means. He favored nuclear power and the use of peaceful nuclear explosions. But he called for the building of nuclear reactors underground as a hedge against accidents or terrorism.

[124] William J. Conyngham, "Technology and Decision Making: Some Aspects of the Development of OGAS," *Slavic Review*, vol. 39, no. 3 (September 1980), pp. 426–445; and Arnold Buchholz, "The Scientific-Technological Revolution (STR) and Soviet Ideology," *Studies in Soviet Thought*, vol. 30, no. 4, Garmisch 80 (November 1985), pp. 337–346.

[125] Dennis Meadows et al., *Limits to Growth: A Report for the Club of Rome's Project on the Predicament of Mankind* (New York: Universe Books, 1972).

of the ruling elite, because they often indicated that the costs of Soviet economic development projects were extensive. The majority of forecasts were classified as confidential or top secret.[126]

The renewed interest in ideas outside the Soviet mainstream contributed to reexamination of the founders of Russian environmentalism who had fallen out of favor or been criticized during the Stalin period, including Vladimir Vernadsky. In the 1950s, specialists acknowledged only his contributions to the natural sciences. But, starting during the Khrushchev thaw, they turned attention to his broader philosophical musings about the place of human beings in nature and the cosmos. The geneticist N. V. Timofeev-Resovsky (1900–1981) significantly contributed to the development of Vernadsky's concept of the biosphere, emphasizing the importance of preserving its balance rather than transforming it. Timofeev-Resovksky wrote,

> ...the Earth's biosphere is a huge living factory, transforming energy and substances on the surface of our planet. It also forms the equilibrium structure of the atmosphere and the structure of solutions in natural waters, as well as the power base of our planet through the atmosphere. The biosphere influences climate. It is enough to mention the huge role of water evaporation by vegetation, and the earth's vegetative mantle in the circulation of water on the globe. Hence, the biosphere forms the entire environment of man. Any negligent attitude toward it, any undermining of its normal work, would mean not only undermining the food resources of people... but also undermining their gas and water environment. In the final accounting, without a biosphere, or with a badly functioning biosphere, people cannot exist at all on the Earth.[127]

The development of diverse views among environmentalists in the Brezhnev era and the rediscovery of the works of such founders as Vernadsky indicate that the Soviet leadership was prepared to listen to expert opinion in formulating rational resource-use policy, even as economic development pressures contributed to further degradation.

Soviet environmentalists were united in their conviction that rational utilization of resources – a kind of Progressive-era "gospel of

---

[126] Richard Vidmer, "Management Science in the USSR: The Role of 'Americanizers,'" *International Studies Quarterly*, vol. 24, no. 3 (September 1980), pp. 392–414. See also Alan McDonald, "Report on the International Institute for Applied Systems Analysis," *Bulletin of the American Academy of Arts and Sciences*, vol. 37, no. 7 (April 1984), pp. 9–13; Harvey Brooks and Alan McDonald, "Report of the Committee for the International Institute for Applied Systems Analysis (IIASA)," *Records of the Academy (American Academy of Arts and Sciences)*, no. 1987/1988 (1987–1988), pp. 24–28; and David Dickson, "Perestroika and Détente Boost IIASA's Prospects," *Science*, vol. 241, no. 4863 (July 15, 1988), pp. 285–286.

[127] N. V. Timofeev-Resovskii, "Biosfera i Chelovechestvo," *Proceedings of Geography Society of the USSR, Obninsk Department*, vol. 1, no. 1 (1968): pp. 3–12.

efficiency"[128] – lay at the heart of their enterprise. Their notions of secular stewardship of natural wealth had much in common with conservation movements elsewhere from the end of the nineteenth century. But they never attempted to develop a philosophy or politics of an "autonomous nature," nor did they question the privileged place of humans in rational-use models of conservation.

The Rise of Environmental Interest and Public Action Groups

In many countries, environmental pressure groups and the general public slowed the headfirst efforts of industries to pursue profit through resource exploitation, especially in modern democracies. In the Soviet Union, conservationists or environmentalists rarely prevented overuse or pollution. They had little influence on the ministries or their officials. Whenever one ministry or organization tried to reduce environmental damage, another, often more powerful one would override their actions, even when industry was clearly breaking the law.

Still, nongovernmental organizations – if one can refer the existence of the forerunners of these bodies in a country whose institutions were all state-run and stated-owned, so that everyone of them was "governmental" – had a long history of environmental protection activities, as we have discussed in previous chapters. VOOP (the All-Union Organization for the Preservation of Nature), in existence since 1924, at its height in the Brezhnev era had nearly 19 million members in the Russian Republic, 6 million in Ukraine, and smaller numbers in other republics, more than even the Communist Party. Most members were enrolled through schools, so they were involved in educational rather than applied work. VOOP members rarely had a high profile in drafting environmental impact statements, let alone on such important issues as Lake Baikal.

By co-opting such organizations as VOOP, the government exerted close control over the broader public environmental aspirations. Yet these broader aspirations indicated growing interest among diverse groups in environmental protection. These groups included the student nature protection movement (the *druzhina* movement); civic initiatives to rehabilitate the urban environment; a new version of the "Garden City" movement in response to growing problems of pollution, garbage, urban blight, and depressing new skyscraper apartment complexes on the outskirts of

---

[128] Samuel Hays, *Conservation and the Gospel of Efficiency: The Progressive Conservation Movement, 1890–1920* (Cambridge: Harvard University Press, 1959).

many cities; the "Ecopolis" program; and the local lore and history study movement (*kraevedenie*).

The druzhinas widened the scope of their activities from the 1960s onward. They gained their impetus and freedom from the political liberalization of the Khrushchev thaw. At first, they focused on fighting poaching and other forms of environmental crime. Next, they shifted to more comprehensive environmental activities, including extensive programs for protecting flora and fauna, raising environmental awareness, and so on. Druzhina activists developed recommendations on improving hunting and fishing laws and establishing new protected territories. Networking among the druzhinas of different universities that crossed the empire also gradually developed, in which the biology department of Moscow State University took a leading role. Students identified, contacted, and educated newly established groups, publicized their activities, and encouraged creation of new groups in provincial universities. The universities provided facilities, equipment, and funding to the druzhinas. In 1975, the druzhinas sponsored the first nationwide workshop on poaching. In another program, "Fact," launched in 1976, students collected, catalogued, and analyzed environmental news and publications. The student druzhinas published their first handbook for organizing the movement in 1978. Six years later, at the movement's conference in Sverdlovsk, the students established a national council to consider methodological issues in depth, not to adopt an activists' platform. The network of druzhinas could not engage in any overt political activity, and vertical organization – the creation of hierarchy within the druzhinas – would likely have been rejected by the groups themselves. But they were comfortable addressing methodological concerns in an informal fashion.

Three other kinds of clubs whose members focused on growing problems of environmental degradation engaged in activism: naturalists' clubs, urban recreational clubs, and civic clubs. The first type is best represented by the Club of Young Biologists at the Moscow Zoo. Such leading figures of a burgeoning environmental movement in the Gorbachev era and beyond as Aleksei Yablokov, the dean of the Russian environmental movement and the druzhina movement's spiritual leader, and Evgenii Shvarts, a major defender of druzhinas, were involved in the club. The club members promoted scientific research on environmental issues. Many of these clubs were run by the old scientific intelligentsia, people who retained professional and family roots in clubs and student organizations even from the tsarist era. Recreational clubs promoted hiking and scouting activities for teenagers, often under such ideologically permissible pretexts as studying the history of World War II, for example, identifying lost soldiers' graves and planting memorial trees. The third type of club was mainly involved in voluntary activity aimed at planting trees and improving the quality of the surroundings. Gradually, they turned to more aggressive campaigns. For example, the Kazan club "Green World" conducted a "Boomerang Operation" in which members gathered garbage from forests and other illegal disposal sites and returned the trash to the enterprises guilty of dumping it. Many other clubs began their activities in the forests and countryside but gradually focused on the urban environment.

The average citizen gained environmental awareness in another way. These individuals, especially those living in workers' settlements and villages, grew concerned about the endemic pollution they saw around them. A long-term crisis in the rural economy had left them destitute and forced them to grow produce in window boxes and courtyard gardens, especially at their small summer homes or dachas. Many of them survived by scrounging for local natural resources – by hunting, fishing, and gathering mushrooms, nuts, and berries. They saw firsthand evidence of the impact of pollution on their gardens and on their scrounging activities.

Scientists, teachers, social leaders, and local authorities in Pushchino, a biological research town near Moscow, embarked on "Ecopolis" in 1980 with the goal of saving the town's environment. Ecopolis attracted experts from as many as a dozen scientific establishments as well as local clubs and associations. Based on environmental and social studies, these activists developed a master plan for the town to protect the natural and cultural environment and conducted numerous environmental education and awareness-raising activities. The group closely collaborated with the Pushchino municipal council. The Ecopolis program stimulated the growth of environmental concern in the local population and helped to create mechanisms through which the residents could influence decision making. Pushchino became a model "Garden City" for the urban environment by leading planners to consider how to preserve green space and discourage litter.[129]

Toward the end of the Brezhnev era, such official or co-opted organizations as VOOP had lost any pretense of objectivity. Castoffs of the party apparatus, men who were either incapable of serious work or guilty of gross misconduct, filled the headquarters of such groups as VOOP and the Fisherman's Society. They also began to infiltrate the druzhinas, which

---

[129] Katy Pickvance, "Social Movements in Hungary and Russia: The Case of Environmental Movements," *European Sociological Review*, vol. 13, no. 1 (May 1997), pp. 35–54.

ceased to challenge official environmental policy. On the contrary, allegiance to the state and its organs was the cornerstone of their ideology. The activities of these groups ultimately had little to do with any attempt to grasp the nature, scale, and pace of environmental degradation. One reason for this was that the emerging civil society was not itself a sui generis phenomenon – few, if any, Soviet institutions could fully escape some degree of incorporation into the existing Party-state machine. Yet, the "machine" metaphor did not mean implacable rigidity – the unwritten motto of the Brezhnev era was "live and let live," which is why corruption rather than violent coercion was always the greater threat to such groups.

Civic groups lacked energy for a second reason. Most people had grown accustomed to self-censorship. Stalinist repression ceased with the dictator's death in 1953, but Communist ideologues had nevertheless achieved the cherished goal of an ideologically uniform system of values that discouraged political participation. Conformity, willful indifference to policy making, and intolerance of political gatherings beyond permitted forms – all of these habits permeated civic society. In fact, the "de-ideologization" of the values of the authoritarian Soviet system long after the end of the Soviet Union remains a major obstacle to the creation of a modern environmental consciousness in the Russian Republic into the twenty-first century.

Political repression had one paradoxical but positive effect. The exiled intelligentsia, and in particular natural scientists, created "oases" of ecological thought, education, and research in many cities remote from Moscow and Leningrad. People who had been exiled because of their anti-Lysenkoist biological leanings, or their reformist politics, or, more likely, simply because of the arbitrariness of the totalitarian system often found life in towns distant from the center that were more open to their activities. They established a network of reserves, protected areas, and experimental testing stations, many of which continue to function. After the mass rehabilitation of political prisoners under Khrushchev, these cells of environmental activity became elements in the state system of nature conservation. Most important, they served as the foundation for environmental research and teaching in such fields as genetics, soil science and so on in provincial universities. They became important organizations of the Gorbachev era, for example, the Socio-Ecological Union under Sviatoslav Zabelin, who won international recognition for his efforts.[130]

---

[130] On the Socio-Ecological Union, see Sviatoslav Zabelin and Olga Berlova, "A Decade of Work, A Decade of Making a Difference," http://gadfly.igc.org/russia/soceco.htm.

Literature, the Press, and Environmentalism[131]

The local lore or kraevedenie organizations also rekindled interest in environmental issues in the Brezhnev era. Foremost among these organizations were writers connected with the conservative journal *Nash Sovremennik* (Our Contemporary) and other nationalist periodicals. *Talaka* was one of the first associations aimed at protecting local lore. Established in Minsk in 1985, its members sought to revive Belorussian language and culture, embracing environmental concerns as a consequence. In the same year, the Historical and Patriotic Union *Pamyat'* (Memory), a reactionary and anti-Semitic group resembling the tsarist-era "Black Hundreds" pogromists, was founded in Novosibirsk (Siberia). One of its goals was to return Russia to an "unspoiled" agrarian lifestyle, free of the influence of Jews and the filth of industry. The central newspaper *Sovetskaia Rossiia* and others also became havens for various forms of local studies. In the mid-1960s, republican-level societies for preservation of monuments of history and culture were established under the aegis of republican or regional authorities across the country. Party ideologues regarded proponents of kraevedenie as much more loyal to the regime than environmentalists of other sorts.

Yet, not all of the organizations or writers were conservative or deeply nationalistic. The eminent writer (and earlier hydrologist) Sergei Zalygin contributed to kraevedenie through his association "Ecology and the World," founded in 1983 under the auspices of the Soviet Committee for Peace. In 1988, Zalygin initiated the joint expedition of writers, journalists, and scientists titled "Aral Sea-88" under the aegis of popular literary monthly *Novyi Mir* (New World), whose liberal politics stood in contrast to those of *Nash Sovremmenik* and *Oktiabr'*. Zalygin used *Novyi Mir*, whose editor he became, as a sounding board for those environmentalists and others who opposed the Siberian rivers diversion project.[132]

What role did the Soviet press play as a "fourth estate" in the environmental movement? How did journalists, employees of the state, see their role next to those other public groups? By the mid-1960s, various media

---

On Sviatoslav Zabelin, winner of the 1993 Goldman Environmental Prize, see http://www.goldmanprize.org/node/176.

[131] Felix Shtil'mark and Roberta Reeder, "The Evolution of Concepts about the Preservation of Nature in Soviet Literature," *Journal of the History of Biology*, vol. 25, no. 3 (Autumn 1992), pp. 429–447.

[132] Ritta Pittman, "Perestroika and Soviet Cultural Politics: The Case of the Major Literary Journals," *Soviet Studies*, vol. 42, no. 1 (January 1990), pp. 111–132.

had become a hub for various environmental groups. Their activities became more strident in the Brezhnev era. Regional or specialized media that dealt with specific sectors of the economy or industry occasionally managed to broaden discussions of environmental issues that drew the attention of reporters in *Pravda* and *Izvestiia*. In this way, the media gradually became an important tactical weapon in the hands of a widely based coalition of activists, including scientists, environmentalists, and other intellectuals. Even when journalists brought environmental issues to public attention, certain themes remained closed for discussion. Always under the shadow of censorship, leading publications on environmental problems took on a didactic character, frequently lecturing the population on the importance of a "communist attitude" toward conservation of nature.

Soviet writers contributed to the opening of environmental discussions not only through letters and articles but also through novels, as Leonid Leonov had in *Russian Forest* during the Khrushchev thaw.[133] A number of Brezhnev-era writers were more strident. The nationalistic Valentin Rasputin, a leader of the village prose genre of writers, spoke out in defense of Lake Baikal, directly accusing some top officials of lying and incompetence for approving the construction of paper mills on the shore of Lake Baikal. He was an active figure in the Baikal movement. He regarded the struggle to preserve the lake and its environs as a fight to save Russian culture. His *Farewell to Matyora* describes the failed efforts of the inhabitants of a historic island town to save their home from inundation of water backing up behind a new hydroelectric power station and their forced removal into ugly new concrete apartments.[134] This fictional account represents the scores of cases when Soviet industrial projects destroyed peoples' lives, culture, memories, and nature itself.

A major test for Soviet environmentalists was the project to divert the flow of Siberian rivers to increase the irrigation area of Central Asia. The gigantic construction project had been the dream of hydrologists for decades but gained official support in the 1960s, and by the late 1970s, several hundred organizations and thousands employees from engineers to workers were engaged in it. The goal of the project was to divert the flow of 7 percent to 10 percent of several Siberian rivers into Central

---

[133] Leonid Leonov, *Russkii Les* (Moskva: Mologaia Gvardiia, 1954), p. 652.
[134] Valentin Rasputin, *Farewell to Matyora* (New York: MacMillan, 1979), p. 227. See also Julian Laychuk, "Conflicts in the Soviet Countryside in the Novellas of Valentin Rasputin," *Rocky Mountain Review of Language and Literature*, vol. 47, no. 1/2 (1993), pp. 11–30.

Asia (with modest amounts of water for northeastern European Russia) through massive transfer canals to support agriculture, industry, and municipalities. By official decree in July 1973, information about the diversion project was classified.[135] Scientists pushing the project claimed after extensive study that the project was environmentally sound, but many others opposed it, correctly noting its impact on ecosystems of the Arctic Ocean and local climatic changes along the rivers. In 1986, a central newspaper published a statement by a group of writers, including Leonov, Rasputin, and others, protesting against another diversion plan in European Russia. The river diversion projects were abandoned under Mikhail Gorbachev only in 1988 after growing protests in the media.[136]

Environment and Society on the Eve of the Gorbachev Reforms

On the eve of Mikhail Gorbachev's accession to power, the Soviet Union had the world's second-largest economy, a remarkable transformation from an agrarian system in less than a half-century. The Brezhnev regime relied on specialists' input to ensure rational policies to continue economic development, including the harnessing of the nation's wealth of fish, forest, oil and gas, metallurgical, and other resources. The government had embraced a series of new laws, standards, and statutes to ensure environmentally sound practices and to bring the country within the standards of the world. A number of citizens' groups actively contributed to environmental improvement and awareness. This may have been more the result of a highly bureaucratized modern state than of Soviet-style socialism, with its emphasis on economic growth on the foundation of seemingly inexhaustible natural resources, oil, gas, coal, and ore.

Developed socialism notwithstanding, the Soviet environment had become increasingly degraded, with only the extensive landmass providing a false sense that a cushion existed. Whether heavy industry, agriculture, or forestry, each sector of the economy contributed to degradation through rapacious and inefficient use of resources, overuse of toxic chemicals in a variety of processes, and haphazard if not simply intentionally poor waste and pollution management, treatment, and disposal practices. As economic growth slowed, managers and planners turned even more to wasteful, inefficient harvest, mining, processing, and production

---

[135] Komarov, *Destruction of Nature*.
[136] The final environmental impact statement for the diversion project is G. V. Voropaev and D. Ia. Ratkovich, *Problema Territorial'nogo Pereraspredeleniia Vodnykh Resursov* (Moscow: IVP AN SSSR, 1985). See also Philip Micklin, "Water Diversion in the Soviet Union," *Science*, vol. 234, no. 4775 (October 24, 1986), p. 411.

practices based on the fiction that the Soviet Union's extensive resources would continue to be the foundation of the economy – no matter how poorly they were exploited – for decades to come. The forests suffered; the fish suffered; the inland waterways and lakes suffered; and Soviet citizens suffered. The evidence indicates significantly higher levels of various acute and chronic diseases related to environmental degradation – emphysema, blood disorders, heart disease – than in other industrialized nations, with levels of diseases actually growing. Although the causality is uncertain, infant mortality had also begun to rise, an anomalous circumstance in an advanced industrial country; the Brezhnev regime responded by classifying infant mortality data.[137]

There are many reasons for this situation, as we have explored in this chapter. They include the fact that, although the government begun to address problems of water and air pollution, acid rain, a hole in the ozone layer, erosion of soils, and so on, its allegiance to economic development overrode concern about the environment. Pollution standards were weak; enforcement was lax; resources devoted to tracking, measuring, and fighting pollution were insufficient; and there was no national environmental protection agency with the power and authority to enforce the law. On top of this, such large-scale projects as river diversion, huge Siberian energy complexes, and Central Asian cotton and agriculture continued to be the backbone of development. These huge, costly projects acquired such great momentum that it was impossible to arrest them.

Still, there were signs of hope. One of the most important signs was the rise of environmental thinking and informal groups of environmentalists who began to challenge the traditional ways of thinking about the economy, natural resources, and development. Although the authoritarian political system put strictures on their activities, they grew in size and activity so that by the mid-1980s they might serve as a voice of the modern environmental movement. In the Soviet media, especially newspapers and journals, various writers, poets, scientists, and others worried openly about continuing degradation and debated other paths. When Gorbachev came to power in 1985 and embraced political and economic reforms, these individuals and groups were ready to participate in environmental activities, especially after the Chernobyl disaster in April 1986.

---

[137] For chapter and verse on the great public health costs of Soviet pollution, hazardous waste, and other environmental problems, see Feshbach and Friendly, *Ecocide*.

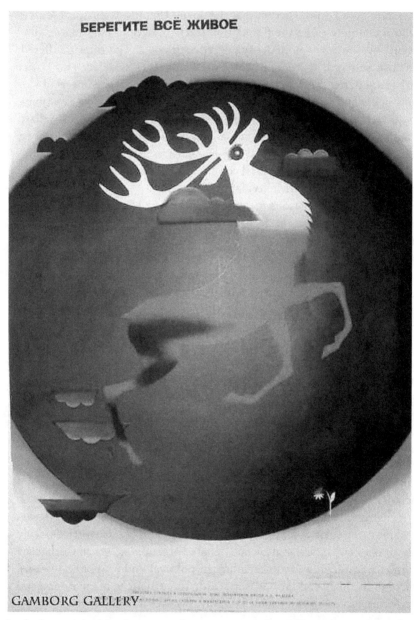

In the last years of Soviet power, the government showed its concern about the natural environment in laws – and posters – and citizens groups grew in size and activity in support of environmentalism. "Protect the environment!"

Under Brezhnev Siberia and the Far East fell to the planners' pencil and builders' machines, but from the point of view of the environment, there was not glorious sunshine but great waste and devastation. Young people were encouraged to join the great building projects "of the Far East and Siberia!"

5

Gorbachev's Reforms, Glasnost, and Econationalism

During the brief era of leadership of the Soviet Union of Mikhail Gorbachev (1985–1991), civil society was born – or reborn – in public participation concerning a wide variety of issues in which environmental concerns began to occupy a central position. Gorbachev's policies of perestroika (restructuring, revolution) and glasnost (openness) at first encouraged public involvement, although he thought he could control its extent; ultimately, he could not shape or control the forces of political change he had supported with the goal of revitalizing the Soviet system. Indeed, openness shed direct light on the nature of that system – the high level of corruption among officials; the waning support among citizens; the shocking environmental costs of the Soviet development model; and the recognition that talk of the advantages of socialism had become empty sloganeering, in particular after the Chernobyl disaster in 1986 that became a symbol for endemic technological failure, pollution, and profligate use of natural resources. Two years later, in connection with campaigns for a new legislative body, the Congress of Peoples' Deputies, political parties, interest groups, and nongovernmental organizations (NGOs) formed that insisted on both investigation of the past and a new path to the socialist future, many of which were concerned with the environment.

After the death in quick succession of three aging party leaders – Leonid Brezhnev, Yuri Andropov, and Konstantin Chernenko – dramatic political, economic, and social changes shook Soviet society in the mid-1980s. Many Communist Party leaders believed that only profound reforms could help the Soviet Union survive. With oil exports leveling off, the domestic economy had few other prospects for expansion. In foreign affairs, the effort to produce thousands of nuclear warheads and equip the Warsaw Treaty Organization troops with conventional weapons had grown extremely costly, and the Soviets had fallen behind in the race to develop new high-tech weapons. The terrible military strain on state budgets could no longer be met by cutting investments in health care, agriculture, or housing. When the Politburo advanced the fifty-four-year-old Mikhail Gorbachev to address these predicaments in March 1985, the Stavropol, Russia, native promptly led the Soviet government in a series of campaigns to reform socialism.

The era of perestroika became synonymous with the final years of the Soviet Union. Because perestroika ultimately failed to save the state, it has also become a marker of an incomplete transition from dysfunctional Soviet socialism to a more dynamic form of rule. The Soviet system grew increasingly unstable, and the economy went into free fall. In the effort to save their perquisites and the Soviet multinational empire, old-style Communist leaders staged a coup in August 1991. The coup failed, which accelerated the dissolution of the empire, and the Soviet Union disappeared in December 1991 when Gorbachev resigned. At the time of its collapse, the Soviet Union consisted of fifteen republics, each of which became a Newly Independent State (NIS), with Boris Yeltsin the president of the Russian Federation. The term NIS is usually used only for the twelve nations that became "newly independent," excluding those of the three Baltic countries – Estonia, Latvia, and Lithuania – whose subjugation to Soviet power such countries as the United States did not recognize. The collapse triggered a period of slow nation building, economic change, and crisis in these new states.

Still, before the collapse, glasnost and perestroika led to a full-throated evaluation of the environmental costs of the Soviet development model – its emphasis on large-scale programs to organize workers, secure resources, and provide the promised but never materialized comforts of socialist life. These programs were costly not only from an environmental point of view, but also from the points of view of inefficient use of resources, and taken together they had a significant impact on the health and safety of the Soviet citizen. Infant mortality had begun to rise, life expectancy had begun to decline, and into this mix such technogenic disasters as Chernobyl had put increasing pressure on the government to prove its legitimacy to citizens. In this chapter, we discuss the impact of the five years of the Gorbachev reforms on environmental policies in the last years of the empire.

Gorbachev, Reforms, and Environmental Issues

Gorbachev's policies triggered extensive environmental awareness and activism. Building on activism surrounding efforts to protect Lake Baikal from further assault, citizens banded together in informal and eventually formal environmental protection organizations. By the mid-1980s, many Soviet citizens had come to realize they lived in a dangerously polluted environment. They read in newspapers and journals extensive exposés about the extent and costs of pollution that had damaged virtually all ecosystems and threatened the public health of all citizens. They began to question whether the roots of the problem were in the Soviet system itself, especially as more and more of them saw with their own eyes that the quality of life in Western democracies was much higher – and much less polluted – than they had been led to believe.

The large-scale projects paradigmatic of the Soviet Union in the Stalin era may have served the important function of the mobilization of society. But, as many commentators have noted, they mobilized capital and labor inputs and were virtually impossible to halt when their costs became clear. Hence, although understandably the product of the ministerial system of economic growth that prevailed in the Soviet Union, they were no longer appropriate for the more modern, urban, educated society that existed by the mid-1980s. These expensive projects – smelters and mines that stretched from the Arctic Circle to Siberia; dozens of nuclear power stations being built increasingly on the outskirts of major cities; canals, hydropower stations, and river diversion projects – faced criticism from a variety of quarters, both within and outside the government. Those ministries connected with mining, metallurgy, water resource development, and so on jealously protected their hold over resources of manpower, equipment, and funding. Ministries and other institutions connected with resource management, with fisheries and forests, on the other hand, were rightly critical of the cost of these projects and their lack of environmental sustainability at the same time that they themselves also pursued policies that had extensive costs.

One major project involved a plan to divert the flow of Siberian rivers through massive transfer canals into Central Asia to develop agriculture. Planners spoke hopefully of using scores of small nuclear devices to help excavate the transfer canals. Supporters produced reams of documents indicating the environmental soundness of the project. Ultimately, open discussion required political change: Gorbachev's accession to power triggered open discussion of Siberian diversion, dozens of smaller projects, and the Soviet development paradigm generally. By 1989, river diversion had been put on hold, and environmentalists began to hold a prominent place at the decision-making table.

Gorbachev's government pursued policies for environmental protection much more actively than any previous Soviet government. Gorbachev sought to displace obsolete political categories with what he called new thinking (*novoe myshlenie*) in Soviet domestic and foreign policy. A prominent feature of this new thinking was the emphasis on "common" or even universal human values as opposed to the earlier Marxist doctrine that all values were class based. New thinking thus enabled people to embrace certain Western values and openly learn from the West in working toward comparable goals. These people recognized that there were global issues of common interest – for example, the environment – that could only be solved through international cooperation. Gorbachev's position was an explicit rejection of Brezhnev-era approaches, to which his administration referred as the "time of stagnation"; of the Khrushchev era as a period of voluntarism; and of the Stalin era as the cult of personality.

The shift in official ideology during perestroika supported environmental reform in three ways. First, securing a clean environment became a central priority, even if authorities did not realize the extent of the problems they faced, how much they would cost to rectify, and what changes would be required. Second, the reforms required significant institutional changes, both within existing institutions and through the formation of new bodies with power, authority, and budgets. Third, as part of the "new thinking," the authorities aimed to learn from and to cooperate with the West in these reforms. In keeping with glasnost, they admitted that environmental protection was not a purely technical matter, but one that required open public debate representing various viewpoints where human values were central and choices were based on normative concerns.

By the eve of the collapse of the Soviet Union, most citizens believed that socialism, even if significantly reformed, was no longer the best system to secure a healthy environment. Quite the contrary, they believed that socialism was the root cause of environmental problems. Murray Feshbach's *Ecocide in the USSR* accurately reflected the widely shared view that the Soviet state had almost deliberately poisoned its own citizens, especially after Chernobyl, when the newspapers and televisions shows were filled with accounts of the high costs – public health, environmental, and financial – of living in Soviet society.[1] Through increasing contacts with the West, people discovered the relatively clean and attractive environment in prosperous countries in Western Europe and North America, quite in contrast with pictures of the dirty and polluted capitalist world portrayed by Soviet propaganda. They widely assumed that dismantling the Soviet regime would almost automatically deliver a better environment similar to that in the West. The proponents of this view argued that economic liberalization would end subsidies to outmoded industries, force factories to use resources more efficiently, and foster the adoption of modern, clean technologies from the West. Moreover, radical political reform would make the government more accountable, transparent, and thus more responsive to the environmental concerns of its citizens. Environmental reform became a synonym for political and economic change. The most trying event for Soviet citizens, especially those of Belarus and Ukraine, was the Chernobyl nuclear explosion. For Gorbachev and his leadership, too, Chernobyl required coming to grips fully with perestroika and glasnost.

The Chernobyl Disaster[2]

The Chernobyl Disaster may be the most famous anthropogenic environmental event in history, along with the chemical disaster in Bhopal, India, in 1984, and one of the crucial factors that led to the collapse of the Soviet

---

[1] Murray Feshbach and Alfred Friendly, *Ecocide in the USSR: Health and Nature Under Siege* (New York: Basic Books, 1992), p. 376.

[2] Sergei Mirnyi contributed to this section on Chernobyl. Of the many fine books and articles on the Chernobyl disaster on which this section is based, see S. A. Amirazian, S. M. Filippova, and M. Iu. Tikhomirova, "Semiotika 'Chornobyl's'kogo Sindromu' ta Rol' Social'no-psihologichnyh Chunnukiv u Jogo Formuvanni ta Rozvytku," *Ukrainskii Radiologicheskii zhurnal*, vol. 5, no. 1 (1997), pp. 13–15; R. J. Baker and R. K. Chesser, "The Chornobyl Nuclear Disaster and Subsequent Creation of a Wildlife Preserve," *Environmental Toxicology and Chemistry*, vol. 19, no. 5 (2002), pp. 1231–1232; V. G. Bariakhtar, ed., *Chernobyl'skaia Katastrofa* (Kiev: Naukova Dumka, 1995); R. Gale and T. Hauser, *Final Warning* (New York: Warner Books, 1998); David Marples, *Chernobyl and Nuclear Power in the USSR* (London: Macmillan, 1987), and *The Social Impact of the Chernobyl Disaster* (London: Macmillan, 1988); Zhores Medvedev, *The Legacy of Chernobyl* (New York, London: Norton, 1999); Sergei Mirnyi, "Chernobyl as a Model Case of Eco-disaster," in *Fifth Conference on Environmental Education. Case Studies in Environmental Education and Research* (Zurich, Switzerland, April 15–17, 1999), and *Chernobyl Liquidators' Health as a Psycho-Social Trauma* (Budapest: Bogar Kiado, 2001), and Mirnyi and A. Yastrzhembska, "Information and Communication Factors' Impact on the Chernobyl Zone (1986–1990) Workers' Health," *New Solutions (Journal of Environmental Occupational Health Policy)*, vol. 11, no. 3 (2000), pp. 229–241.

Union. Its scale, long-term impact, and transnational significance may be unrivaled. The accident grew out of a series of human errors, and Soviet officials blamed the station operators. Yet, the accident had more sinister roots which were not solely the responsibility of the operators at the station but reveal the inadequacy of the entire safety philosophy of the Soviet nuclear industry as well as the inherent danger involved in operating the specific technology. The reactor design was, in fact, inherently unsafe, and the personnel of the industry had been overcome by hubris that led them almost inevitably toward disaster. On April 26, 1986, as a result of a poorly designed and foolishly-carried-out experiment, the fourth of four reactors (a total of ten were planned at this reactor "park") at the Chernobyl nuclear power station exploded, releasing large quantities of radioactive materials into the environment.

Considering Hiroshima and Nagasaki, Chernobyl is not the world's largest nuclear catastrophe. Considering, for example, the accident at the Bhopal, India, Union Carbide chemical plant in 1984 that released poisonous gas that killed more than three thousand people and seriously injured more than twenty thousand, Chernobyl has rivals for the world's greatest technological accident. It led to more extensive radiation contamination than from the catastrophe at the Kyshtym nuclear waste dump in the Urals in 1957 connected with a plutonium production effort; this accident was shrouded in Cold War secrecy deep within Soviet borders; and when it was uncovered in the United States, it was kept silent lest American citizens lose faith in a nascent nuclear power program.[3] Chernobyl's significance relates to a combination of factors: its impact within and beyond the Soviet Union to Europe; its symbolism as a failure of indigenous technology; its occurrence at the beginning of perestroika; and the failure of Soviet authorities initially to understand and reveal the extent of the crisis.

On the morning of April 28, workers at a Swedish nuclear power station first detected an anomalous increase in airborne radiation. The search for its origin started. When it was determined to be the Soviet Union, the Soviet authorities reluctantly admitted an accident had occurred. But because of their tradition of secrecy and reticence, and because they themselves did not initially fathom the extent of the accident, rumors spread wildly throughout the nation and beyond that perhaps two thousand people were killed in the initial explosion, although thirty-three

---

[3] Zhores Medvedev, *Nuclear Disaster in the Urals*, trans. George Saunders (New York: Norton, 1979).

individuals actually perished, mostly firemen arriving on the scene to put out the highly radioactive flaming graphite that was ejected from the reactor core during the explosion. Granted, hundreds of thousands of people were exposed to excessive levels of radiation, both residents downwind, especially in Belarus and Ukraine, and the "biorobots" – soldiers – and others ordered into the exclusion zone to eradicate the disaster.

Some individuals argue that Chernobyl's impact has been exaggerated in a mass-media frenzy of misleading stories about birth defects, cancers, and death. The scientific literature has been much more extensive and measured in its evaluation, although a wide gulf separates various estimates of total excess mortality worldwide from the accident. The International Atomic Energy Agency (IAEA) has estimated that, over time, five thousand excess deaths will occur, while others argue the total will be closer to fifty thousand deaths.[4] Given the failure of the Soviet authorities to inventory and track those individuals who may have been exposed to radiation after the accident, it seems fair to assume that the IAEA estimate is too low because it is based on incomplete information. Several critics have pointed out that it is also in the interests of IAEA personnel to arrive at a lower estimate because they support the further development of nuclear power and argue that it is safe.

Nothing indicated that the small, quiet, provincial Ukrainian town of Chernobyl would become world famous when the nuclear industry ordered the construction of a nuclear power station twelve kilometers to the northwest of the town on the Pripiat River that flows into the Dnieper River and directly to the capital of Ukraine, Kiev, only ninety kilometers away. Construction commenced in 1970; the first reactor generated electricity in 1977, the second unit in 1979, the third in 1981, and the fourth in 1983. All were RBMK (channel graphite) reactors, as opposed to the safer pressurized water reactors that the Soviets and other nuclear nations had built as well. To many people, nuclear power stations were safe, reliable, and an undoubted symbol of technological progress. Yet, the Chernobyl accident was nearly inevitable. Its principal cause was a combination of the drawbacks in the design of the RBMK, namely a positive void coefficient that made the reactor unstable at low power. A positive void cofficient means that as the void content inside the reactor grows because of increased boiling or loss of coolant (in most reactors, water), the reactivity increases. Because the coolant acts as a neutron

---

[4] Fred Mettler, "Chernobyl's Living Legacy," *IAEA Bulletin*, vol. 47, no. 2 (March 2006), pp. 4–6.

absorber, the less coolant, the fewer neutrons are absorbed by the water, and the more they lead to fissions of nuclear fuel. The fissions in turn release more neutrons – and vast quantities of heat. This becomes a positive feedback loop that can quickly – seemingly instantaneously in an RMBK reactor – boil all the coolant in the reactor, and a runaway chain reactor results. This happened in the Chernobyl disaster. The construction of power reactors with a positive void coefficient was not pursued in any other country. Operators at the Chernobyl station did not have a complete understanding of these drawbacks. Instead, the operators forced the reactor into a positive void to save time rather than lose time on a poorly designed experiment.

The accident happened during the cooling down of reactor four before scheduled maintenance. Operators intended to permit the turbines to spin from their own momentum after the shutdown to see how long they would continue to generate electricity. In the middle of the shutdown, the Kyiv grid called for more electricity. But rather than bring the reactor online again, a timely and costly process that would prevent the experiment, the operators disabled various safety systems and removed control rods from the reactor core in order to keep electricity production up – but without any safety margin. Already at low power and without safety or control systems in place, the positive void coefficient came into play. In this case, the Chernobyl reactor began an exponential surge in power, the reactor core overheated, the cooling water boiled out of the core (increasing the power further), the core melted down, and a chemical reaction of steam with metal and/or graphite yielded an explosive mixture of hydrogen and oxygen. Two powerful explosions ripped through the reactor, destroying it and lifting its lid – at 2,000 tons – into the air and down on its side, destroying the roof of the standard factory building. One hundred to two hundred megacuries of radioactive substances poured into the environment over the next ten days, falling onto the land and entering the water around the station and also into the atmosphere, where it spread through the northern hemisphere. Fuel rods, burning graphite, and other material scattered on the ground and the roof of reactor unit three next door, which, against the regulations, had a flammable bitumen cover, and instantly caught fire. Inside several other areas caught on fire, through the heroic – and mortal – actions of the firefighters, the most dangerous fire spots were extinguished by 5:00 A.M. But the core of the uncontrolled nuclear reactor was open and its graphite burned, emitting visible fumes and invisible radiation into the environment. The base of the reactor was forced down by four meters, the explosion having demolished the supporting structure. Highly radioactive lava of the melted nuclear fuel and construction material flooded the lower corridors and rooms of the building.

The authorities ordered the evacuation of the 45,000 residents (including 17,000 children) of Pripiat, the town just two kilometers from the reactor built for employees and their families to serve the station, only thirty-six hours after the explosion. Evacuation of such heavily contaminated settlements as Chernobyl (with 20,000 inhabitants) and the Gomel region of Belarus followed days later. Sometime in August, the evacuation of 166,000 people from one hundred and eighty-eight towns and villages in Ukraine and Belarus, was complete. In addition, 60,000 cattle were transported from the zone, some of which made it into sausage sold through the Soviet Union – except for Moscow province.[5] The authorities used 8,500 vehicles, including 2,500 buses, in the evacuation – many of which were buried in a vehicle "graveyard" near Chernobyl. Eventually, they established a thirty-kilometer–radius exclusion zone surrounded by barbed wire and protected by armed guards. Soldiers were sent in to shoot all animals, including pets, lest they escape the zone.

To prevent a spontaneous nuclear explosion should the fuel somehow gather in sufficient quantity and shape, and to suppress fire and prevent radiation from escaping into the environment, over the next two weeks military helicopters dropped about 5,000 tons of sand, clay, dolomite, lead (2,400 tons), and boron-containing compounds into the core. Already, thousands of people were working around the clock to mitigate the accident. Eventually, they put out the reactor fire, the reactor began to cool, and other stabilization measures commenced.

In order to prevent further contamination of the aquifers, Donbass coal miners, who were ordered into dangerous working conditions along with tens of thousands of soldiers, drilled a horizontal tunnel under the reactor and pumped concrete into it, creating a massive concrete dish. They drilled water-table–depressing wells at a larger distance and connected them into a system and constructed an earthen embankment on the right bank of the Pripiat. For months, trucks washed down the main roads constantly; dosimetric checkpoints and vehicle decontamination depots were employed to prevent spread of radioactive beyond the exclusion zone. Workers scraped the most contaminated topsoil in the vicinity, loaded it into steel boxes, and transported it to nearby, newly established burial grounds. Neither Soviet nor foreign remote control electronic equipment

[5] Paul Josephson *Red Atom* (Pittsburgh: University of Pittsburgh Press, 2005), p. 166.

could handle the intense radiation, so that soldiers wearing little more than cotton masks – some lucky individuals had respirators – and lead vests did most of the remedial work in the zones of highest contamination. Over three years, perhaps as many as nine hundred thousand soldiers and volunteers labored in dangerous conditions. To prevent the spread of radioactive dust and protect the damaged reactor from the elements, the authorities determined to build an enormous enclosure or sarcophagus. This project involved ten thousand workers employed daily in round-the-clock shifts over a five-month period from August 1986. They poured 6,000 tons of concrete daily, with trucks arriving at two-minute intervals. In all, they poured 360,000 tons of concrete and used 5,000 tons of metal girders and plates. Although the sarcophagus contains more than 180 tons of the spent nuclear fuel with more than twenty megacuries of radioactive substances, the releases from it have been negligible.

Even in these conditions and with this kind of RBMK reactor, the authorities pushed to begin producing electricity as soon as possible from the remaining reactors. They restarted unit one on October 1, 1986; unit two on November 5; and unit three, just next to the destroyed unit and sharing some of its equipment, on December 3, 1987.[6] To operate the station, the authorities hurriedly built a new town, Slavutich, only fifty kilometers from the station, and a special train to bring workers to and from the station. They promised higher salaries and coveted apartments to entice workers. The residents of Slavutich tended to be young people in pursuit of higher salaries and apartments. In Slavutich, they met people of the opposite sex, married, and had children – with more birth defects than in the normal population.[7]

Radiation Contamination

The explosion expelled roughly 4 percent of the reactor's radioactive content (100 to 200 megacuries) into the environment, about 40 percent of which fell within the thirty-kilometer zone. The nature of the spread of radioactive material depended on its form, size, and the weather conditions (for example, wind and rain) and therefore was deposited unevenly across soil, buildings, machinery, vegetation, roads, and bodies of water throughout large areas of Ukraine, Belarus, Russia, and beyond. But because of the nature of this radiation, some of which can enter the body

---

[6] Bariakhtar, *Chernobyl'skaia Katastrofa*, p. 19.
[7] Marples, *The Social Impact of Chernobyl*, pp. 225–230.

TABLE 1. *Areas of ^{137}Cs Contamination and Their Population, 1995*

Zone of:	Obligatory	Guaranteed Resettlement	Resettlement	Enhanced Radiological Control
Contamination (Ci/km^2)	>40	15–40	5–15	<5
Area (1,000 km^2)	3.25	7.15	14.1	~116
Population (1,000s)	33.8	234	544	3,070

Source: V. G. Bariakhtar, ed., *Chernobyl'skaia Katastrofa* (Kiev: Naukova Dumka, 1995), p. 27; and United Nations, *The Human Consequences of the Chernobyl Nuclear Accident. A Strategy for Recovery* (Minsk: United Nations, 2002).

only through contaminated ingestion as opposed to through the skin and all of which decays over time, it has been difficult to gauge its impact on ecosystems and people. Many officials and scientists have used this complex picture to downplay the impact, whereas others have generated a kind of "radiophobia" through an exaggerated picture.[8] One of the differences between the Chernobyl disaster and nuclear explosions is the presence of radioactive iodine (^{131}I) in the former, which is characterized by the relatively short physical half-life of eight days and a rather long period of biological half-elimination from the body (120 days). Yet, deposited in the thyroid gland, the radioactive iodine represented a substantial hazard, especially for children, and the authorities were inadequately prepared with iodine pills to mitigate this hazard. This led to a sharp increase in childhood thyroid gland cancer in areas of Belarus.

The fallout affected primarily rural areas, mostly covered with forest, wetlands, pastures, and arable lands. The main dose-forming radionuclides are caesium, strontium, and plutonium.[9] The degree of the contamination usually is expressed in the average radioactivity of ^{137}Cs (caesium) per square kilometer, and scientists consider the area contaminated if the value exceeds one curie per square kilometer (see Table 1). The total area of significant contamination was 140,400 square kilometer, with 37,600 square kilometers in Ukraine (5 percent of the country's surface area with 5 percent of the country population), 43,500 square kilometers in Belarus (23 percent and 19 percent, respectively), and 59,300 square kilometers in Russia (1.5 percent and about 1 percent, respectively).[10]

---

[8] Zhores Medvedev, *The Legacy of Chernobyl*, p. 89; *Atlas Chernobyl'skoi Zony Otchuzhdeniya* (Kiev: Kartografiia, 1996), pp. 22–25.
[9] Bariakhtar, *Chernobyl'skaia Katastrofa*, p. 26.
[10] United Nations, *The Human Consequences of the Chernobyl Nuclear Accident. A Strategy for Recovery* (Minsk: United Nations, 2002).

TABLE 2. *Land Removed from Various Uses Because of Chernobyl Contamination (square km)*

	Ukraine	Belarus	Russia	Total
Agricultural land	5,120	2,640	170	7,930
Forest	4,920	2,000	20	6,940
Total	10,040	4,640	190	14,870
Percent of the country area	1,66%	2.24%	0.001%	

Source: United Nations, *The Human Consequences of the Chernobyl Nuclear Accident. A Strategy for Recovery* (Minsk: United Nations, 2002).

Officials ordered that substantial radiation-affected areas outside the zone be removed from productive use (Table 2). In Belarus, the exempt land constitutes the largest share of the affected countries' area and is the best arable land of the country. This continues to put a strain on the economy into the twenty-first century.

In the areas affected by the fallout, radioactive contamination continues to pose health risks and economic constraints. However, the direct health risk for the majority of the inhabitants is insignificant and may be important only for a small group of rural dwellers, who depend for their nutrition on their own milk and wild berries, mushrooms, and game.[11] The indirect impacts – a dramatic decline in economic activities and social life because of restrictions on the use of land and other resources, and migration of skilled and young people away – have been profound. The authorities have been unable or unwilling to invest sufficient resources toward the recovery of these affected areas. Part of the problem is that the costs of the Chernobyl disaster are difficult to assess. In addition to mitigation and recovery costs, direct losses (abandoned and destroyed equipment, buildings, and infrastructure), indirect losses (removal of the contaminated land from use, closure of enterprises, and diversion of resources away from productive use), and health and social costs must be taken into account, and there is also the loss of electrical energy production to consider. Beyond the perhaps $19 billion spent on mitigation, other costs have been huge. Belarus estimates its losses will be $235 billion for 1986 to 2016, whereas Ukraine estimates its losses at $148 billion for 1986 to 2000. To help pay for some of these costs, both Ukraine and Belarus introduced a special Chernobyl tax; in Belarus, in 1994, it was 18 percent of wages for all nonagricultural firms. Analysts

[11] United Nations, *The Human Consequences of the Chernobyl Nuclear Accident.*

TABLE 3. *Victims of Chernobyl, Thousands through December 2002*

	Ukraine	Belarus	Russia	Total
Invalids and Acute Radiation Sickness Cases	89	9	50	148
Liquidators, 1986–1987	62	70	160	292
Liquidators, 1988–1989	489	37	40	566
Relocated	163	135	52	350
Contaminated Areas' Inhabitants	1,141	1,571	1,788	4,500
Total	1,944	1,822	2,090	5,856
Percent of Country's Population	4	18.4	1.4	

Source: United Nations, *The Human Consequences of the Chernobyl Nuclear Accident. A Strategy for Recovery* (Minsk: United Nations, 2002).

considered the tax as stultifying to economic development, and it has been progressively reduced.

Officially, victims of the Chernobyl accident have been subdivided into several categories, with children included. In Ukraine, the number of Chernobyl "invalids" (*invalidy*) has increased from only 200 in 1991 to 64,500 in 1997 and 91,000 in 2001 because of both the health of the individuals directly affected and changes in requirements a citizen needs to meet to be formally recognized as an invalid. Ukraine recognizes fifty different categories of people, whereas Russia and Belarus recognize seventy (see Table 3). The total is nearly 7 million people, some of whom get several forms of compensation, but a number large enough to raise the question again of whether the IAEA estimate of total excess deaths of five thousand is too low. The data also reveal that Belarus, where nearly one-fifth of the people are entitled to some compensation as victims of Chernobyl, bore the brunt of the disaster.

Chernobyl compensation takes such forms as welfare payments, free medicine and public transport, access to sanatoria, and pensions. The compensation program has become a severe burden on the national economies of the NIS.

On top of this, the Chernobyl welfare system is flawed and likely inadequate. One problem is that it largely compensates for the estimated level of exposure to radiation-connected risks rather than actual injury. Another flaw is that compensation is low, and many payments have limits or have been cut. In Belarus, for example, 500,000 people – more than 400,000 of them children – were entitled to access to sanatoria, but only 294,000 received this benefit in 2000. In Ukraine, this benefit was cut in half between 1994 and 2000 and then eliminated. Several people argue that the system also encourages some individuals who are less severely disabled than they claim – or are not disabled at all – to secure benefits for psychosomatic or economic reasons. Still, the traumas of exposure and relocation are real; the physical and psychological damage of having to leave one's home hurriedly is great; and many people have not adapted well to new homes in cities, having come from forest or agricultural regions. High levels of mental illness, post-traumatic stress syndrome, and crime affect these people. This may explain why thousands of people, mostly elderly, have returned to the exclusion zone to live, and the authorities have left them alone.[12]

Uncertainties dating to the dawn of the nuclear era about whether low-dose exposures are sufficient to cause adverse effects or if some threshold of total exposure must be surpassed have also made evaluating the public health impact of the Chernobyl disaster difficult. If an individual develops cancer, was the cause genetics, smoking, exposure to industrial particulate, or radiation from Chernobyl? Still, evidence that has accumulated since Hiroshima and Nagasaki indicates that low doses are sufficient to cause cancers. And, of course, there is no doubt that acute radiation sickness (ARS) is directly related to exposure. For Chernobyl, of 237 ARS cases, 29 died within three months. This is a small number given the number of liquidators involved, for whom the mortality rate seems to be within the normal range. But in any event, ARS is a horrific way to die. Also, the increase in thyroid cancer among children of Belarus will be scores, perhaps hundreds of times higher because of Chernobyl, with perhaps eight thousand cases; thyroid cancer is not necessarily lethal but requires early detection, surgery, and lifelong attention. No matter the debate over threshold or low dose, Soviet officials ordered an increase in temporary dose limits after the Chernobyl disaster in 1987 and 1988. According to one source, cancers increased by 32 percent in Ukraine in the first five years after the disaster, childhood cancers by 92 percent, and thyroid cancers by 82 percent.[13]

Beyond death, cancer, destruction of nature, destruction of homes and family history; beyond elimination of local cultures and ways of life, perhaps the most unexpected impact of the Chernobyl disaster was on the process of perestroika. If Gorbachev was to demonstrate his faith in perestroika and his belief in glasnost, then he and his advisors, and all leading Communist Party officials and administrators throughout the

---

[12] Bariakhtar, *Chernobyl'skaia Katastrofa*, p. 67.
[13] Janie Brummond, "Liquidators, Chornobylets and Masonic Ecologists: Ukrainian Environmental Identities," *Oral History*, vol. 28, no. 1 (Spring 2000), pp. 52–62.

Soviet empire, would have to report openly on the nature of the disaster, its extent, and efforts at eradication of the serious human and environmental impact, no matter the cost, in full view of Soviet and world society. After a few halting steps – the initial reports out of Chernobyl were terse and lacked an honest evaluation of the extent of the greatest technological disaster of human history – the authorities were much more forthcoming. Perhaps they had no choice: mitigation took place on the world's stage, and many nations offered help in the process, which the Soviet leaders gratefully, if with embarrassment, accepted. Given the fact that the Soviet system was a closed political system, Gorbachev seems to have risen to the level of openness required. And given his attempts to advance reforms in the sphere of environmental protection generally, Gorbachev indicated his faith in perestroika. Finally, Chernobyl contributed to a general questioning of the Soviet development model and the technologies, factories, and agricultural systems at its core.

The nuclear industry in the Russian Federation has been rebuilt and has embarked on an ambitious program to recapture the nuclear enthusiasm of the 1970s and 1980s. That enthusiasm – as manifested in the Soviet period by the design and construction of Chernobyl-type reactors; the construction of a factory, Atommash, in Volgadonsk to "mass produce" eight pressurized water reactors and associated equipment annually; and the use of nuclear explosives to build canals and dams – has been rediscovered in a program to build up to 100 reactors by 2030 and to manufacture floating nuclear power stations that can be moved around on barges. Granted, the Russian Ministry of Atomic Power (Rosatom) fully subscribes to international safety standards. But the nuclear legacy of Soviet power – 50,000 tons of spent fuel rods, nuclear waste whose disposition is as yet unclear, such disasters as Kyshtym and Chernobyl, and so on – remains to be solved.[14]

Perestroika and the Formation of New Environmental Institutions

Public discussion about the need to create a single, nationwide environmental protection agency intensified after Gorbachev's accession to power in early 1985 and gained impetus from Chernobyl. A new generation of Soviet leaders was eager to demonstrate their commitment to the

---

[14] On the rebirth of nuclear enthusiasm in the Russian Federation in the twenty-first century, see Paul Josephson, *Would Trotsky Wear a Bluetooth?* (Baltimore: Johns Hopkins University Press, 2009), chap. 4, "Floating Reactors."

environment – a popular, "modern," and, they mistakenly believed, politically neutral issue. They approached the problem through the means they knew best: they set up a new government bureaucracy. Such a step would bolster not only the domestic but also the international image of Soviet authorities, an area in which Gorbachev excelled, and would contribute to the overall objectives of perestroika. Support for the creation of an environmental protection ministry was thus transformed from a dissident idea into mainstream policy. Yet, leading Soviet environmentalists understood that creating one more bureaucracy alone would not solve environmental problems. They pointed out that every Soviet ministry had been ineffective in achieving its social and economic objectives, and there was no reason to believe that the new ministry would break with this tradition and contribute to environmental improvements. Many asserted that more comprehensive and radical solutions, including effective legislation, greater penalty and enforcement capability, and anticorruption measures, were needed to address systemic causes of environmental degradation.

Any new institution requires new tools to succeed. Toward this end, an important change to environmental policy in the Gorbachev era was the rise of the environmental impact statement based on sound data, almost all of which had been classified. The State Statistical Administration (Goskomstat) led the way in 1987 with the publication of a statistical yearbook devoted to the seventieth anniversary of the October Revolution. The 760-page book, published in 25,000 copies, contained a 12-page section on "Natural Resources and Environmental Protection."[15] Soon afterward, the State Committee on Hydrometeorology (Goskomgidromet) declassified its publications on air and water pollution and background radiation.

The Soviet government struggled to find institutional solutions that would work within the context of perestroika. Many of the important proposals were inspired by successful environmental policies used in Europe and the United States. An Interministerial Council on Environmental Science and Technology was formed in the mid-1980s to explore the options for improving environmental management in the Soviet Union. In a classified report issued in 1987, the Council identified the so-called Environmental Impact Statement (EIS) procedure as an important step. EIS – introduced in the West in the 1970s and 1980s – were used to analyze potential environmental impacts of proposed projects and

---

[15] *Narodnoe Khoziaistvo SSSR za 70 Let. Iubileinyi Statisticheskii Ezhegodnik* (Moscow: Finansy i Statistika, 1987).

to stop or force the modification of those projects that might have significant negative consequences. Environmentalists embraced the EIS, which they believed was a useful tool to stop such gigantic, environmentally damaging projects as Siberian river diversion that were still the mainstay of Soviet economic development programs.

However, when they adopted this tool, the members of the Interministerial Council translated EIS as *ekologicheskii ekspertiz* (expert ecological assessment). This inaccurate rendition of EIS, whether intentional or not, led to an ingenious solution of adapting a Western policy tool to the Soviet administrative context that shaped environmental policy in the Soviet Union and post-Soviet states for years to come: it placed a premium on the supposed objectivity of engineers and scientists trained within the Soviet context and limited public participation as not scientific. The expert ecological assessment was a fully Soviet mechanism of centralized economic management that involved the review of projects and plans by special expert committees affiliated with such centerpieces of the state with a vested interest in big, environmentally dubious projects as the State Planning Committee (Gosplan), the State Construction Committee (Gosstroi), Minvodkhoz, and other industrial ministries and bureaucracies. Altogether, the government established 900 expert committees, each responsible for the appraisal of specific types of planning documentation.[16] No project, plan, or program could proceed without the authorization of the appropriate expert committee.

Officials initially overlooked the fundamental differences between the EIS in the West and expert assessments in the Soviet Union. Whereas the EIS involved various scientific studies and consultations that encouraged but did not replace democratic decision making, the Soviet assessments were part of decision making and perceived as largely technical procedures. When a multidisciplinary panel of experts was employed to analyze the impacts of a proposed project, the participants' diverse backgrounds were intended to represent different scientific perspectives rather than different group interests. In fact, the committees did not recognize the very existence of different interests. A multidisciplinary panel therefore was normally expected to arrive at a "conclusion" that summarized the results of the scientific investigation of different aspects of the proposed activity, with only marginal, if any, public participation. Still, the

---

[16] Iu. L. Maksimenko, I. D. Gorkina, and A. P. Cherdantsev, "Proizoidet li Reforma Gosudarstvennoi politiki?" in *Otsenka Vozdeistviia na Okruzhaiushchiu Sredu. Materialy 2-oi Konferentsii Sordruzhestva Nezavisimykh Gosudarstv* (Moscow: Znanie, 1992).

government believed that a tool similar to EIS had merit; on July 3, 1985, the Supreme Soviet issued a decree, "On the Implementation of Nature Protection and Nature Resource Use Legislation," that directed the government to elaborate environmental impact procedures for new technologies and projects.[17] The Chernobyl disaster made the need for environmental management reform all the more apparent.

A watershed in this reform process was Decree No. 32, "On the Radical Perestroika of Nature Protection" (see **Box 1**). The 1988 decree introduced a mandatory procedure for State Environmental Review of all new plans, programs, projects, materials, and technologies. The review, a hybrid of socialist planning and the Western EIS, became the central environmental policy tool in the Soviet Union and the post-Soviet countries. The decree resolved wrangling between the State Planning and State Construction administrations through the creation of a new agency, the State Committee for Nature Protection (Goskompriroda), a kind of Soviet Environmental Protection Agency. On September 20, 1988, Goskompriroda issued instructions to all developers to submit planning documentation for all state environmental reviews. Developers were specifically required to conduct an "assessment of environmental impacts." The review procedure was strengthened on November 27, 1989, when the Supreme Soviet passed a decree, "On Immediate Measures of Environmental Improvement," that prohibited the financing of projects and programs until the review was completed.[18]

The principles of these measures reflected not only environmental policy priorities but also the concerns of non-state environmental organizations that the Soviet-style EIS be independent, objective, open, multidisciplinary, and focused on human and environmental health as the lead criteria for judging the soundness of a project. In an attempt to resolve endemic conflicts of interest, an agency independent from the one developing the project would carry out the review. The experts were to base their decisions on rigorous scientific premises rather than on political and economic considerations. To ensure openness, citizens were granted access to the review. In this way, the review became one of few mechanisms through which Soviet and later NIS citizens could influence decisions

---

[17] Postanovlenie Verkhovnogo Soveta SSSR ot 3 Iuilia 1985 g., "O Sobludenii Trebovanii Zakonodatel'stva ob Okhrane Prirody i Ratsional'nom Ispol'zovanii Prirodnykh Resursov," in *Vedomosti Verkhognogo Soveta SSSR*, 1985 (no. 27), p. 479.

[18] Oleg Cherp and Norman Lee, "Evolution of SER and OVOS in the Soviet Union and Russia, 1985–1996," *Environmental Impact Assessment Review*, vol. 17 (1997), pp. 177–204.

> Box 1. Decree No. 32, January 7, 1988: "On Radical Perestroika of Nature Protection Throughout the County"
>
> The Central Committee of the Communist Party of the Soviet Union and the Council of Ministers issued Decree No. 32 on January 7, 1988. The decree emphasized the importance of environmental protection and referred to many of the deficiencies of the existing environmental management, explaining the latter by reference to the negligence of "certain ministries and departments" that did not effectively implement Party and government decisions. The decree directly stated the need to "eradicate the notion of unlimited availability of natural resources and consumerist primitivism in relation to nature." The decree stipulated the formation of a Soviet State Committee on Nature Protection (Goskompriroda) and similar state committees in the Soviet Republics and their branches in territories, provinces, districts, and cities. Goskompriroda was designated the central authority, with primary responsibility for protection of the environment and "rational use" of natural resources.
>
> The functions of the national Goskompriroda included coordinating environmental protection activities of other ministries and agencies, establishing and enforcing environmental standards, proposing environmental strategies to Gosplan, conducting state environmental reviews for national and sectoral development, issuing environmental and some resource use permits, overseeing the system of protected territories, promoting environmental education and awareness, and participating in international environmental cooperation. Within its mandate, Goskompriroda could issue decisions obligatory for other ministries, order cessation of activities that violated environmental norms, and sue and fine the violators. The decree also established a public council that included prominent scientists, parliamentary deputies, representatives of public organizations, and industry leaders whose task was to discuss key environmental issues and provide policy advice to Goskompriroda.
>
> The decree outlined the policy for charging enterprises for utilization of natural resources and for environmental pollution. Moreover, environmental fines were to be subtracted from wages (in order to avoid indirect subsidies by the state of environmental payments). This introduction of the "polluter pays principle" was meant to encourage efficiency of resource utilization and stimulate environmental

protection activities by linking their effectiveness to employee remuneration. Environmental charges and fines were to be accumulated in special funds exclusively designated to finance environmental protection activities.

Finally, the decree addressed environmental education, international environmental policy, funding research and development for the environment, and environmental review of new products. The decree recognized the role of the public in environmental protection and proposed the creation of a national environmental protection society and publication of a specialized newspaper called *Priroda* (Nature).[19]

affecting their lives. The review was explicitly based on the premise that "natural norms" – based on the belief that these norms for ecosystems, independent somehow of human needs and environmental rights, could be determined – could not be violated under any circumstances, even for achieving economic objectives, and recognition of the complex nature of most environmental problems and their intricate links with social and economic matters. Finally, in contrast to past practices, the review was to be applied to all policies, plans, programs and, projects. This generated controversy in the 1990s, because it ensured that hundreds of thousands of projects were checked to verify if they met environmental standards and resulted in significant improvements in projects while fostering the environmental education of developers. Yet, the requirement was applied indiscriminately, with reviews a significant burden on both state officials and businesses. The much larger number of reviews in many cases meant lower quality and invited discretionary application and even corruption.[20]

Officials debated who might direct Goskompriroda. The new leader was supposed to combine several incompatible qualities: he would be an experienced Soviet bureaucrat, an excellent manager, and a Gorbachev supporter; have a good international reputation; and be familiar with environmental protection. Eventually, Gorbachev appointed Fedor Morgun, the first secretary of the Poltava Provincial Party Committee, in May 1988. Morgun was indeed an experienced Party activist who shared many of Gorbachev's ideas. Yet, his only environmental experience, if one could call it that, was in managing agriculture in Poltava province

---

[19] Adopted from Larin et al., *Okhrana Prirody Rossii: Ot Gorbacheva do Putina*.
[20] Cherp and Lee, "Evolution."

in a more environmentally friendly way than traditional practices; collective farms in Poltava province used fewer pesticides and herbicides than those in other provinces. Still, by Soviet standards, the appointment of a reformer like Morgun was a significant achievement. During his short time as the chairman of Goskompriroda, Morgun established the main administrative structures of this body and initiated its operations in key areas. However, Goskompriroda required a more active hand with knowledge of the biological world. Nikolai Vorontsov, a former biology professor specializing in theoretical evolution, replaced him. Vorontsov's knowledge of biology and his firsthand understanding of the sad history of the environmental sciences in the Soviet Union made up for his lack of administrative experience. Viktor I. Danilov-Danilyan, the future Russian minister of the environment, served ably as his deputy chairman.[21]

Finding qualified staff for Goskompriroda proved as difficult as identifying its leader. Most of the leaders of provincial nature protection committees were former party functionaries with no environmental experience at all. The central apparatus of Goskompriroda was largely staffed with former employees of branch industry ministerial environmental protection departments. Because the environment was never a priority for such ministries, these individuals lacked the skills and world view required to enforce Decree 32 and new legislation. However, the reach and public visibility of the new agency increased until 1991, when, with the breakup of the Soviet Union, its headquarters was merged with the Russian Ministry of the Environment, and its branches in the republics were re-formed into the national ministries of the Newly Independent States.

Civil Society and Environmentalism in the Gorbachev Era[22]

The evolution of informal environmental organizations during perestroika was perhaps more dramatic than the changes in formal governmental institutions. A strong environmental movement managed to organize politically and, together with other civil movements, helped bring about the demise of the socialist political system and the collapse of the Soviet empire. Glasnost and tolerance of civil activism resulted in the

---

[21] V. B. Larin et al., *Okhrana Prirody Rossii: Ot Gorbacheva do Putina*.
[22] This section is drawn largely from V. Larin et al., *Okhrana Prirody Rossii: ot Gorbacheva do Putina* and Oleg Yanitsky, *Ecologicheskoe Dvizhenie v Rossii* (Moscow: Institut, 1996).

emergence of thousands of grassroots organizations and networks. By the late 1980s, environmental citizens' groups and NGOs had between thirty thousand and fifty thousand members.[23] In addition to research, education, lobbying, and cooperation with government environmental services, they organized protests that had previously been illegal and engaged in electoral politics, which had not existed. In one of their first national political campaigns in 1987 and 1988, they focused on preventing the adoption of the All-Union Water Resources Management Program, which, like the Stalinist Plan to Transform Nature (1948), envisioned a series of gigantic dams, canals, and a capstone project to "redistribute" Siberian river flow to burgeoning Central Asia for cotton and fruit industries. Having contributed to the scuttling of the "All-Union" program, the emboldened environmentalists set out to block a variety of other environmentally insensitive strategies and projects at the national, republican, and local levels.

At the beginning of perestroika, the Communist Party elite allowed only those forms of activism that did not threaten their primacy in establishing policies or the large-scale projects that had become the major form of economic development. But they quickly lost control of the situation as various environmental, cultural, religious, and humanitarian civil society groups organized for action with the expressed aim of political change. Several "Popular Fronts for Defense of Perestroika," especially strong in the Baltic countries, united various civil initiatives at the city, regional, and republican levels. Public committees to save such threatened rivers as the Volga and Neva from further development and pollution formed, as they did for self-management of urban neighborhoods. As soon as political parties were permitted, several environmental groups actively participated in forming green parties and anarchist organizations.

The peak of political activity of Soviet environmental movement was in 1989–1990, in the run-up to the first multicandidate elections to the Soviet and republican parliaments. The national parliament consisted of a Congress of People's Deputies elected directly by the voters and a Supreme Soviet elected from the congress. Gorbachev had sought the parliament as a way to invigorate perestroika and to push intransigent, anti-reform, conservative Communist Party officials aside – or into action. The elections guaranteed that individuals would be elected both at large and from

---

[23] Oleg Yanitsky, "The Environmental Movement in a Hostile Context: The Case of Russia," *International Sociology*, vol. 14 (1999), pp. 157–172.

slates of candidates in various organizations and institutions controlled and endorsed by Communist Party officials. Gorbachev did this to ensure he had a workable Party majority to achieve other ends. Yet, he underestimated the extent to which the parliament both invigorated the public and encouraged intransigence. In any event, in the first democratically elected congress in 1989, roughly 300 of 2,200 deputies included environmental issues in their electoral programs, and more than 40 of them were leaders of various environmental initiatives.[24] Especially prominent among these was Alexei Yablokov, the de facto leader of the parliament's environmental committee, who subsequently served as the environmental advisor to the first Russian president, Boris Yeltsin. Yablokov used the parliament as a tribune to raise environmental awareness and inspire local environmental groups and movements.

Glasnost encouraged not only party officials but also citizens to question Communist ideology. This questioning led many people to state openly that the system had failed; the glorious Communist future was nowhere to be seen, and not the Soviet state, but its ideology was the first thing to "wither away." In its absence, an ideological vacuum arose that was filled by a profusion of philosophies, world views, and ideologies, some of which were mystical and pseudoscientific. Non-Marxist environmentalists gained legitimacy. Scientists, philosophers, writers, and a variety of numerous informal groups and individuals advanced divergent views, philosophies, and policies on the environment that openly challenged official views. How did state views of the environment change under Gorbachev? What were the ideologies of these non-state environmental actors?

Given the Soviet tradition of strict state control of public initiatives, a number of NGOs formed surprisingly quickly outside state organizations. At first, the NGOs were called "informal organizations" (*neformaly* in Russian). The number of informal organizations increased rapidly and reflected a variety of positions (see **Box 2**).

During perestroika, environmentalists had to define their relationship with other civil movements for the first time. This was not a simple task, because the environmental movement had taken shape long before perestroika, whereas most other groups emerged only as a consequence of the reforms. Some environmentalists viewed these newcomers with a certain detachment. They believed that the green movement was fundamentally

---

[24] Marat Khabibullov, "Crisis in Environmental Management of the Soviet Union," *Environmental Management*, vol. 15, no. 6 (1991), pp. 749–763.

> **Box 2. Informal Environmental Movements in the Soviet Union in the Late 1980s**
>
> Oleg Yanitsky, who has studied environmentalists in the former Soviet Union, identified seven main groups or approaches among these individuals: conservationists, alternativists, traditionalists, members of civil initiatives, ecopoliticians, ecopatriots or econationalists, and ecotechnocrats. Initially, the *conservationists* were the core of the Russian Greens. Their starting point was bioscientism ("Nature knows best") and the idea that ecological catastrophe is inevitable. The key values of the conservationists' ideology were the creation of a world brotherhood of Greens and construction of a society of modest material needs. In autumn 1988, the group's leaders and patriarchs of the druzhina movement founded the Socio-Ecological Union, one of the Greens' biggest umbrella organization up to that point.
>
> The *alternativists* were the most ideologically-oriented group in the movement. Its founders were professional ideologists of ecoanarchism. In those times, the majority of alternativists were simultaneously members of green parties and of the anarcho-syndicalist movement. The alternativists were principled adversaries of the state as a political institution. In their opinion, an ecological "turn" could be carried out only via an alternative project for the whole society of which the key ideas would be decentralization of power and economic activity, self-provision, and self-organization.
>
> The *traditionalists* represented the humanistically-oriented layer of the Russian intelligentsia, with its eternal ideals of good, tolerance, nonviolence, and the desire "to understand and to help." The traditionalists looked to the past with respect in the sense that they appreciated the culture of the nineteenth century, with its ideals of serving and enlightening the people and of nature protection. The main core of the group was composed of academics, writers, journalists, and scholars. The traditionalists adhered to pastoralist values and opposed both Russification and westernization of the Russian ethnic minorities.
>
> The various *groups of civil initiatives* shared four critical types of values. The first one was the responsibility for the condition of an environment ("If not us, then who?"). The second one was the self-organization of their creative activity. The third set of values comprised the need for self-realization and for fellowship with like-minded
>
> *(continued)*

> Box 2 (*continued*)
>
> people. The fourth set of values consisted of those related to preservation of a safe and clean immediate milieu. Taken together, these values helped to provide a sense of social protection, of emotional comfort and mutual support.
>
> From the very beginning, the *ecopoliticians* were the most heterogeneous group. It included druzhina patriarchs who, though not formal members, had a great influence on the group; theoreticians, who imparted the already well-developed ideological doctrines to different groups of the moment; leaders of numerous Russian green parties; former politicians (the deputies of the Soviets as a political institute); and practicing politicians, who originated from the milieu of civil initiatives and combined the role of professional politician with membership in an NGO. The group was united by the idea that environmental protection should be at the top of the national agenda.
>
> The ideology of *ecopatriots* was characterized by left radicalism, the idea of forceful ecologization of society and explicit sympathy for socialism, that is, for state ecological patriotism, for a strongly regulated market, for restriction of private ownership, and for the economy – whether state or private – and social justice to reflect ecological values. The ecopatriots stressed that the powerful Soviet modernization program had annihilated the natural landscape of the Russian core (the Volga River region).
>
> Until the late 1990s, the *ecotechnocrats* were the smallest group in the movement. They believed that the solution of all environmental problems lay in the adoption of ecologically sound technologies – they were naively technocratic, but also the least politicized segment of the Russian environmental movement.[25]

different from other ones because it did not strive for power and was not based on the interests of a particular social group. Other environmentalists embraced all civil society movements that shared democratic values; they believed that only democratic reforms could ensure environmental improvements and actively supported most popular fronts and the "Democratic Russia" movements. In all of these ways, environmental movements and individual activists played an important role not only in

[25] Oleg Yanitsky, *Ekologicheskoe Dvizhenie*.

reshaping Soviet environmental institutions during the perestroika period, but also in the rejection of the Soviet political and economic system as a whole.

As a sign of the vitality of informal groups, mass awareness of Chernobyl and other disasters often resulted in environmentally inspired protests that attracted tens of thousands of people. Strikingly new for the Soviet Union, these protests were covered by the media and in several cases were backed by local authorities, as it happened in Kazan, the capital of Tatarstan, in 1988 (**Box 3**).

Public enthusiasm died away in the early 1990s when the average citizen faced tremendous economic hardships. Consumerism, which had existed in a Soviet form, although officially derogated, became a new mass ideology and also a significant obstacle to spreading environmental awareness. The nouveau riche adopted a mode of highly conspicuous consumption with strong normative overtones that was emulated even by a largely impoverished population. Environmentally motivated consumption and lifestyle choices that are accepted ethical norms in many parts of Western Europe have made little headway in the NIS. Still, the average Soviet and post-Soviet citizen became more environmentally astute, and the consumer frenzy of the wealthy had little in common with the social glorification of the destruction of nature seen under Stalin or Khrushchev. They also had important environmental concerns: housing, land privatization, traffic congestion, lack of recreation areas and opportunities, and food safety were all on the agenda. In the Baltic countries, environmental concerns were more widespread, with residents of Estonia, for example, valuing a clean environment as highly as health, freedom, justice, and peace in a 1995 survey.[26]

The Rise of Econationalism in the Republics

In most of the republics, environmental concerns were intermeshed with nationalist ones. Citizens perceived environmental degradation as both a systemic fault of socialism and a direct result of Moscow's desire to weaken a particular nation by destroying its natural base, exploiting its resources, and poisoning its people while preserving Russia for the Russians. For example, the Lithuanian and Ukrainian independence movements drew great strength from environmental activism, especially

[26] Yanitsky, *Ekologicheskoe Dvizhenie*.

> **Box 3. Mass Environmental Protests in the Soviet Union in the Late 1980s**
>
> In 1987, *Pravda* published an article about air pollution in Ufa, the capital of the Bashkir autonomous province. This article was perhaps one of the most grim to appear in Soviet press. *Pravda* reported that it started its investigation into pollution in Ufa by 400 local enterprises after receiving a collective letter signed by 3,000 residents of the city. The letter constituted an unprecedented move by thousands of Soviet citizens acting independently of authorities.
>
> Also in 1987, more than 30,000 people in Latvia sent written protests to state bodies in opposition to the construction of the Dagavpils hydroelectric power station on the Dagava River. The Soviet news agency, TASS, indicated that environmental protests had stopped work on the plant in order "to avoid loss of arable lands and preserve the proper ecological environment."[27]
>
> In 1988, the Soviet government had to close a ferrous metal plant in Nizhny Tagil, an industrial city in the Ural Mountains, after 10,000 people demonstrated against its pollution. The efforts to organize an anti-pollution demonstration were initially blocked by local authorities, but the Komsomol (Communist Youth League) managed to organize the protest.[28]
>
> In 1988, journalists reported that the government of the city of Kazan, the capital of Tatarstan, had sided with its citizens against a Soviet ministry. Rallies and demonstrations took place when a local newspaper published plans of the Ministry of the Medical and Microbiological Industry to build a biochemical plants in a Kazan suburb. The central news agency, Novosti, which highlighted the situation, stressed that in the past, such a plant would have been built without any preliminary notification, and any protests by local officials would have been ignored.[29]
>
> In May 1988, the newspaper *Komsomolskaia Pravda* published a story that the town of Kirishi, southwest of Leningrad (St. Petersburg), was about to "explode" with popular protests because of the harmful effects of a local biochemical plant whose pollution had caused

---

[27] TASS, November 9, 1987.
[28] Reuters, April 7, 1988.
[29] Novosti, February 20, 1988.

> allergies. A total of 12,000 workers participated in the spontaneous meeting.
>
> In April 1988, between 12,000 and 15,000 people, according to unofficial sources, protested against planned construction of a subway system in Riga, Latvia.[30]

from an antinuclear stance because of the presence of Chernobyl-type reactors in both nations. Dawson calls this linking of environmental and nationalistic concerns "econationalism."[31] Such sentiments arose even in Russia, where ecopatriots accused the Soviet government of deliberately destroying the landscape that supported the core of the Russian nation, for example, the Volga floodplains, through extensive industrial and agricultural development.[32]

Econationalism was most prominent in the Baltic countries. In Estonia, plans to construct a large phosphorus production plant near Rakvere sparked mass protests in 1987. By early 1988, the protesters attracted media attention and gained the attention of national politicians. During an environmental television debate, the economist Edgar Savissaar, who later served in the government in various environmental and economic functions, proposed creating a popular front in support of perestroika in Estonia that drew in part on environmentalism. The Estonian Popular Front was established the next morning, and within a few months it became the most influential political force in the drive toward Estonian independence. In his report to the Congress of the Popular Front of Estonia in October 1988, Savisaar pointed out that approximately 10 percent of the Front's supporters were connected to environmental groups.[33]

In Lithuania, many of the public concerns focused on the safety of the Ignalina nuclear power plant, with its Chernobyl-type design. The Zemyna public club, led by the nuclear physicist Zigmas Vaishvila, organized a number of public meetings and press conferences. (Zemyna is the goddess of the earth in Lithuanian mythology.) The anxiety and mistrust

---

[30] *Sovetskaia Molodezh'*, April 30, 1988.
[31] Jane Dawson, *Eco-Nationalism. Anti-Nuclear Activism and National Identity in Russia, Lithuania, and Ukraine* (Durham: Duke University Press, 1996), p. 221.
[32] Yanitsky, *Ekologicheskoe Dvizhenie*.
[33] Matthew Auer, "The Historical Roots of Environmental Conflict in Estonia," *East European Quarterly*, vol. 30 (1996); and Eestimaa Rahvarinne, *The Popular Front of Estonia: Charter, General Programme, Resolutions, Manifesto* (Perioodika, 1989), adopted at the Congress of the Popular Front of Estonia on October 2, 1988.

toward Ignalina was fueled by the fact that the majority of Ignalina's operating personnel were mostly Russians who had been sent from nuclear power stations in other regions of the Soviet Union. Lithuanian ecological groups were among the founders of Sajudis, formed in summer 1988 as the Lithuanian analogue of the Estonian Popular Front. In one of its first and most prominent actions, Sajudis organized the "Ring of Life" on September 17, 1988, with more than 20,000 people holding hands in a living chain surrounding the power plant. In Latvia, the ecological movement focused on opposition to a hydroelectric power station on the Daugava River, a pulp and paper mill in Sloka, and an oil terminal in Ventspils, all three near crucial ecosystems. Econationalism in the Baltics had a profound impact at a national level, because the members of the Supreme Soviet from the Baltic countries lobbied for environmental reforms and spread environmental ideas in Moscow and the Kremlin.[34]

In other republics, both environmental and nationalist movements were somewhat weaker yet crucial. In Ukraine and Belarus, they largely united around the issue of providing adequate information on the consequences of the Chernobyl disaster and compensation for its victims. The journalist Vasil Yakavenka, who gained notoriety for publishing critical articles on Moscow's handling of Chernobyl, was among the founders of the Belarus Popular Front in 1989. Chernobyl Memorial Day (April 26) remains important to nationalist political parties in Belarus. (At the initiative of Yakavenka in 1994, the Socio-Ecological Union of Belarus held an essay competition for schoolchildren seeking essays on Chernobyl. The result was a book titled *Sled Chornaha Vetru*, published in English as *The Trace of the Black Wind*.)[35] In Armenia, the consolidation of the national and reform movements occurred in connection with protests in 1988 against the Metsamor nuclear power plant, located not far from the capital, Yerevan, and built in a seismically active region. In 1988, a catastrophic earthquake occurred at Spitak in Armenia that claimed thousands of lives. The earthquake fueled fears that an earthquake might trigger an accident at Metsamor and contributed to growing radiophobia in the Soviet Union. Not only was the Metsamor nuclear plant shut down, but a research reactor near Tbilisi was mothballed because of mass protest, and Azerbaijan rejected a project to build its own nuclear plant.

---

[34] Dawson, *Econationalism*, pp. 51–56.
[35] Vasil Yakavenka and Mark Bence, *The Trace of the Black Wind: Through the Eyes of Children* (Minsk: Belarusian Socio-Ecological Union, "Chernobyl," 1996).

In Kazakhstan, the poet Olzhas Suleymenov, later Kazakh ambassador to Italy, established a powerful NGO called Nevada-Semipalatinsk in 1989 to oppose nuclear testing at the Semipalatinsk test site where the Soviet Union conducted hundreds of nuclear weapons tests. Nevada-Semipalatinsk achieved its goal when the new government of Kazakhstan banned all testing, demanded withdrawal of Russian troops from Semipalatinsk, declared Kazakhstan a non-nuclear state, and saw to the removal of all nuclear weapons to Russia.[36]

Similar to the advocates of radical political and economic reforms, econationalists assumed that environmental degradation would be automatically reversed on achieving independence from Moscow. This assumption proved to be premature. Government officials respected environmental priorities as long as they were in line with mainstream political and economic reforms. By the mid-1990s, various nation-building and geopolitical priorities – economic growth, security, independence – that have so often dominated emerging national ideologies pushed environmental concerns into the background. In dealing with international aspects of environmental problems, for example, the Russian elite shifted from Gorbachev's "new thinking," with its emphasis on common international values and the need for cooperation, to the need to defend "Russian national interests." This was especially clear in the Russian debate over signing the Kyoto Protocol to the global treaty on combating climate change, as well as in Russia's position with respect to other environmental treaties (see Chapter 6 on the resolution of Russian debates about Kyoto; Russia eventually ratified the treaty).[37]

The Lithuanian and Ukrainian independence movements, so closely tied to anti-nuclear sentiment, gave up their demands for immediate closure of Ignalina and Chernobyl, respectively. The Chernobyl disaster merged with deep-seated and long-term desire for independence during perestroika in the Ukrainian Popular Front (Rukh) and led to the creation of societies, foundations, and other movements both to raise awareness of the true impact of Chernobyl and to push independence.[38] After the breakup of the Soviet Union, the situation changed radically, with the

---

[36] See, for example, Pauline Jones Luong and Erika Weinthal, "The NGO Paradox: Democratic Goals and Non-Democratic Outcomes in Kazakhstan," *Europe-Asia Studies*, vol. 51, no. 7 (November 1999), pp. 1267–1284.
[37] Dmitri Efremenko and Paul Josephson, "Lego li Byt' Kiotskim Mogil'shchikom?" *Nezavsimaia Gazeta*, February 11, 2004 (no. 27).
[38] Janie Brummond, "Liquidators, Chornobylets and Masonic Ecologists."

growing priority for most people a concern for national independence, not environmental safety. Both Ukraine and Lithuania began to insist on continuation of the operation of their dangerous nuclear plants despite pressure from the European Union. Independent Lithuania continued to operate the Ignalina nuclear power station, which supplied 80 percent of the country's electrical energy and generated revenue from sales of electricity to Russia.[39] Ultimately, under strong pressure from the European Commission as a condition to join the European Union, Lithuanian leaders agreed to shut the station; one unit was closed in 2005 and the second was closed in 2009.[40] Ukraine agreed to shut down the Chernobyl nuclear power station's remaining two operating reactors only in 2001 – in exchange for Western donors financing two new nuclear plants in Rivne and Khmelnitsky. Ukraine's government sought Western subsidies to build a new "sarcophagus" around the increasingly decrepit one built in difficult circumstances in 1986. In 2007, the project contract was finally signed with the French consortium Novarka, with the new structure yet to be completed.[41]

On the Eve of the Breakup of the Soviet Union

Such events as the Spitak earthquake and other natural disasters at first glance had little to do with environmental considerations, but were simply human misfortunes. Yet closer examination indicated that they, too, were often closely connected to the Soviet model of economic development, if not in their immediate cause then in their technological underpinnings and human cost. In this way, the Spitak tragedy also contributed to the belief that the very foundations of the system – and not only homes and apartment buildings – had collapsed. On December 7, 1988, at 10:41 A.M., an earthquake in Transcaucasia in the northern region of the Armenian Soviet Socialist Republic – the Spitak earthquake – killed at least 25,000 people. Except for the Ashkhabad Earthquake of 1948, when the number of dead was between 110,000 and 130,000 people, this was the largest natural disaster in the Soviet Union. Since the beginning of the century,

---

[39] D. Elliott and T. Cook, "Symbolic Power: The Future of Nuclear Energy in Lithuania," *Science as Culture*, vol. 13, no. 3 (2004), pp. 373–400.
[40] Susan Houlton, "Lithuania Shuts Down Last Reactor," *Deutsche* Welle, December 31, 2009.
[41] "Chernobyl to be Covered in Steel," *BBC News*. 18 September 2007, http://news.bbc.co.uk/1/hi/world/europe/6999140.stm.

there have been about 200 devastating earthquakes in the Soviet Union (fortunately, mostly in sparsely populated regions).[42]

The town of Spitak, which was nearly leveled, was less than five kilometers from the fault break. Four minutes after the 6.8-magnitude mainshock, a major aftershock of 5.8 magnitude struck that collapsed many buildings that had been weakened by the mainshock. In addition to bad luck, Soviet-style mass production construction techniques were at fault. According to structural engineer Loring Wyllie, of the more than fifty frame buildings with precast components attached to column and beam construction, fewer than a dozen remained standing, and even these were heavily damaged. "There was very little reinforcing to tie some buildings together. The buildings basically came apart the way they were put together," another specialist noted.[43] Mikhail Gorbachev immediately visited the region to demonstrate once again how the new leadership during perestroika would publicly recognize disaster, whether natural or technological or a combination of the two. Gorbachev also did not hesitate to ask the United States for humanitarian aid, a sign of compassion that past Soviet leaders would have seen as weakness.

In the Spitak quake, in 11 percent of the settlements, more than 70 percent of the buildings were destroyed; in 18 percent of the settlements, 30 percent to 70 percent of the buildings were destroyed; in 22 percent, between 5 percent and 30 percent were destroyed; and in 48 percent, fewer than 5 percent were destroyed. Study of the structures indicated that the collapses – and resulting deaths and injuries – had more to do with the quality of buildings than with distance from the epicenter. Low-quality mass production techniques were used to provide housing quickly to satisfy pent-up need. One such technique was "large-panel construction" – large squares of prefabricated concrete blocks. In Spitak, these structures collapsed like a deck of cards. On top of the disaster, reconstruction was seriously hindered by political factors. Ethnic conflict between Azeris and Armenians led to blockades of railway traffic from Azerbaijan to Armenia. Furthermore, reconstruction planning did not take into account the distinct characteristics of the rural areas. A study group concluded that "the approach 'one settlement – one (two) construction organization(s)'

---

[42] A. K. Borunov, A. V. Koshkariov, and V. V. Kandelaki, "Geoecological Consequences of the 1988 Spitak Earthquake (Armenia)," *Mountain Research and Development*, vol. 11, no. 1, Transformation of Mountain Environments (TOME). Part 2 (February 1991), pp. 19–35.

[43] Richard Kerr, "How the Armenian Quake Became a Killer," *Science*, vol. 243, no. 4888 (January 13, 1989), p. 170.

that disregards the size, site, damage to a settlement, and capacities of a construction unit that was spontaneously adopted in the first days after the catastrophe, has led to unequal rates of reconstruction."[44]

Gorbachev's arrival in Armenia, though a sincere expression of his concern that was welcomed by the survivors, of course, could do nothing to slow the centripetal forces rocking the Soviet system. In this atmosphere, environmental movements connected with nationalist independence movements ("econationalism") formed in the Baltic states, Ukraine, and several autonomous republics. In the next chapter, our conclusion, we evaluate the environmental legacy of the Soviet Union against the backdrop of environmentalism in the successor states to the Soviet Union.

---

[44] A. K. Borunov, A. V. Koshkariov, V. V. Kandelaki, "Geoecological Consequences of the 1988 Spitak Earthquake (Armenia)."

Conclusion

After the Breakup of the Soviet Union
Inheriting the Environmental Legacy

Symbolic of the Soviet environmental legacy, the sarcophagus that encases Chernobyl reactor number 4 is crumbling. Built heroically yet hastily under dreadfully dangerous circumstances, it is unlikely that it will outlive typical aging processes of any concrete. In spite of efforts to strengthen it, it may collapse, leading to the release of tons of radioactive dust. The European Union (EU) and the United States agreed to provide the perhaps $2 billion needed to build a new sarcophagus for the reactor – 325 feet tall, 800 feet wide, and 475 feet long. Construction has been delayed several times and it is unclear when it will be completed. This would permit the first entombment to be dismantled and all remaining nuclear fuel to be removed from the site. The entire station, closed for years, with unfinished cooling towers of additional planned reactors, serves as a ghostly reminder of the costs of pushing ahead to produce electricity without circumspectly considering the social and environmental costs of development. Yet around the station and in the exclusion zone, wild horse, boar, lynx, and wolf populations have returned to the area, and birds are even nesting in the reactor building without any obvious ill effects.[1] This suggests how the zone has become a vibrant ecosystem when sealed off from general human use. But, of course, nature exists only in interaction with human beings.

---

[1] http://www.chernobylee.com/blog/sarcophagus/2010/02/. In *Wormwood: A Natural History of Chernobyl*, Mary Mycio describes her multiple visits to the exclusion zone, where, while wearing protective gear, she was shocked to find a rebirth of plant and wildlife that suggested to her an optimistic, not postapocalyptic landscape. See Mycio, *Wormwood: A Natural History of Chernobyl* (Washington, DC: Joseph Henry Press, 2005).

This chapter is a discussion of the environmental history of the successor states to the Soviet Union that have undergone a transition from Soviet socialism to new forms of government – from democracy in Georgia to a new kind of authoritarianism in Belarus. This discussion of post-Soviet environmental issues will illustrate both the legacy of the Soviet Union and the continuity of environmental issues, actors, ideas, and institutions in Russia, the major successor state to the Soviet Union and in such other nations as Ukraine, Belarus, Georgia, and Armenia, the Baltic states and the Central Asian states. The Soviet model of development, with its emphasis on breakneck economic growth centered on large-scale programs; with its closed political system that limited public involvement in the policy process and its ideological underpinnings that considered nature malleable and believed it desirable to create socialist nature, continues to have a strong influence on the environmental history of the former Soviet Union.

What is the Soviet environmental legacy? During late imperial Russia, environmental thinking was associated with trying to understand and make better existing forestry, agricultural, and fishery practices. Among scientists and their institutions, ideas about "nature" emerged that remained powerful throughout the Soviet period. The ideas largely reflected the beliefs of an increasingly self-aware scientific community whose members argued that natural resources should be "better" managed to improve the people's lives. The limited yet important results of ecological thought before 1917 included the advancement of ideas for the scientific management of fish, forest, and water resources; the establishment of the nation's first nature preserves; and the first, halting steps of cooperation between tsarist officials and scientists. Many of the leading ecologists of the Soviet period received their training before the revolution.

The Russian Revolution of 1917 created hardships for millions of people and disrupted environmental activities. Civil War and famine followed the Revolution. The new Bolshevik state, seeking to control the economy and polity, instituted so-called War Communism to militarize labor and control capital. Economic production plummeted, to recover only in the mid-1920s after the introduction of the New Economic Policy, a policy that permitted small-scale capitalist enterprises to operate. In spite of this turmoil, scientists managed to establish dozens of new research institutes with state support and were permitted to found scores of professional organizations to promote their research and social programs. These nascent efforts in environmental science and policy included the expansion of the network of nature preserves; the study of natural resources, from surveys to analysis of data in support of efforts at scientific management; and the development of ecological ideas unique to the Soviet setting, for example, Vladimir Vernadsky's concept of the noosphere.

When Joseph Stalin took power in 1928, he abandoned the New Economic Policy. If during the New Economic Policy of the mid-1920s, the state focused more on economic recovery than on large-scale mining, metallurgy, forestry, fisheries, and geological engineering projects, by the late 1920s the Bolsheviks were prepared to subject nature itself to plans, no less than industry or labor. New policies mirrored the outcome of the struggle to succeed Lenin with the coming to power of militant young Communists impatient to create a socialist economy and society. Stalin's self-proclaimed Great Break with past social, political, and economic policies involved – among other things – programs for rapid industrialization and collectivization of agriculture that had long-term, extensive environmental impacts. Engineers who suggested a circumspect approach often faced charges of subversion or wrecking. The result was large-scale programs whose human and environmental costs continue to play out in the twenty-first century.

During the Stalin era, party officials, economic planners, and engineers joined in the effort to master the empire's extensive natural resources toward the end of economic self-sufficiency and military strength. At their order, armies of workers built giant dams and reservoirs on all major rivers. Irrigation systems spread across vast arid and semiarid areas of Central Asia. The workers erected massive chemical combines and metallurgical works across the empire, paying little attention to the pollution they produced or the scars they left behind. They erected entire cities to house the laborers, whom they exhorted to meet plans and targets irrespective of the environmental costs and the risks to the workers' own health and safety.

Stalinism did not tolerate any rest. World War II (1941–1945) involved complete ruination of vast regions of the country, its agricultural and forest lands, on top of the deaths of more than 20 million people. But rather than focus on investment and policies to rebuild human lives, in the postwar years Stalin redoubled industrialization efforts in the fourth Five-Year Plan (1946–1950). At the same time, millions of people lived in rubble or barracks, and at least 1 million people starved. Stalin then advanced a bold Plan for the Transformation of Nature, promulgated in 1948 to turn the European Soviet Union into a well-functioning machine based on massive reclamation, afforestation, irrigation, and other projects. This plan failed completely. On top of this, the Soviet Union and United States engaged in the Cold War, with its significant environmental costs of the production of nuclear, chemical, and biological weapons.

Nikita Khrushchev (1954–1964) triggered an era of reformism in Soviet politics and society with his Secret Speech to the twentieth Party Congress of the Communist Party. In this speech he condemned the excesses of Stalin and Stalinism, although he himself had contributed to them as a member of the Party's inner circle. Reformism extended from art and literature to the economy and science. What were the impacts of the reforms on environmental science and policy? The increasing autonomy of scientific specialists and the greater freedoms given to various civic groups led to a reevaluation of the Soviet economic development paradigm. This in turn contributed to a more circumspect attitude among planners, policy makers, scientists, and citizens toward issues of environmental concern. The government strengthened environmental protection laws and increased fines for factory managers who failed to follow them. It sought greater participation of specialists in policy making, which suggested a reevaluation of notions of scientific management of resources. Mass and scientific environmental groups that dated to the 1920s grew much larger in the Khrushchev era and beyond.

Because of the fits and starts of reforms, the continued emphasis on heavy industry, the pressures of the Cold War, and perhaps even the essential nature of Soviet Marxism, in which increased output in gross terms remained the major indicator of success or failure, the quality of life deteriorated from public health and environmental perspectives. The centralized management of resources, the primacy of the annual and five-year plans based on planners' preferences, and the unwillingness of the regime, even during the Thaw, to tolerate dissent, all ensured continued environmental degradation.

Because of the prevailing emphasis on heavy industry, problems with pollution grew still worse; and because of the embrace of inappropriate and often outrageously faulty agricultural programs, farmlands were also degraded. The plowing of the Virgin Lands to increase arable land was an environmental and social disaster resulting in losses of vast areas of pasture lands. Ultimately, Party officials were unable to abandon the large-scale, centralized approach to resource management. Rejuvenated citizens' and scientists' environmental groups and associations began to actively fight against several state development programs. These included the (ultimately unsuccessful) efforts to protect Lake Baikal, the "jewel" of Siberia and by volume the largest body of freshwater in the world, from the construction of paper mills on its shores.

Under Leonid Brezhnev (1964–1984), Party officials increasingly sought the involvement of scientists and engineers in economic planning, resource management, and evolving regulatory and legal activities. This was a period of self-satisfied proclamations about great social and economic achievements, including on the environmental forefront. Officials pointed to successes in advancing environmental protection laws that rivaled those in the capitalist West. The life sciences experienced a kind of flowering once Trofim Lysenko, a quack biologist, was finally removed from power. A kind of environmentalism burst forth that paralleled that in the West after Rachel Carson published *Silent Spring* in 1962. Millions of Soviet citizens joined clubs and societies of "admirers" of nature. But limited discussion of how Soviet ocean trawling, forestry, riverine engineering, and other practices were among the most rapacious in the world did little to temper environmentally disastrous practices, and the truth about many of these practices remained classified information. Furthermore, the effort to increase output of crops through the "chemicalization" of agriculture accelerated poisoning of land and water with pesticides, herbicides, and fertilizers and led to significant erosion of topsoil. Haphazard disposal of chemical, nuclear, and metallurgical wastes; the pumping of particulate, heavy metals, and various other air pollutants into the atmosphere; and the poisoning of water was the rule, not the exception. Large-scale projects to develop Siberia and divert vast quantities of water through massive canals along with clear cutting of forests of forests contributed to "ecocide." The many contradictions of Brezhnev-era environmental policies and practices ultimately did little to improve the situation.

Through his policies of glasnost and perestroika, Mikhail Gorbachev (1986–1991) triggered the rebirth of reformism in the Soviet Union. Such taboo subjects as the role of the Party in its sordid past of prosecuting of citizens and destroying the environment now received open treatment in exposés that filled the press. Environmental movements were reborn, especially after the Chernobyl disaster. With Chernobyl, the Communist Party leadership was confronted by the need finally to address directly and openly the costs of the Soviet economic development paradigm. Ecology grew as an established and respected science. The government ultimately created an environmental protection agency, although this lacked the bureaucratic clout of other ministries that were dedicated to economic growth. Yet, the bugaboo of environmentalists – the Siberian rivers diversion project – was finally mothballed. In this atmosphere, environmental movements connected with nationalist independence movements ("econationalism") formed in the Baltic states, Ukraine, and several autonomous republics.

The breakup of the Soviet Union and Russia's transition from a planned economy and authoritarian political system commenced in the 1990s. Many observers argued that dismantling the centrally planned economy would automatically deliver environmental improvement and that the environmental legacies of socialism would fade away. The reality has proved dramatically different. There have been new threats to sustainability, including the fire sale of resources, the restructuring of the economy that drastically reduced resources available for environmental protection, and President Putin's decision to ultimately disband the Russian Federation's Environmental Protection Agency in 2000. As for the panacea of public participation in environmental movements leading to improvements in the situation, citizens have lost interest in the environment because they have been distracted by political and economic problems. Nationalist movements based on environmental leanings contributed to the breakup of the Soviet Union, but econationalism has faded. Moreover, in Ukraine, Armenia, and even Lithuania, citizens reversed their attitude toward Soviet nuclear plants because the plants provided jobs and electricity. (Lithuania has closed its Ignalina reactors, and its leaders now plan to build state-of-the-art reactors to replace them. Armenia has restarted one of its two pressurized water reactors, a reactor very close to an earthquake fault and built without any containment vessel.) If some environmental improvements took place (mostly associated with decreasing air and water pollution), they were due to the collapse of industry, rather than to the implementation of sound environmental policies.

Post-Soviet Reforms in the Newly Independent States

Perestroika prompted significant reforms in state environmental institutions, whereas glasnost resulted in a rapid growth in environmental concerns shared by the general public and a proliferation of informal environmental organizations. Many of the organizations, institutions, and ideas born during perestroika still form the basis for environmental management in the countries of the former Soviet Union; paradoxically, the Soviet Union collapsed before the environmental reforms of perestroika led to any tangible results. This created completely new challenges to the emerging nations. Living standards plummeted everywhere, and anxieties over political and economic problems overwhelmed environmental concerns of ordinary citizens and leaders alike. The newly constituted national bureaucracies of the new countries suddenly had to deal with poverty, the collapse of the social safety net, the lack of basic services, and generally dysfunctional infrastructure typical for developing countries. By the mid-1990s, the economic output in such nations as Moldova, Ukraine, and Georgia had declined by as much as two to three times compared with pre-1985 levels. The market economies and consolidated democracies that emerged in the former Soviet Union in many case still face daunting political, economic, and environmental challenges, whereas such nations as Turkmenistan, Uzbekistan, and Belarus have retreated to authoritarianism.

If in the late 1990s, the economies of the new nations began to rebound, then it will be many years before economic performance returns to pre-1985 levels. Moreover, the main driving forces behind this renewed growth have been resource- and energy-intensive industries exporting raw materials to developed countries. The share of the economy devoted to these potentially environmentally damaging sectors has increased significantly in comparison with the Soviet period. Many Newly Independent States (NIS) are heavily dependent on one or two commodities; for example, Russia, Azerbaijan, and Kazakhstan base economic growth on oil and gas exports; Ukraine on steel, Armenia on diamond cutting; and Kyrgyzstan on gold mining.

Another result of the disintegration of the Soviet Union has been the increasing economic inequality between and within its former republics. For example, in 2003, the gross domestic product per capita in Russia exceeded that of Tajikistan by almost eight times and that of Moldova by almost six times. Within Russia, income inequality has risen dramatically from levels similar to those in Nordic countries in the late 1980s to levels exceeded today perhaps only by those in Brazil and the United States. Poverty, inequality, and other social problems have essentially displaced environmental priorities.

The collapse of the Soviet Union has also exacerbated ethnic and religious conflicts, with their own direct and indirect social and environmental consequences. Civil wars erupted in Moldova and Tajikistan in 1993. Armed ethnic conflicts engulfed the Northern and Southern Caucasus (in Chechnya, Abkhazia, Nagorno-Karabakh, and South Ossetia). Simmering tensions in the overpopulated Ferghana Valley and along the Afghan border in Central Asia persist. Most of these conflicts are intricately linked with environmental issues because they are motivated by the desire to control natural resources.

TABLE 1. *Dates of Adoption of Framework Environmental Protection Laws in the Former Soviet Union*

1991	Armenia, Kazakhstan, Kyrgyzstan, Russia, Turkmenistan, Ukraine
1992	Azerbaijan, Belarus, Uzbekistan
1993	Moldova, Tajikistan
1996	Georgia

Initially, the collapse of the Soviet Union had a positive impact on the environment. Giant industrial complexes supported by centralized state subsidies closed and ceased polluting air and water. Others were forced to introduce cleaner technologies and to use materials and energy more efficiently in response to higher resource pricing. Although transnational corporations have accelerated exploitation and exports of natural resources (oil, gas, timber, and so on), they have also imported modern technologies and managerial practices at refineries, smelters, and paper mills. Of course, as independent international actors, the new governments of the successor states not Russia, control their own natural resources, and they consider environmental issues a matter of the state. For the most part, they choose to participate actively in international environmental organizations and processes such as the "Environment for Europe." In response, Western governments and international organizations like the World Bank and United Nations Development Programme provide financial resources and support for environmental policy reform and some environmental projects.

Post-Soviet Environmental Institutions

Will Russia, Ukraine, and the other new nations be able to clean up Soviet environmental legacies and prevent the emergence of new environmental problems? To answer this question for the post–Soviet Union, we focus primarily on developments in the Russian Federation because these are largely representative of the situation in the rest of the former Soviet Union. Yet, we provide examples and insights from the other fourteen former republics. After the breakup of the Soviet Union in December 1991, the Soviet republics inherited its environmental institutions, including Goskompriroda and its branches. The twelve countries of the NIS (excluding the Baltic countries) established national environmental ministries based on these branches and passed framework environmental laws soon after or just before achieving independence (Table 1).

Just before the dissolution of the Soviet empire in November 1991, the Russian Ministry of the Environment was amalgamated into a huge organization with the State Committee on Geology and Mining, the Ministry for Forestry, Goskomgidromet, and the Committee for Water Management. Viktor I. Danilov-Danilyan, an economist and specialist in mathematical modeling, became the head of the new ministry and held this position until its dissolution by President Putin in May 2000. Soon after its formation, the new ministry appropriated the buildings, assets, and personnel of a number of disbanded all-union agencies (including Goskomprirody, the State Committee on Forests, the Commission on Mineral Resources, the Committee on Geodesy and Cartography, and the Commission on the Arctic and Antarctic Affairs). This massive new environmental ministry incorporated virtually all agencies dealing with environment and natural resources except for the Chief Directorate on Hunting and Nature Reserves Management and the Department of Fisheries, both of which were parts of the Ministry for Agriculture. This mega-bureaucracy did not exist in such an inflated state for long. In a process that begin in March 1992, the government stripped the resource- and data-amassing units from the ministry (the Committee on Geology and Mineral Resources, the State Committee on Forests Committee that became the Federal Forest Service, Goskomgidromet, and the Committee on Geodesy and Cartography), with the renamed Ministry of Environmental Protection and Natural Resources appearing later in the year. Officials pursued this and other reorganizations to blunt the effective scope of state environmental institutions while exploiting natural resources more aggressively, because natural resources were one of the few income-generating possibilities left to Russia during the economic collapse of the Russian economy in the 1990s.

One indication of the importance of resource exploitation concerns the merging of the Committees on Geology and Mineral Resources and on Water Management into a new Ministry for Natural Resources in 1996. At the same time, the Ministry for Environmental Protection became the State Committee on Environmental Protection. This change in nomenclature signaled a decline in its status, including the fact that its head was no longer a member of the president's cabinet. Reflecting its lower status – and also rampant inflation and recession – the number of staff members shrank from 1,500 people in 1991 to 630 in 1993, 421 in 1996, and 483 in 1998. Yet the State Committee on Environmental Protection established a strong track record, in part because it raised the qualifications of its inspectors. If in the early 1990s its personnel inspected roughly 20,000 annually, then in 2000, just before its disbandment, roughly 360,000 enterprises were subject to inspection.[2]

Russia established a number of other relatively high-level bodies with environmental responsibilities that indicated confusion over how to proceed, worries that regulation would handicap economic development, and the challenges of creating a new government. One such body was the Commission on the Environment and the Exploitation of Nature chaired by a deputy prime minister; the commission's rocky existence indicates the fragility of environmental concerns during the Yeltsin and Putin administrations. This commission was established in 1993 to coordinate implementation of national environmental policy. It included the chairpersons of other relevant state committees, key deputy ministers, the vice president of the Academy of Sciences, and Alexei Yablokov, at that time the presidential advisor on the environment and health. In 1998, the new prime minister, Sergei Kireenko, disbanded the commission together with thirty-three other bodies to streamline the burgeoning central bureaucracy. Then the government reestablished the commission in 2000 with a wider mandate that included protection of biological resources of the Russian continental shelf, territorial waters, and the Azov and Caspian Seas, all of which were significantly threatened by overexploitation and pollution.[3]

Another high-profile environmental agency in Russia was the Inter-Agency Commission on Ecological Security attached to the Security Council. The Commission was created in 1993 and initially chaired by Yablokov, who was subsequently replaced by the vice president of the Academy of Sciences, Nikolai Laverov, a geologist by training who later served on a presidential commission to monitor a program to generate income through the importation of spent fuel rods from other countries' civilian nuclear power programs. Laverov's sense of the environment was that resource use was a crucial activity to be supported by government policies. This commission largely duplicated the activities of the Commission on the Environment and Exploitation of Nature, although its focus on security issues reflected the concerns of major security, intelligence, and defense agencies, and therefore the belief that defense concerns usually trumped environmental ones. This commission reported directly to the president. The commission analyzed and published reports on

---

[2] V. B. Larin et al., *Okhrana Prirody Rossii: Ot Gorbacheva do Putina* (Moscow: KMK, 2003).
[3] Larin, *Okhrana Prirody Rossii.*

many contentious and high-profile environmental issues ranging from the sinking of the *Komsomolets* nuclear submarine in April 1989 to dioxin contamination.[4] Although the commission could not directly influence government decisions, its reports and deliberations led to the promulgation of roughly forty environmental decrees and resolutions.

The two major environmental policy tools introduced in the Soviet Union in the late 1980s were employed by the post-Soviet states: the system of expert assessments, a kind of environmental impact statement, and environmental charges. Most NIS attempted to reform the expert assessment system in response to external and internal pressures for environmental responsibility and transparency in decision making. More than fifty legislative measures related to the assessments were adopted in the various nations during the 1990s. By the end of the decade, parliaments in all twelve of the new republics had passed legislation regulating the system, and it had been applied to thousands on thousands of projects, resulting in the rejection of thousands of environmentally dangerous projects, encouraging conformity to environmental standards, and increasing environmental awareness of developers. Reflecting on this vast experience, several nations introduced a second generation of expert assessment legislation to bring the systems into conformity with internationally accepted standards of transparency, focus, and relevance to decision making. Yet, many observers question how effective expert assessments were because they rarely influenced policy outcomes or waylaid environmentally costly, large projects.

Even less encouraging, in 2000 the Putin administration effectively destroyed the autonomy and authority of the State Committee on Environmental Protection, first by dissolving it in 2000 while transferring its responsibilities to the provinces without either resources or personnel, and then by essentially dismantling the system of expert assessments. Administration officials based their "revision" of the system on the claim that it hindered investment activities; the administration of George W. Bush in the United States made a similar argument that the U.S. Environmental Protection Agency employed several policy tools that were a barrier to economic growth. Fortunately, Russian environmental nongovernmental organizations (NGOs), government officials, and businesses managed to organize to deflect this threat. They enlisted the support of the World Bank, which suspended signing off on several projects in Russia until the government provided assurances that a working system of environmental

---

[4] Larin, *Okhrana Prirody Rossii*.

impact statements had been restored. A joint study of the World Bank and the Russian government subsequently demonstrated numerous benefits of expert assessments and the absence of any negative effects on the investment climate.[5]

The second major policy tool involved a system of charges. The Russian framework Environmental Protection Law introduced in December 1991 stipulated collection of pollution charges and their accumulation in special environmental funds that covered as much as 95 percent of all government environmental expenditures in Russia. The pollution charges varied depending on whether the enterprise could meet the established "emission limit values." The pollution charges were set at very low levels in comparison with what businesses from Western or even Eastern Europe were required to pay. Consequently, neither charges nor fines influenced the enterprises' behavior. Introducing higher pollution charges proved to be politically – and economically – impossible during a time of economic free fall. At the same time, the Russian government introduced payments for utilization of water; extraction of mineral resources; and use of forests, land, and biological resources. These payments, in limited total amount, were accumulated in specialized funds often used to exploit these same resources.

In a number of ways, however, the system was a marked improvement over its Soviet predecessor. The collection and dissemination of data became more regular and open. The complex system for collecting and organizing environmental information that was established in the Soviet Union in the 1970s was overcentralized, secretive, and ineffective, resulting in low-quality data that hardly supported rational policy making. Its performance improved under Gorbachev and in the 1990s, with monitoring of air, water, and soil quality both at reference points (away from major human settlements) and in more than 220 population and industrial centers. Although the monitoring network shrank after the collapse of the Soviet Union, it continued to collect and store vast quantities of information in Obninsk, south of Moscow, and the public gained access to the data. The 1991 Environmental Protection Law explicitly prohibited classifying environmental data, and from that time on almost all environmental information collected by public agencies was declassified. Similar provisions exist in most of the other NIS.

Since the early 1990s, annual "state of the environment" reports have been published and widely disseminated in all NIS; the first such reports

---

[5] Larin, *Okhrana Prirody Rossii*.

(1988) for the Soviet Union and Russia were published in 1989. In contrast to most other NIS, however, Russia did not sign the Aarhus Convention (1998), which stipulates public access to environmental information at reasonable costs. Many agencies still tried to classify environmental information. In a series of instances, federal prosecutors have brought charges of espionage against Russian citizens who have disseminated reports based on already published but damaging information. For example, Alexander Nikitin was arrested, imprisoned, and only after a five-year battle found innocent of all charges for publishing extant information on the practice of the Soviet, not Russian, navy of dumping radioactive waste in the Arctic Ocean.[6]

With the Russian Federation inheriting the Soviet embassies, diplomatic staff, and other outposts of foreign policy after 1991, many post-Soviet states found it extremely difficult in practical terms to play the role of independent international actors. Even Russian politicians were ill prepared to conduct coherent international environmental policy in this new atmosphere, as evidenced by Russia's ineffective participation in the Earth Summit, the United Nations (UN) Conference on Environment and Development in Rio de Janeiro in June 1992 (see **Box 1**).

The lack of diplomatic infrastructure did not prevent active participation by other NIS environmental officials in international environmental regimes during the 1990s. Representatives of all NIS took part in the Earth Summit in Rio in 1992. By joining the "Environment for Europe" process initiated by the EU in the early 1990s, the NIS committed to assess regularly the state of the environment and to developing national environmental action plans. Although the plans rarely enjoyed adequate financial or political support, they were important steps in identifying and evaluating key environmental tasks. The fourth Ministerial Conference of the Environment for Europe process, which took place in Kiev, Ukraine, in May 2003, further strengthened cooperation between the NIS and the EU. The NIS also acceded to a number of key environmental conventions at both the global and the European levels, including the Convention on Transboundary Air Pollution, Transboundary Water Courses, the Environmental Impact Assessment in a Transboundary Context, Protection of Biodiversity, and the UN Framework Convention on Climate Change; with Armenia, Azerbaidzhan, Belarus, Kazakhstan, Moldova, and Ukraine the most active in pursuing these international accords.

---

[6] On Nikitin, see http://www.goldmanprize.org/node/139 and http://hrw.org/english/docs/1998/10/15/russia1356.htm.

> **Box 1. Russian Participation in the Rio Earth Summit in 1992**
>
> The Rio Conference was the largest international forum of the 1990s, involving delegations from more than 150 countries. The American delegation had 200 members; the Indonesian about 60. Only ten officials and two scientists independently funded by the Academy of Sciences represented Russia, while several Russian NGOs also took part in nonofficial meetings. This was insufficient staff to participate in the two dozen or so committees that worked at the summit.
>
> Russian President Boris Yeltsin did not participate in the summit, although the presidents of other major countries attended. The head of the Russian delegation, Alexander Rutskoi, infamous for his participation in the 1991 and 1993 coups against Gorbachev and then Yeltsin, attended for only two of the eleven days of the meeting. He also missed the heads of state round table for sightseeing in Rio de Janeiro, endangering the emerging mutual understanding between Russia and the largest Western countries. The chair of the Russian delegation remained conspicuously empty at the enormous negotiation table. The absence of the Russians at the final meeting alarmed the U.S. delegation, which suspected that Moscow's policy had suddenly changed and that secret instructions had been sent to Rio.
>
> After returning to Russia, delegate Victor Danilov-Danilyan briefed President Yeltsin about the conference. He noted that the Earth Summit had been attended by nearly all the major heads of state. On hearing the report, Yeltsin thoughtfully asked, "Maybe it was wrong for me not to attend the conference?" Mr. Danilov-Danilyan diplomatically answered that, had Yeltsin been in attendance, Russian participation could have been "more effective."[7]

The environmental movement was weak in the Soviet Union and is still weak in modern Russia. The story of the ratification of the Kyoto Protocol illustrates that environmental neglect persists into the twenty-first century. The Kyoto agreement, drafted in 1997, requires industrialized countries to cut their greenhouse gas emissions to 5 percent below 1990 levels by 2012. Russia signed the protocol in 1999. By 2003, the protocol was ratified or accepted by 124 countries, which collectively account

---

[7] V. Larin, Interview with Danilov-Danilyan, 2001 (adapted from V. B. Larin, *Okhrana Prirody Rossii*.

for 44 percent of the industrialized world's emissions. But 55 percent was needed for the treaty to take effect, and after the United States withdrew its support in 2001 because of the ignorance of the administration of George W. Bush, Russia, with its 17.5 percent share of emissions, became the treaty's last remaining hope. The World Climate Change Conference in Moscow in 2003 would have been a very good place to announce the ratification of the Kyoto Protocol, and many Western politicians put their hope in it. However, the atmosphere of the forum was tense. President Vladimir Putin's statements in his welcoming speech were ambiguous and ironic. Andrei Illarionov, Putin's top economic adviser and a staunch opponent of any government interference in the economy, described the problem of climate change as nonexistent in a speech not included in the official papers of the conference. The chairman of the forum, Yuri Izrael, the 74-year-old director of the Moscow-based Institute of Global Climate and Ecology who had served as the head of Soviet delegations in all international environmental forums since the early Brezhnev era, sought to put aside any discussion of the ratification of the Kyoto Protocol by Russia during the forum. His efforts regarding Kyoto earned him the name "the communist fossil fighting for fossil fuel."[8]

What makes things worse is that Russia was under virtually no pressure from its scientific community to take steps to avert climate change. Although the majority of members remained silent, a small group within the Russian Academy of Sciences headed by Izrael himself spoke with nationalistic fervor about the need to avoid restrictions on the Russian economy. Among developed countries, Russia, perhaps, has the smallest portion of scientists who evince concern about any global environmental issues. In the end, the story was a happy one, as Russia surprisingly ratified the protocol in November 2004. Journalists speculated that Russia agreed to ratify the treaty only in exchange for agreement by the EU on terms for Moscow's admission to the World Trade Organization.

By the end of the decade, Russia had determined not to accept new Kyoto commitments. Russia supported Canada's decision to pull out of the Kyoto Protocol. A government spokesman at the Durban, South Africa, 2011 Kyoto Meeting noted that the treaty does not cover all major polluters, and thus cannot help solve the climate crisis. He said, "This is yet another example that the 1997 Kyoto Protocol has lost its effectiveness in the context of the social and economic situation of the 21st century."

---

[8] Q. Schiermeier and B. MacWilliams, "Crunch Time for Kyoto," *Nature*, vol. 431 (September 2, 2004), pp. 12–13.

He observed that protocol does not cover the world's largest polluters, China and the United States.[9]

The Public and the Environment in the Post–Soviet Union

Given the changing political atmosphere of the government's lagging commitment to the environment and its determination to base economic growth on resource development, environmental organizations and movements have had to adopt new strategies. The first major challenge was the decline in the publicly perceived importance of environmental issues vis-à-vis acute social, political, and economic problems. The second one was the difficulty of organizing voluntary efforts amid the economic hardships facing most families. During the Soviet period, the intelligentsia who formed the core of the environmental movement had guaranteed, if paltry, salaries. They could give their free time (and sometimes even work time) to environmental or other "legal" or permitted activism.

In the 1990s, employment and income were no longer secure; many professionals had to work at two or three jobs to secure a minimal living. They had little time for voluntary work. Environmental NGOs had to acquire grants or donations from external, often foreign, sources. Furthermore, after the initial infatuation with democracy, the leaders of several NIS had returned to more familiar authoritarian methods of government. In some post-Soviet states, it became as dangerous to criticize the regimes as it had been in the Soviet Union. In this atmosphere, environmental activists faced an increasingly hostile atmosphere. Still, post-Soviet environmentalists have dramatically expanded their activities and organizational scope. Because biodiversity in Russia and several other NIS has attracted increasing international attention, such conservation NGOs as the Center for Wild Nature Protection have become integrated in international networks and actively cooperate with such major international organizations as the World Wildlife Federation and Greenpeace.

Many of the individuals associated with civil initiatives that emerged in the late 1980s changed their social affiliations and core values. In the late 1980s, they had closely collaborated with local Soviets. When the Soviets were largely disbanded in 1991–1992, civil initiatives naturally entered into a phase of decline. They began to attract marginalized and vulnerable communities who had often lost their faith in "democracy" and who

---

[9] http://www.usatoday.com/news/world/story/2011-12-16/russia-kyoto-climate-change/52003796/1.

sought to protect public health and secure a safe environment through any available political means. They viewed official science with suspicion, although they were prepared to trust those whom they viewed as independent experts. They increasingly collaborated with global and national human rights organizations, but their main efforts have been directed at changing local and regional policies. In many cities, for example, civil initiatives have opposed the surge in poorly constructed apartment houses, office buildings, shopping malls, and garages. They also seek to protect parks and historic landmarks from development.[10]

The druzhina movement experienced a significant decline in the early 1990s, partially as a result of reduced contact between students from universities in the different republics, now new nations each with its own passport. By the end of the decade, many former druzhina activists resumed their environmental work and encouraged revival of student organizations. For example, in 1993, a former druzhina leader, Vladimir Boreiko, founded the Kiev Ecological-Cultural Center to hold local, national, and international meetings of environmentalists; train young activists in working with mass media; design nature reserves; and publish monographs and a journal, *Humanitarian Environmental Magazine*. The latter provides one of the few post-Soviet forums to address environmental ethics and philosophy.[11] Activists at the EcoCenter in Karaganda, Kazakhstan, founded a conservationist group, "Berkut" (Golden Eagle). In Dushanbe, Tajikistan, druzhina members resumed activities after nine years of civil war and political turmoil. Former activists organized through such umbrella organizations as the Socio-Ecological Union and the Taiga Rescue Network and through locally established groups.

Although most post-Soviet environmentalists lack political clout, some have engaged in both civil initiatives and political action. In Russia, a group of influential ecopoliticians is affiliated with Yablokov's Center for Environmental Policy. Many of them are natural scientists and NGO organizers. They aim to promote scientifically based state environmental policy in Russia by lobbying for effective environmental laws. They have formed green parties, although the parties never became a significant political force. Three green parties existed in Belarus in the mid-1990s, but they only managed to secure a few seats on local councils. Compounding the dismal picture, the traditional environmental movements

---

[10] Larin, *Okhrana Prirody Rossii*.
[11] Vladimir Boreiko and Oleg Listopad, "Zelenye na Ukraine. Vchera, Segodnia, Zavtra?" *Tretii Put'*, no. 36 (1995), pp. 1–7.

and the newly emerged green parties were often suspicious of each other. NGOs accused green parties of selfishness and pragmatism, whereas green party leaders considered NGO activists to be idealists unable to understand political realities. The mutual hostility between the Green Party of Ukraine and the Zelenyi Svit umbrella NGO is a typical example.[12]

The diversification and fragmentation of post-Soviet environmental movements have made any coordinated activity, like the mass protest actions of the late 1980s, highly unlikely. The Greens managed to organize a united front only once – to protest a Russian parliamentary vote in 2000 to permit the importation and storage of spent nuclear fuel from abroad to earn money for the Ministry of Atomic Energy (Rosatom).[13] The Greens organized a mass protest and collected sufficient signatures to call a national referendum on the issue; the vast majority of Russian citizens rejected the new law. Unfortunately, the courts disallowed the signatures, and the referendum was not held. In most other cases, environmentalists have little influence over public opinion and lose out to the government's well-financed and professional public relations machinery.

In several nations, increasingly authoritarian governments have forced environmentalists essentially to close down their activities. In Belarus, many environmental groups stopped functioning because of increasingly tough registration laws and the state's insistence on controlling all grants and donations. Some environmental activists in Central Asia were arrested under various political pretexts. Generally, most national, regional, and local authorities have found it difficult to break with Soviet political traditions, and they refuse to involve various social groups meaningfully in governance. Endemic corruption has also hindered transparent interaction between citizens and the state. Environmental NGOs have not helped their cause by keeping their distance from the general population, preferring to organize actions on their own rather than spend time and effort recruiting and teaching new activists. By and large, the Greens have failed to engage with the problems of everyday life to make environmentalism meaningful to citizens. According to Yanitsky, a sociologist with the Russian Academy of Sciences, so-called ecotechnocrats often have contempt for the inability of citizens to think and act rationally, whereas "deep ecologists" have acted too paternally.[14]

---

[12] On Zelenyi Zvit, see http://www.zelenysvit.org.ua/, accessed October 23, 2009.
[13] On the rejuvenation of the Russian nuclear industry and the importation of spent fuel for money, see Paul Josephson, "Technological Utopianism in the Twenty-First Century: Russia's Nuclear Future," *History and Technology*, vol. 19, no. 3 (2003), pp. 279–294.
[14] Yanitsky, *Ekologicheskoe Dvizhenie* (Moscow: Institute of Sociology Press, 1996).

Many environmentalists held cosmopolitan, universal attitudes that the environment knows no borders and democratic values strikingly different from those of nationalists. The conflict between those Ukrainian environmentalists who believed that an oil terminal in Odessa was essential for securing the country's independence from oil imports from Russia and those who believed that it was dangerous and unnecessary to build the facility illustrated the clash of these values; Greens protested vigorously after a March 1997 spill at the terminal. Similar protests occurred in Lithuania over the Butinge Oil Terminal after an oil spill in 1999.[15]

Because of these philosophical and organizational difficulties, representatives of other social movements and open-minded government officials pushed many post-Soviet NGOs to look for support and partnerships abroad, especially in democratic, stable, and prosperous Western countries whose citizens were largely environmentally conscious. After 1991, international cooperation (both within and beyond the former Soviet boundaries) was the central aspect of NGO activities. Almost 50 percent of Russian NGOs surveyed in 1995 worked closely with their Western counterparts and received foreign donations or grants, another 30 percent exchanged information, and 35 percent joined networks of foreign NGOs. A number of international organizations established offices in the NIS. These included the World Wildlife Federation, the International Union for the Conservation of Nature, and Greenpeace. Western foundations also supported various forms of civil activity, including the environmental movements. Among the major actors are the British Charities Aid Foundation, the Open Society Institute funded by George Soros, the MacArthur and Ford Foundations, the Eurasia Foundation, and the Initiative for Social Action and Renewal in Eurasia (formerly the Institute for Soviet-American Relations), which established offices throughout the NIS in Azerbaijan, Belarus, Kazakhstan, Russia, and Ukraine and supported a variety of educational, nature protection, and informational programs.[16] The most frequent form of cooperation involved financial support from international or bilateral development agencies or large international NGOs to carry out specific projects.[17] In the process of applying for funding and implementing projects, the NGOs members

---

[15] http://www.kyivpost.com/nation/1957/print and http://www.planetark.org/dailynews story.cfm/newsid/13580/newsDate/5-Dec-2001/story.htm.
[16] Larin, *Okhrana Prirody Rossii*.
[17] Regional Environmental Center for Central and Eastern Europe, *New Regional Environmental Centers: A Feasibility Study on Establishing New Regional Environmental Centers for Countries Beyond the Mandate of the Regional Environmental Center*

learned new skills and expanded activities. At the same time, some of them learned that approaches advocated by foreign partners were not necessarily feasible or effective in their home countries.

Another aspect of internationalization of the environmental movement was growing interaction with, and opposition to activities of, multinational corporations (MNCs). Environmentalists believe that MNCs, whose presence in the former Soviet Republics dramatically increased during the 1990s, bring both risks and opportunities for the environment. NGOs from the Caspian region (Azerbaijan, Georgia, Iran, Kazakhstan, Turkmenistan, and Russia) held a workshop in 2000 on tactics to force changes in production activities of MNCs that threatened the sea basin.[18] This cooperation among ideologically different NGOs across the region was triggered by proposals to build pipelines to transport Caspian oil across the Caucasus. In another case, the Russian NGO "Forest Club" criticized several large MNCs involved in the timber trade in Russia for violating "Principles of Responsible Trade of Russian Timber." Yet, a number of other forestry MNCs largely met the principles. They have cooperated with Forest Club and the World Wildlife Federation-Russia to introduce a system of forestry certification that is widely used in Scandinavia and North America, aiming at ensuring environmental and social sustainability in the forest sector.[19]

Many individual NGOs successfully expanded international contacts on their own without substantial international support. The Rainbow Keepers (Ukraine) cooperated with leftist green organizations as well as anti-fascist and peace movements in the United States, Europe, and South America. In 1991, the Rainbow Keepers initiated an international protest campaign against a pitch distillation shop located at a metallurgical plant in Zaporozhe, and in 1994 they protested the construction of the Odessa oil terminal. The EcoDefense NGO, based in the western Russian enclave of Kaliningrad, organized several international campaigns, including one to protest potential exploration of oil in the Baltic Sea.[20]

In summary, during the 1990s in the former Soviet Union, many environmental NGOs matured and broadened their activities. They often shifted from critical and idealistic approaches based on assumed shared civic values to pragmatic and utilitarian environmentalism based on

for Central and Eastern Europe (Budapest: Regional Environmental Center and Aqua, 1995), p. 111.
[18] Larin, *Okhrana Prirody Rossii.*
[19] Ibid.
[20] Ibid.

exchange of information and paid for largely through the mobilization of Western financial aid. Leaders of the NGOs recognized that if a post-Soviet environmental movement was to secure cohesion and influence, it had to redefine its relations with such new social actors as political parties, private businesses, mass media, and other social movements. Yet, this would be a major challenge. According to a pessimistic view of the situation offered by Oleg Yanitsky, Russia is exhausted in terms of its human capital, particularly regarding political activity. In a typical democracy, when one layer of activists and dissidents becomes "coopted" and bureaucratized, another layer of activists takes its place. But Russia lacks people with the skill, resources, and interests to act as vigorously as those earlier activists who became politicians and bureaucrats. There is, in short, no fund of "human resources" to renew the movement.[21]

Environmental Change in the Former Soviet Union

What were the costs of economic development in the former Soviet Union? How extensive was environmental degradation? Did the situation improve after the collapse of the former Soviet Union? In most NIS (except Uzbekistan, Ukraine, and Turkmenistan), environmental pressures have dropped significantly. For example, total air emissions in the Russian Federation declined by 47 percent, from 67.0 to 35.3 million tons between 1991 and 1999.[22] Carbon dioxide emissions dropped in all NIS except Turkmenistan, but especially dramatically in Georgia, Ukraine, and Moldova, with declines continuing through 2010.[23]

Yet, as the following litany of challenges and problems concerning air, water, land, pollution, and preservation indicates, any good news must be treated with caution. These reductions have not always led to improvements in air quality, especially in urban areas where automobiles have replaced industry as the main source of pollution. Second, these reductions may be reversed once economies recover if the structure and the mode of operation of industrial, transport, and power generation

---

[21] Ernest Partridge, "A Conversation with Oleg Yanitsky," http://gadfly.igc.org/russia/yanitsky.htm.
[22] A. Cherp, R. Mnatsakanian and I. Kopteva, "Economic Transition and Environmental Sustainability: Effects of Economic Restructuring on Air Pollution in the Russian Federation," *Journal of Environmental Management*, vol. 68 (2003), pp. 141–151.
[23] European Environment Agency (EEA), http://www.eea.europa.eu/publications/92-9167-087-1/page014.html.

sectors do not change for the better. In fact, air pollution remains the most serious environmental problem faced by cities of the NIS; urban air pollution is again on the rise. In Russia, the annual average environmental quality standards have been exceeded in seventy-two cities and the acute short-term limits have been exceeded in ninety-five cities of Russia for at least one substance.[24] Unsatisfactory air quality also is a problem in the capital cities of Tbilisi, Dushanbe, Bishkek, Kiev, Chisinau, Almaty, and Ashgabad; in such industrial centers as Ust-Kamenogorsk, Ridder, and Temirtau in Kazakhstan; and in Donetsk, Lutsk, and Odessa in Ukraine. Environmental quality standards were exceeded in 30 percent of Russian and 40 percent of Ukrainian cities in 1998 and 2000, respectively.[25]

In another example, an international team of researchers plotted different research points in the Pechanganikel industrial area and confirmed continued, astronomically high heavy metal devastation of birch, moss, lichen pine, and fir into the twenty-first century. The devastation began in the 1940s, and even if the releases of heavy metals and sulfur dioxide have dropped significantly in the last twenty years (the latter from 380,000 to 120,000 tons since the breakup of the Soviet Union), the nickel smelter in the city of Nikel, Russia, alone emits five to six times the entire annual Norwegian output of sulfur dioxide.[26]

Most of the increase in urban air pollution is due to increasing emissions from motor vehicles resulting from the growing numbers of cars and their poor states of maintenance. Many of these cars are secondhand vehicles from Europe, Eastern Europe, and Japan. Even in the worst recession years of 1990 to 1994, the number of private cars in Armenia, Russia, and Ukraine doubled. In Moscow, automobiles are responsible for more than 80 percent of all air emissions. According to the Russian State Committee on Nature Protection, the annual cost of environmental damage due to the transport sector has grown to 1.5 percent of the Russian GDP.[27] And the situation is getting worse – as gridlock in Moscow, Petersburg, and elsewhere for six or seven hours a day indicates.

---

[24] EEA, http://www.eea.europa.eu/publications/environmental_assessment_report_2003_10/kiev_chapt 05.pdf.
[25] Ibid.
[26] Tor Myking et al., "Effects of Air Pollution from a Nickel-Copper Industrial Complex on Boreal Forest Vegetation in the Joint Russia-Finish-Norwegian Border Area," *Boreal Environment Research*, vol. 14 (April 2009), pp. 279–296. See also Tor Norseth, "Environmental Pollution Around Nickel Smelters in the Kola Peninsula (Russia)," *Science of the Total Environment*, vol. 148, nos. 2–3 (June 1994), pp. 103–108.
[27] EEA, Environmental Assessment Report (2003), "Transport" http://www.eea.europa.eu/publications/eea_report_2007_3/07_Transport.pdf, p. 13.

Similar to air emissions, registered discharges of waste water have significantly declined in all the NIS as a result of the decrease in industrial and intensive agricultural activities. For example, in Georgia, the use of fertilizers declined from 240 to 250 kilograms per hectare in the late 1980s to about 10 kilograms in 1994; the usage of mineral fertilizers in Armenia declined to 3 percent of its previous levels. Yet, declining water discharges have not always led to notable improvements in water quality. The reasons might be that unorganized (diffuse) pollution of rivers and lakes may have increased, and many wastewater treatment facilities have stopped functioning normally. Water extraction, especially for industrial and agricultural purposes, also declined after the breakup of the Soviet Union. The Central Asian countries and Azerbaijan still face considerable problems because vast expanses of their territories are under severe water stress; water use is high in relation to the renewable water resources available to them.

Respiratory diseases linked to air pollution are perhaps the next most significant health effect of environmental pollution. Although the precise effects of air pollution on health in NIS cannot be precisely known because of lack of data, it is likely that respiratory diseases in highly polluted cities such as Kyiv and Tbilisi occur more frequently than in other cities.[28]

Land use issues remain critical in every country. In the Soviet Union, as elsewhere, modern measures to bring more land into agricultural production (for example, plowing under the "Virgin Lands" in Kazakhstan under Khrushchev, the construction of massive irrigation systems in Central Asia under Brezhnev, and drainage of marshes in Belarus) resulted in long-term land degradation in the form of soil erosion, bog formation, salinization, and desertification. These negative trends accelerated during the 1990s because of the spread of previously banned logging, grazing, and water extraction practices; inadequate maintenance of drainage and irrigation systems; and poor law enforcement. About two-fifths of the territory of Armenia, one-third of Georgia, and 50 percent of Azerbaijan face severe problems because of soil erosion, itself a result of excessive use of chemicals in agriculture, clear cutting, and so on. The damage from soil erosion amounts to 7.5 percent of the agricultural GDP of the South Caucasus countries; erosion increases by 1.9 percent every year. At the same time, the area of irrigated lands in Armenia declined from 311,000

---

[28] EEA, Environmental Assessment Report (2003), "Air Pollution," http://www.eea.europa.eu/publications/environmental_assessment_report_2003_10/kiev_chapt_05.pdf, p. 126.

hectares in 1985 to 280,000 hectares in 1995 and 217,000 hectares in 2000.[29]

Another form of pressure on land is waste disposal. Although the quantities of industrial waste have generally fallen (in Armenia, nonhazardous industrial waste generation fell from 35.2 million tons per year in 1985–1990 to only 251,000 tons per year in 1995–1996[30]), the amount of domestic waste has increased, and its composition has become more complicated with the spread of consumerism, especially in metropolitan areas. Municipal waste disposal infrastructure and Soviet-era landfills are often ill equipped to deal with this problem, which has been complicated by the widespread practice of illegal disposal of residential and commercial waste. Pressure on natural ecosystems declined significantly at the beginning of the 1990s, largely because of a significant increase in the forest cover. This increase resulted not only from artificial reforestation efforts, but also from a reduction in agricultural activities and the decline in forest industries. For example, in the late 1990s, total removals of forest in the NIS amounted to only between one-quarter and one-third of amounts extracted in the 1970s and 1980s.[31]

Yet this increase in the total forest cover conceals several negative trends, including a rise in illegal forest cutting, with severe local deforestation in some areas. According to Greenpeace and the Forest Club, illegal practices, which account for almost 20 percent of total logging in Russia, range from individuals cutting wood for home heating and cooking to companies exceeding sanctioned quotas and criminal groups logging in nature reserves.[32] Whereas in Russia most illegal logging is for profit, in the Caucasus and Crimea, which have faced severe energy shortfalls due to pipeline failures and embargoes from Russia over political disputes, valuable woodlands have been felled near to villages for individual use. Poaching hurt such valuable forest types as beech in the Caucasus; forests on slopes vital for erosion control; and even urban forests, parks, nature preserves, and botanical gardens, where poachers have been active.

[29] Ron Witt et al., Caucasus Environment Outlook (Tbilisi: GRID, 2002), http://www.grid.unep.ch/product/publication/CEO-for-Internet/CEO/index.htm.
[30] Witt Caucasus Environment Outlook, http://www.grid.unep.ch/product/publication/CEO-for-Internet/CEO/index.htm.
[31] Food and Agriculture Organization (FAO), Global Forest Resources Assessment 2000.Main Report, FAO Forestry Paper 140, Rome, ftp://ftp.fao.org/docrep/fao/003/Y1997E/FRA%202000%20Main%20report.pdf.
[32] A. Morozov, Survey of Illegal Forest Felling Activities in Russia (Forms and Methods of Illegal Cuttings) (Moscow: Greenpeace, 2000), http://www.forest.ru/eng/publications/illegal/index.html.

Corruption of officials and poor law enforcement exacerbated the problem. A decline in *quality* of forests reflects a drastic reduction in state-supported reforestation and forest management programs. And the threat of pollution remains; more than 500,000 hectares of Russian forests have been damaged by pollution in the Siberian region of Norilsk alone.[33] Chernobyl affected nearly 1 million hectares of forests in Russia, Belarus, and Ukraine that are likely to be excluded from use for decades.

Some threats to biodiversity have receded, whereas many new ones have emerged. In the Soviet era, the system of nature preserves protected some biodiversity. Although not extensive in terms of total land area, the preserves were additionally sheltered by their remoteness and restrictions on entry. As these restrictions have been lifted and budgets for the protected areas significantly cut, poaching of rare and endangered species has acquired unprecedented dimensions. Armed conflicts have been another serious threat to biodiversity. A number of nature preserves have suffered from military conflicts that occurred in their proximity, for example, Tigrovaia Balka in Tajikistan and the entire Northern Caucasus because of the war in Chechnya. Armed conflicts directly harm flora and fauna in the military arena, prevent maintenance of zapovedniki, and displace people.

Illegal fishing and hunting have hit all regions. Poaching in the Caspian Sea triggered the application of the CITIES (Convention on International Trade in Endangered Species of Wild Fauna and Flora) to prevent the trade of caviar from endangered Caspian sturgeon. Commercial fishing in the Sea of Azov declined tenfold since the 1980s; the sturgeon catch in the Caspian Sea dropped by 90 percent; and the fish catch in Georgia was reduced to essentially zero.[34] Regarding hunting (and bribes to officials to permit it) and loss of habitat, the number of critically endangered species has grown. According to the 2004 *IUCN Red List of Threatened Animals*, the critically threatened or endangered animals include the evorsk vole, the Mediterranean monk seal, the Muisk vole, the wrangel lemming, the blue whale, the European bison, European mink, fin whale, long-tailed birch mouse, sea otter, snow leopard, Steller's sea lion, tiger, and Ussuri Tube-nosed Bat. Among the vulnerable are listed the West Caucasian tur, argali, Asiatic black bear, Bechstein's bat, Caspian seal, dhole, East

---

[33] R. Mnatsakanian, *Environmental Legacy of the Former Soviet Republics*, (Edinburgh: Centre for Human Ecology, University of Edinburgh, 1992), p. 220.

[34] For example, see Wallace Kaufmann, "Caviar Wars," March/April 2003, http://www.orionmagazine.org/index.php/articles/article/117/, and http://www.cites.org/eng/news/press/2004/pdf/1008_sturgeonquota.pdf.

Caucasian tur, Eurasian otter, a number of different bats and mole-rats, humpback whales, harbor porpoises, desman, Musk deer, spotted souslik, Steppe pika, white whale, wild goat, and wolverine.[35]

The situation is endemic in the former Soviet Union. Armenia's Ministry of Nature Protection plans to submit a new edition of the *Red Book of Endangered Species* to the government for approval, hoping to replace as a foundation for protection of biodiversity a book published in the Soviet era; as of 2009, independent Armenia had yet to replace the *Armenia SSR Red Book* published in 1989. That book listed 387 rare species of plants and 99 species of animals that are endangered. Since independence, several factors have had a negative impact on the nation's ecology, and a number of valuable and endemic species have found themselves on the verge of extinction as their natural environments have degraded and their habitats have shrunk. An updated study of biodiversity, an assessment in accordance with international criteria, and a new edition of the *Red Book* to organize the protection of species have become urgent. On top of this, although laws on hunting rare animals and fines for violations are becoming tougher, the number of poaching cases has not decreased.[36]

Kyrgyzstan reports that wolf populations have increased or become stable since independence. Yet most evidence indicates this is not the case, but rather that wolf populations are likely dropping.[37] And even that symbol of the Arctic, the polar bear, is at great risk. Efforts to save polar bears and preserve their natural habitat commenced belatedly in the first years of the twenty-first century through a large-scale effort to fit the bears with satellite tracking sensors and thereby learn more about their whereabouts, how and when they sleep and move, how much time they spend in the water, and so on, as global warming and encroachment of civilization press them into narrower and narrower ranges.[38]

Decoupling Economic Growth from Environmental Security

The former socialist countries of Eastern Europe and the Baltic states have largely managed to reform their economies and political systems to engage

---

[35] http://www.animalinfo.org/country/russia.htm#ixzzotJr4zHG3.
[36] Gayane Mkrtchyan, "Red Book Concerns: Armenia Upgrades Soviet-era Species Protection Code," http://www.armenianow.com/news/19731/red_book_concerns_armenia_upgrades.
[37] C. J. Hazell, "The Status of the Wolf Population in post-Soviet Kyrgyzstan," findarticles.com/p/articles/mi_qa444/is_4_18/ai_n28871869.
[38] "Russian Scientists Trying to Save Polar Bear from Extinction," http://rt.com/Top_News/2010-04-30/scientists-save-polar-bear.html.

in sustained economic growth while establishing the legal framework to ensure environmental sustainability according to European standards. In the NIS, however, most environmental improvement has occurred mainly because of economic decline and may be reversed by accelerated economic growth unless accompanied by other reforms. Analysis of decoupling between economic growth and environmental pressures reveals positive decoupling in the Baltic states and Eastern European nations. For one, economies are growing faster than environmental pressures. This is not the case in most of the NIS, including Russia.[39] Expected or ongoing economic recovery or reform is likely to bring environmental pressures back to levels experienced in Soviet times. The main reason for this negative trend is the recovery of the economy based on resource- and energy-intensive industries. (See **Box 2**.)

Yet, decoupling from the past is not a simple matter. Lithuania reluctantly closed the Ignalina power station on December 31, 2009, a precondition to entering the EU in 2004. A twenty-year decommissioning project, at billions of dollars, will soon commence. More immediately, Lithuanians are worried that the country must rely to a much greater extent on fossil fuels for power, including buying electrical energy from Russia. Many Lithuanian people have an uncomfortable feeling that they were once a colony of Russia whose energy sector was built by Russians, and they risk being energy dependent again, although many of the nuclear workers from Ignalina have returned to Russia to staff existing and new reactors in contemporary Russia. Lithuanian President Dalia Grybauskaite said, "After the decommissioning of Ignalina, 2010 will be the start of our energy independence. [Yet] Lithuania's energy system was and continues to be dependent on Russia because all supplies of gas and electricity are related to that country. Dependence on one country was caused by the old power plant. The issues around it tied Lithuania to Russia."[40] Ironically, Lithuania has decided to build a new nuclear plant at Visaginas, the site of the Chernobyl-type Ignalina station, in partnership with Latvia, Estonia, and Poland, in a project likely to be built by a European vendor.

---

[39] EEA, Environmental Assessment Report (2003), "Air Pollution," http://www.eea.europa.eu/publications/environmental_assessment_report_2003_10/kiev_chapt_05.pdf, p. 126.

[40] A. Ozharovsky and C. Digges, "Lithuania Shuts Down Soviet-Era NPP, but Being a Nuclear-Free Nation Is Still Under Question" (Bellona, January 12, 2010), http://bellona.org/articles/articles_2010/ignalina_shut_down.

Box 2. Decoupling Economic Growth and Environmental Pressures in the NIS in the 1990s

Decoupling describes the link between economic and environmental indicators and is often expressed in the pollution-, energy-, or resource-intensity of the economy, that is, the amount of pollution released or energy or resources consumed per unit GDP. Positive decoupling means that this intensity is declining such that the economy is growing faster (or declining more slowly) than environmental pressures. Positive decoupling is a precondition for sustainable environmental improvements. In the late 1990s, several studies contrasted positive decoupling in East Central Europe and the Baltics with negative or neutral decoupling in the CIS, especially Russia and Ukraine.

According to Environmental Sustainability Index indicators, nine of eleven NIS are in the lowest 10 percent of all countries in the world in the energy intensity of the economy, with an average ranking 134 of 142. Kyrgyzstan and Armenia are notable exceptions, most likely because of the high ratio of hydroelectricity in their energy balance. The new EU member states (except Latvia and Slovenia) are also performing very poorly, but still better than NIS, with an average ranking 111 of 142. Similarly, carbon dioxide emissions (primarily arising from energy production) per unit GDP in Russia, Azerbaijan, Kazakhstan, Turkmenistan, and Ukraine are among the worst in the world. In contrast to most Central European countries, these high energy and pollution intensities do not show signs of declining.

If the new EU member states achieved better environmental results in the process of transition, what are the causes of the trend in positive decoupling? One of the causes of positive decoupling is the shift from manufacturing to the service sector in the overall structure of economies. This shift has been more profound in Central Europe and the Baltic states. Service is usually environmentally benign, although such branches of the service sector as tourism and transport can inflict severe environmental damage.

The second force for positive decoupling has been efficiency improvements within individual industrial branches. This is the most widely cited benefit of economic transition for the environment. Reasonably, the nations pursue policies promoting efficient use of energy and resources, thus achieving both environmental and economic benefits. Analysts believe that the slow pace of economic reform in the NIS, particularly in liberalizing energy prices, will result in large inefficiencies in industries. Still, studies of individual industrial branches in the NIS show decreases in their pollution intensities and increases in their energy efficiency comparable to that of Central European countries; the energy intensity of Russian industry improved by 17 percent from 1995 to 2001, whereas in Slovakia it has improved by 21 percent over the same time period. Indeed, in those sectors that were economically profitable and managed to attract foreign investments, such resource gains occurred and environmental impacts per unit of production decreased.

What explains the overall negative trends in decoupling in the NIS? An important but often neglected factor associated with economic liberalization is the structural change that *increases* rather than decreases pollution and energy intensity. The proportion of more pollution- and energy-intensive industries in Russian industries at certain periods of time has increased, all but counteracting the efficiency gains in individual branches. Thus, the positive trend associated with increases in efficiencies in some NIS countries was unfortunately counteracted by the shift from less to more resource- and pollution-intensive branches of industry. As a result of free-market reforms combined with globalization, many of the manufacturing industries subsidized by the government in the past proved to be less competitive than more typically polluting branches associated with extraction of raw materials.[41]

Not surprisingly, human vulnerability to environmental threats in the NIS remains a greater problem than in Eastern Europe. This relates particularly to residents of zones of ecological crisis, for example, Ural metallurgical regions, Chernobyl, and so on. The people and ecosystems carry

[41] Organization for Economic Co-operation and Development (OECD), *Environment in the Transition to a Market Economy: Progress in Central and Eastern Europe and the New Independent States* (Paris: OECD, 1999); A. Cherp, R. Mnatsakanian, and I. Kopteva, "Economic Transition and Environmental Sustainability: Effects of Economic Restructuring on Air Pollution in the Russian Federation," *Journal of Environmental Management*, vol. 68 (2003), pp. 141–151; M. Olshanskaya, "Comparative Analysis of Energy Intensity in Russia and Slovakia During Economic Transition of the 1990s," Master's thesis, Department of Environmental Sciences and Policy, Central European University, Budapest, Hungary; and http://sedac.ciesin.columbia.edu/wdc/geonetSearch?geonetService=wdcmetadata.show&id=2457&currTab.

a disproportionate burden of environmental risks, as sharp declines in economic activity and budgetary resources and inadequate state capacity have left NIS governments unable to deal with some of the worst Soviet environmental legacies. It does not help that the environmental institutions in the NIS are weaker than those in other countries in transition and may be unable to ensure environmental sustainability in the future.

The political and economic changes that have resulted from the collapse of the Soviet Union have not, so far, significantly altered the systemic causes of environmental problems. Some progress has been made in increasing public participation, entering into agreements for international cooperation, and helping to integrate environmental concerns into policy making overall. Significant opportunities for reversing environmental degradation may result from public administrative reform, decentralization of bureaucracies, social mobilization, formation of democratic institutions, and reduction of subsidies to the military and other heavy industries. And many NIS nations seem willing to accept international assistance, learn from other countries' experiences, and participate in global environmental conventions.

The environmental threats of the transition period of the 1990s and the present include the shortening of time horizons, as decision makers focus more on immediate economic and social problems than on longer-term environmental issues; such undesirable environmental side effects of globalization as economic pressures that overwhelm sustainable practices and marginalization of poor and indigenous peoples and their resources that are threats to traditional lifestyles; and the growth of resource- and pollution-intensive industries that often exacerbate political and social instability and lead to the marginalization of certain regions and groups.

Even more, in the Russian Federation, the largest and wealthiest of the NIS in terms of resources and population, the administrations of Presidents Putin and Dmitry Medvedev weakened considerably the power of national environmental protection agencies in the name of economic growth; have prioritized the development of oil, gas, and mineral resources; and xenophobically see the specter of foreign influence in every NGO. They have cracked down on the rights of NGOs; raiding their officials, confiscating their computers, and closing some of them outright. Oil and gas across the entire Arctic region have become increasingly important to Russia's chosen path of remaining a superpower on the basis of resource exploitation. Yet, NGOs seem powerless in the battle to get the Russian government to adopt sustainable approaches. For example, activists have struggled to ensure that a new Sakhalin energy complex will develop with minimal impacts on gray whales and their habitat.[42]

In some areas, environmental degradation has become the main factor of poor and declining human well-being. In these cases, environmental issues are becoming more difficult to disentangle from economic and social ones. Finding the resources needed to transform such environmental disaster zones and support sustainable development of communities living there may require generations. Sustainable development of resources remains a paramount problem not only within the NIS, and not only for such traditional natural resources as oil and other minerals, timber, or fisheries, but also for freshwater (in Kyrgyzstan), for potential carbon sinks (the Russian taiga and tundra), and so on, all of which become increasingly important on the global scale. Wise management of such resources may enhance the prospects of sustainable human development of the NIS. The responsibility of ensuring such management lies with environmental institutions that have also undergone profound transformation during the 1990s, and whose future policies, prospects, and strengths and weaknesses remain unclear.

In the mid-1990s, many NIS environmental agencies did little more than create regulations that increased red tape. Ten years later, they are gradually converging toward conformity with international standards and serving the goals of sustainable development. This achievement alone was insufficient to ensure that environmental issues remained effectively integrated into development policies. Such contemporary and proven approaches to environmental regulations as semivoluntary agreements, environmental product policies, strategic environmental assessments, and so on are largely unknown in the NIS. These lacunae jeopardize the prospects for sustainable development. What can the citizens of NIS expect? Will they galvanize to force their governments to promote policies to protect the environment even in times of economic uncertainty? Will governments begin to address at long last the threats of ecocide? Given the fact that environmental issues are global ones that nearly always transcend national borders – especially fifteen new national borders – we must hope that a repeat of the failed environmental policies of the Soviet past has no opportunity to reappear.

---

[42] Jessica Graybill, "Places and Identities on Sakhalin Island: Situating the Emerging Movements for 'Sustainable Sakhalin'," in J. Agyeman, E. Ogneva-Himmelberger, eds., *Environmental Justice of the Former Soviet Union* (Cambridge: MIT Press, 2009).

Environmental Concerns in the Twenty-First Century

The political and economic changes in the former Soviet Union during the 1990s and 2000s did not significantly alter systemic causes of environmental problems, although the countries all made some progress in increasing public participation, international cooperation, and the integration of environmental concerns into decision making. Given administrative reforms and decentralization, social mobilization, and reduced subsidies to military and other heavy industries, the successor states were more prepared than the Soviet Union to accept international assistance, to learn from the experience of developed and developing countries, and to adhere to international environmental agreements. Many of these states, particularly Moldova, Russia, and Ukraine, sought to align their environmental legislation with that of the EU.

Yet, a series of environmental threats associated with the transition period to new regimes included the shift of decision makers in their foci from environmental issues to the resolution of immediate economic and social problems; dependence on external markets and a drive to overexploit natural resources; and political and social instability.

Developments in the region during the 2000s clearly indicate that the environment is far from being a priority for most governments. On the one hand, on several occasions former President Vladimir Putin of Russia (2000–2008) observed that environmental considerations were blocking attractive investment opportunities. He stated that environmental considerations had to be respected unless they were used as a means for economic competition. This explains several actions taken by his government, for example, the closing down of the Ministry of Environment and the transfer of its functions to the Ministry of Natural Resources and to the oblasts in 2002, and the disbandment of the Russian Forest Service, which had operated continuously since the century, in 2007. On the other hand, Russia, after long deliberation, ratified the Kyoto Protocol, thus bringing it into force. Most likely, this was done under pressure from Russia's main trading partner, the EU.

Diversification of the economy and a shift toward energy- and resource efficiency should have been much easier to achieve during the years of unprecedentedly high oil and gas prices in 2003–2008, when Russia was flooded with "petrodollars." But the government took no significant steps along this route. The situation in many of the former Soviet republics was even worse, because many of these states do not have oil revenue to invest in new technologies and will thus be hit much harder by the crisis.

Conclusion 319

As the world economic crisis deepens and continues to affect post-Soviet economies, it is becoming clear that decisions that were made in the beginning of the decade to limit the power of nature protection agencies in order to boost economic growth have backfired. The lack of a state forest service in Russia after the adoption of the new Forestry Code in 2007 and the transfer of functions of that service to local authorities meant that maintenance of the forests became a regional responsibility. In some regions, authorities were up to the challenge, but in others they had no resources to maintain sometimes very vast forest areas. As a result, the system of firebreaks deteriorated and the quantity of combustible material grew; in the stiflingly hot summer of 2010, forest fires spread with ferocity.[43] Of course, extreme climate conditions and drought played their part in the disaster, but the absence of routine preventive fire measures contributed significantly.[44]

One of the most characteristic features of the Putin era is the clear trend of the state to put many of its previous functions on the shoulders of local authorities and citizens. Under the slogans of liberalization and privatization, the state is abolishing some of its vital duties (such as nature protection, nature monitoring, and the forestry service). The consequences of such short-sighted policies have yet to be seen, but the apocalyptic images of smoke in Moscow and forest fires in Russia may become too common a sight in the future.

---

[43] http://www.bbc.co.uk/news/world-europe-10815176.
[44] http://www.bellona.org/articles/articles_2010/repeat_fires; and http://af.reuters.com/article/worldNews/idAFTRE6723BL20100803.

Index

A Little Corner of Freedom, 112
Aarhus Convention, the, 299
Abkhazia, 158, 293
acid rain, 196, 219, 220, 251
acute radiation sickness, 267
Afghanistan, 1, 20
　Soviet invasion of, 195
agriculture, 7, 23, 27, 41, 43, 45, 47, 55, 73, 78, 86, 94, 95, 97, 98, 102, 119, 120, 124, 127, 132, 137, 146, 147, 148, 164, 180, 185, 230, 250, 256, 288
　and famine, 26
　and forests, 18
　and irrigation, 18, 98, 129
　chemical inputs and, 38, 184
　chemicals inputs and, 309
　collectivization of, 5, 12
　ecological change, 250
　ecological change in, 49, 58
　improvement of, 4, 24, 25, 33
　in the taiga, 17
　modernization of, 23
　pollution, 223, 241
　slash and burn, 24, 38, 43
　Soviet modernization of, 2
　subsistence, 24
　subsistence farming, 29
agronomy, 5, 23, 24, 26, 33, 44, 45, 46, 97, 148
air pollution, 85, 115, 177, 178, 179, 184, 185, 189, 214, 215, 218, 219, 221, 269, 280, 308, 309

air quality standards, 198
　control of, 190
　from industry, 178
Alaska, 195
Aldan River, 101
All-Russian Society for the Promotion and Protection of Urban Green Planting, the, 93
All-Russian Society for the Protection of Nature (VOOP), 65, 68, 108, 110, 111, 112, 128, 138, 172, 173, 174, 218, 244, 246
All-Ukrainian Zoological Congress, 107
All-Union Academy of Agricultural Sciences (VASKhNiL), 45, 107, 124
All-Union Conference on Air Pollution Control, the, 91
All-Union Institute of Community Sanitation and Public Health of the People's Commissariat of Health, the, 91
Almaty, 215
　air pollution, 308
Alpatov, V. V., 107
Altai Mountains, the, 158, 159, 175
　forest exploitation, 158, 159
　logging, 159, 238
　nature preserve, 175, 176
　timber, 156
Amtorg Trading Corporation, the, 95
Amu Darya River, the, 19, 20, 81, 98, 99, 120, 232, 233
　hydroelectricity, 164

Amur River, the, 1, 101, 129
Anadyr River, the, 105
Andrei Pervozvannyi, the, 53
Andropov, Yuri, 201, 203, 254
Angara River, the, 1, 14, 128, 129, 131, 226
  hydroelectricity, 167
Angara-Enisei Industrial Belt, the, 186
Angarastroi, 131
animal husbandry, 26, 32
Apatity, 105
Aral Sea, the, 19, 20, 81, 98, 99, 224, 232
  and river diversion, 232
  environmental degradation, 20, 81, 172
Arctic basin, the, 190
Arctic Circle, the, 1, 14, 130
  pollution in, 299
Arctic explorers, Russian, 16
Arkhangelsk, 17, 30, 90, 105, 124
  forest industry, 35, 160
Armenia, 205, 212, 282, 284, 285, 286, 288, 293, 299, 308, 309, 314
  degradation in, 309, 310
  environmental movements, 282
  Ministry of Nature Protection, 312
  nuclear power, 292
  protection of biodiversity and, 312
Armenia SSR Red Book, the, 312
Army Corps of Engineers, the, 8, 119, 131, 167
Arseniev, Vladimir, 56, 57
Asbest, 73
asbestos, 75, 178
Ashgabad
  air pollution, 308
Askania-Nova zapovednik, 56, 66, 174
Association of Russian Naturalists and Physicians, the, 61
Astrakhan
  fisheries, 230
Atommash, 268
autarky, 5, 13, 141, 209, 212
automobiles, 86, 308
  and pollution, 214, 307, 308
Azerbaijan, 212, 230, 282, 285, 293, 299, 305, 306, 309, 314
Azov Sea, the, 63, 76, 80, 81, 170, 171, 190, 224, 233, 234, 296

  fisheries, 82, 171, 180, 224, 233, 234, 311
  pollution abatement, 192
Azovo-Sivashsky nature reserve, the, 174

Baikal Pulp and Paper Combine, the, 227, 229
Baikal-Amur Labor Camp, the, 101
Baikal-Amur Magistral Railroad (BAM), the, 78, 101, 186, 235, 236
Balkash River, the, 19
Balkhash, 79
Baltic Sea, the, 190, 306
  pollution abatement, 192
Barents Sea, the, 51, 53
Bashkir Province, 280
  nature preserves, 175
Basic Law on Forests, the, 121
Batumi, 53, 223
Bay of Baku, the, 230
beekeeping, 27, 29
Belarus, 248, 258, 265, 266, 267, 282, 288, 293, 299, 304, 305, 309. *See also* Chernobyl
  authoritarianism, 288
  Belarus Popular Front, the, 282
  environmental destruction, 265
  evacuations, 262
  fisheries, 50
  forests, destruction of, 311
  Green Party, 303
  nature conservation, 174
  radiation, 260, 263, 265
  Socio-Ecological Union of, 282
Belgium, 126, 239
Belgorod, 28
Belomor-Baltic Canal, the
  and forced labor, 102
Belovezhsky nature reserve, the, 174
Bender, Ostap, 76
Berdiaev, Nikolai, 56
Berezniki, 103
Beria, Lavrenty, 136
Berzin, E. P., 103
biodiversity, 9, 10, 14, 49, 56, 80, 83, 115, 138, 139, 154, 302, 311, 312
biologists, 3, 5, 47, 57, 67, 80, 126, 171, 173, 175
  and environmentalism, 45, 58
biosphere, 67
  concept of, 243

Bishkek
  air pollution, 308
Black Earth Region, the, 58, 119, 124
Black Hundred group, the, 59
Black Sea, the, 14, 59, 63, 80, 81, 202, 234
  fisheries collapse, 82
  pollution, 76, 171, 192, 223, 235
  salinization, 234
Bodaybo River, the
  gold mining, 101
Bolshevik period, the
  and environmental thought, 66
Bolsheviks, the, 2, 9, 26, 61, 62, 63, 71, 77, 119, 121
  agricultural policies, 61
  and nature control, 125, 289
  and science, 67
Bolshevik Revolution, 26, 63
Bondarev, Yuri, 219
Bonhomme, Brian, 122
Bonneville Power Administration, the, 8, 80
boreal forest, 1
Boreiko, Vladimir, 303
botanical gardens, 53, 310
Botanical Journal, 125
Boulogne, 114
Brain, Stephen, 12, 123, 124
Bratsk Hydroelectric Power Station, the, 167, 170
Bratsk reservoir, the, 187
Bratskgesstroi, 131
Brazil, 24, 80, 130
  income inequality, 293
Brezhnev Constitution, the, 186, 198
Brezhnev, Leonid, 11, 101, 146, 152, 162, 184, 185, 188, 195, 198, 201, 226, 235, 236, 237, 240, 246, 247, 248, 249, 250, 251, 254, 257, 291
  and environmental issues, 184, 185, 189, 196, 197, 201, 222
  conservation efforts, 244
  death of, 203
  environmental issues, 239, 240, 242, 243, 291, 301
  Food Program, 185, 209, 210, 211, 212, 233
  human rights, 192, 197
  irrigation projects, 309

Bulgakov, S. N., 56
Bulgaria, 221
Bureau of Reclamation, the, 8, 119, 167
Buriatia, 238
Bush, George W., 297
  policy failures, 301
Butinge Oil Terminal, the, 305

Canada, 81, 152, 166, 199, 220, 227
  Kyoto rejection, 301
canals, 19, 129, 144, 163, 164, 187, 201, 250, 256, 275, 291
Carpathian Mountains, the, 14
  forests, 156, 157, 158, 159, 179, 180, 237
Carson, Rachel, 179, 191, 226, 291
Caspian Sea, the, 18, 63, 64, 81, 82, 164, 171, 190, 222, 224, 229, 230, 231, 233, 296. *See also* caviar
  degradation, 179, 230
  fisheries, 52, 171
  hydroelectricity, costs of, 170
  poaching, 311
  pollution, 222
Catherine II, 31, 33, 40
Catherine the Great. *See* Catherine II
Caucasus, the, 14, 47, 54, 65, 113, 126, 155, 158, 293, 306, 311
  environmental issues, 151, 156, 179, 237, 309, 310
caviar, 63, 171, 230, 311
censorship, 142, 145, 181
  and environmental issues, 142
  of environmental information, 181
Center for Environmental Policy, the, 303
Center for Wild Nature Protection, the, 302
Central Asia, 14, 98, 166, 187, 256
  and agriculture, 19
  and hydroelectric power, 186
  climate of, 19
  collectivization of agriculture, 79
  environmental activism, 304
  ethnic conflicts, 293
  geoengineering, 172
  industrial production, 116
  irrigation, 71, 137, 309
  irrigation systems, 289
  mountains of, 14
  population growth, 188

Central Asia (*cont.*)
  reclamation projects, 232
  river basins, 19
  river diversion, 249, 275
  river drainage in, 19
Central Asian Water Trust, the, 96, 97
Central Black Earth region, the, 42, 43, 154
  drought in, 46, 151
Central Bureau for the study of Local Lore (TsBK), the, 65, 68, 108
Chechens, 149
Chechnya
  ethnic conflicts, 293
  war in, 311
Chekhov, Anton, 39
Cheliabinsk, 142
  industrial pollution, 216
  reforestation programs, 238
Cheliabinsk Tractor Plant, the, 95
Chemical-Bacteriological Institute, the, 90
Cherepovetsk
  pollution, 199
Chernenko, Konstantin, 201, 254
Chernobyl, 80, 204, 212, 251, 254, 255, 257, 258, 259, 260, 261, 262, 263, 264, 265, 266, 267, 268, 271, 279, 281, 282, 283, 284, 287, 291, 311, 313, 315
China, 1, 3, 19, 123, 130, 131
  electrification projects, 163
  Kyoto Protocol, rejection of, 302
  pollution, 302
Chisinau
  air pollution, 308
cholera
  and famine, 48
Chou River, the, 19
Churchill Falls Hydroelectric Power Station, the, 166
Cikhote-Alin' Mountains, the, 156
cities, 13, 73
  environmental effects, 32, 88
  socialist vs. capitalist, 87, 92
civic clubs, 277
civic culture, 3, 4, 93
civic groups, 247
civil war (1918–21), 26, 33, 60, 62, 63, 64, 77, 121, 288, 303

climate, 4, 14, 24, 28, 29, 46, 49, 88, 120, 138, 143, 149, 154, 243
  and agriculture, 105
  and deforestation, 39
  continental, 14
climate change, 39, 172, 196, 283, 301, 319
  19th century views on, 39
  and agriculture, 126
  and reforestation, 123
Club of Rome, the, 242
Club of Young Biologists, the, 245
coal, 27, 28, 73, 75, 76, 83, 89, 104, 105, 113, 116, 129, 130, 136, 165, 186, 216, 235, 250
  and pollution, 69, 241
Cold War, the, 72, 74, 184, 240, 290
  environmental costs, 290
collectivization, 4, 5, 12, 48, 57, 68, 71, 72, 74, 75, 77, 78, 79, 94, 95, 97, 98, 99, 100, 109, 110, 118, 126, 150, 209, 289
Columbia River, the, 80
Commissariat of Agriculture, the, 62
Commissariat of Enlightenment, the, 62
Commission on Conservation and Rational Use of Natural Resources, the, 206
Commission on Ecological Security, the, 296
Commission on Nature Reserves, the, 176
Commission on the Arctic and Antarctic Affairs, the, 295
Commission on the Environment and Exploitation of Nature, the, 296
Committee for the Study of the Productive Forces (KEPS), the, 55, 59, 60, 129
  governmental support, 60
Committee for Water Management, the, 295
Committee on Nature Protection, the, 308
Communist Party of the Soviet Union, the, 3, 72, 94, 95, 99, 129, 136, 139, 147, 148, 201, 239, 244, 275, 290, 291
  and natural resource use, 74, 119
  Central Committee of, 75, 77, 93, 148, 171, 175, 176, 190, 206, 208, 211, 229, 254, 267, 272
  and natural resource use, 209
  Food Program, 209
  under perestroika, 275

Communist Youth League, the, 108
Congress of Peoples' Deputies, the, 254
Congress of Soviet Writers, the, 219, 225
conservation, 1, 3, 5, 11, 12, 25, 34, 36, 37, 38, 48, 51, 61, 62, 64, 65, 66, 68, 103, 107, 110, 121, 122, 123, 157, 159, 162, 180, 190, 192, 195, 206, 213, 232, 277
  attacks on, 110
  definition of, 38
  under Khrushchev, 155
  under the Bolshevik regime, 26
conservation groups and movements, 23, 24, 58, 106, 244
consumption, 214, 279, 310
  and environmental ethics, 279
contruction
  and environmental degradation, 78
Convention on International Trade in Endangered Species (CITES), the, 193, 311
Convention on Transboundary Air Pollution, the, 220
Copet-Dag Mountains, the, 233
corn campaign, the, 149, 153, 154, 180
  environmental consequences, 154
cotton, 186
  environmental impacts of, 98, 233
  irrigation and, 98
Council for the Study of the Productive Forces (SOPS), the, 60
Council of Ministers, the, 91, 147, 171, 177, 190, 206, 229, 234, 272
  pollution control, 217
Council on Environmental Science and Technology, the, 269, 270
Crimea, 54
  illegal logging, 310
Crimean nature reserve, the, 174
Crimean War, the, 59
crop failure, 26, 43, 119, 150, 151
Cuba, 239
cybernetics, 143, 242
Czechoslovakia
  air pollution, 221

Dagava River, the
  hydroelectric station, 280
Dalstroi organization, the, 101

dams, 8, 25, 81, 115, 222, 231. *See also* hydroelectricity
Danilov-Danilyan, Viktor I., 274, 295, 300
Danube River, the, 40, 81, 82, 210, 222
Daugava River, the, 282
Dawson, Jane, 281
DDT, 211
Debeda River, the, 223
deep ecologists, 304
deforestation, 191, 310
  acceleration of, 184
  among peasants, 46
  consequences of, 106
de-kulakization. *See* collectivization
deltas, 16, 19, 105, 144
  and salinization, 81
Denmark, 52, 54
Department of Fisheries, the, 295
Dersu Uzala, 57
deserts, 8, 14, 15, 18, 19, 78
  desertification, 162, 309
  irrigation systems, 164
de-Stalinization, 126, 141, 149
  policies, 136
Dickens, Charles, 56
Dirty Thirty, the, 85
diseases, 119, 233, 267. *See also* public health
  acute radiation sickness, 267
  cancer, 85, 233, 264, 267
  cardiovascular, 251
  cholera, 90
  hepatitis, 233
  malaria, 90
  plague, 90
  pneumonia, 119
  respiratory, 85, 309
  scurvy, 79
  smallpox, 79, 90
  typhoid, 90
  typhus, 79
Dneprodzerzhinsk, 217
  pollution, 217
Dneprodzerzhinsk GES, the. *See also* hydroelectricity
Dnieper River, the, 1, 71, 80, 81, 82, 120, 164, 169, 210, 217, 222
  pollution, 223

Dnieprostroi Hydroelectric Power Station, the, 76, 77, 79, 83, 113
Dniester River, the, 81, 82, 202, 210, 222
　industrial accidents, 201
　Novo-Dnister dam, 202
　pollution and, 202
Doctor Zhivago, 142
Dokuchaev, Vasily Vasilievich, 36, 46, 47, 48, 49, 51, 57, 123
　expeditions of, 54
dolphins, hunting of, 194
Don River, the, 1, 14, 71, 119, 120, 121, 163, 172, 222
　damming of, 233
　discharge, 81
　pollution, 177
　water use, 233
Donets River basin, the
　and coal production, 28
Dostoevskii, Fyodor M., 56
Dresden, 114
　bombing of, 114
drought, 15, 49, 97, 119, 180, 319
　and famine, 15, 26
　in 1946, 15
　of 1873, 48
druzhina, 244, 245, 246, 303
　activists, 303
Du Pont, 96
Dudinka, 105
Duma, 55, 59
Dushanbe, 166
　air pollution, 308
Dvina River, the, 30

earthquakes, 236, 292
　Spitak, 282, 284, 285
Eastern Europe, 3, 148, 241, 312, 314, 315
　automobiles, 308
　emissions limits, 298
ecoanarchism, 277
ecocentrism, 57
ecocide, 12, 291, 317
Ecocide in the USSR, 257
ecology, 1, 2, 4, 5, 6, 8, 12, 32, 45, 53, 64, 66, 67, 68, 74, 107, 109, 110, 143, 173, 174, 185, 192, 247, 288
econationalism, 14, 277, 278, 281, 282, 283, 286, 292, 318

economic development, 2, 3, 5, 56, 76, 111, 145, 180, 191, 226, 240, 243, 251, 275, 290
　and environmental awareness, 250
　and environmental degradation, 144
　costs, 307
　costs of, 73, 284, 291
　criticisms, 255
　environmental consequences of, 184
　impacts of, 179
　published statistics, 137
　Soviet, 190
　under Khrushchev, 139
　under the Bolshevik regime, 26
economic growth, 12, 68, 139, 140, 156, 184, 226, 250, 283, 293, 297, 316
　Communist predictions, 75
　nature conservation and, 109, 319
　rapid plans for, 5
　Soviet, 288
　sustainability, 313
　under socialism, 250
　under Stalin, 77
economic planners, 13, 24
economic policy
　and famine, 26
economic reforms, 136, 137
economics, 2, 45, 122
　agricultural, 2
economy, 2, 6, 42, 62, 123, 139, 254, 290
　agrarian, 11
　and fossil fuels, 187
　and propaganda, 86
　and Siberia, 24
　changes in, 11, 96
　collapse of, 255
　energy intensity of, 314
　investment, 165
　management of, 188
　of indigenous people, 16
　planned, 13, 121
　planned projects, 137
　problems with, 201
　restructuring of, 13
　Stalinist, 99
　transformation of, 76
　under Stalin, 77
Ecopolis program, the, 246
ecopoliticians, 277, 278, 303

ecosystems, 6, 10, 12, 14, 24, 119, 131, 144, 188, 210, 264
  and dams, 81
  and pollution, 256
  and settlement, 40
  change of, 12
  conservation of, 190, 193
  destruction of, 106, 113, 164, 234
  food webs, 228
  improved health of, 310
  industrialization of, 282
  management of, 75
  transformation of, 88, 163
  understanding of, 35
  wartime destruction, 114
ecotechnocrats, 277, 278, 304
Ehrenburg, Ilya, 136
electricity, 113, 115, 216
  in rural areas, 164
emissions regulations, 85, 92
endangered species, 193, 311
energy production, 130, 138, 140, 180, 206
Engel'gardt, Alexander, 54
Engels, Friedrich, 5, 240
engineering, 76
engineers, 12, 24, 79, 80, 84, 95, 109, 128, 130, 131, 138, 141, 166, 168, 172, 185, 202, 222, 233, 239, 270, 291
  and environmental issues, 47, 71, 138, 164, 180, 181, 231, 289
  and urban planning, 89
  river diversion efforts, 249
  views on hydroelectric power, 170
England. *See* United Kingdom, the
Enisei River, the, 14, 100, 101, 105, 129, 222
  hydroelectricity, 129, 168
Enlightenment thought, 33, 74
environmental charges, 297, 298
environmental costs and degradation, 3, 6, 39, 48, 54, 60, 72, 73, 78, 81, 86, 97, 101, 112, 127, 132, 133, 138, 139, 145, 168, 180, 184, 185, 187, 189, 191, 192, 199, 201, 203, 207, 214, 236, 247, 250, 251, 257, 269, 279, 283, 287, 289, 294, 307, 309, 318
  and central planning, 138
  and industrialization, 68
  ethnic conflicts and, 293
  of collectivization, 48
  of the gulag, 99
  public discussion of, 197
Environmental Impact Statements (EIS), 163, 269, 270, 271
  expert assessments, 297, 298
environmental movements, 4, 12, 13, 93, 185, 256, 291, 292, 305
  and state oppression, 111
  during perestroika, 274
  during WWII, 111
  post-Soviet, 304
  role of literature and, 249
environmental planning, 88
Environmental Protection Agency, disbandment of, 292, 318
environmental protection and policy, 11, 13, 201, 213, 272
  incentives for, 207
Environmental Protection Law, the, 298
environmental restoration
  under the Bolshevik regime, 67
environmentalism, 3, 45, 56, 143, 245, 251, 291, 304, 306. *See also* conservation
  and social activism, 240
  under Stalin, 12
  under the Bolshevik regime, 26
erosion, 49, 73, 120, 152, 179, 180, 209, 212, 234, 291, 309
  acceleration of, 184
  and logging, 237
  prevention of, 190, 310
Estonia, 175, 189, 255, 313
  econationalism, 281
  environmental legislation and perestroika, 189, 281
  environmentalism, 279
  Estonian Popular Front, the, 281, 282
  Congress of, 281
Eurasia Foundation, the, 305
Europe, 1, 2, 4, 12, 15, 16, 31, 44, 54, 80, 81, 108, 113, 114, 132, 144, 215, 234, 258, 279
  air pollution, 214, 219, 220
  automobiles, 308
  environmental issues, 191, 269, 306
  impact of Chernobyl, 259

European Union, 299, 301
　Chernobyl aid, 287
　environmental concerns, 39, 191
　membership conditions, 284
expeditions, 32, 53, 55

F. F. Erisman Institute of Hygiene, the, 90
factories
　pollution, 56, 68, 220
　zoning of, 216
famine, 48, 63, 74, 97, 115, 118, 123, 132, 153, 180, 288
　in 1946, 136
　in Kazakstan, 79
　in Ukraine, 79
　of 1891–92, 11, 26, 38, 48, 49
Far East, the, 4, 105, 111, 188, 237, 238
　and hydroelectric power, 186
Far North, the, 4, 73, 118, 128, 188
Farewell to Matyora, 249
farms and farming, 28, 48, 73, 113, 114, 120
　and deforestation, 38
Fedorov, Evgenii, 232
Fedorov, N. F., 56
Fergana Valley, the, 98
fertilizers, 38, 42, 46, 73, 82, 138, 146, 186, 202, 211, 212, 223, 233, 291, 309
　chemical, 146, 208, 209, 211
Feshbach, Murray, 257
Finland
　emissions, 221
First All-Russian Congress for Conservation, the, 109
First General Congress, the, 55
fish and fisheries, 5, 10, 11, 16, 25, 26, 49, 50, 51, 52, 53, 62, 63, 65, 67, 71, 81, 82, 120, 170, 180, 188, 193, 224, 228, 229, 230, 250, 256, 288, 289, 317
　and hydroelectricity, 80
　decline of, 224, 230, 234
　depletion of, 251
　development of, 24
　fish farming, 52
　fishing, 17, 30, 51, 311
　habitat destruction, 56, 76, 80, 82, 163, 199
　in the Russian empire, 50
Fisherman's Society, the, 246

Five Year Plans, 72, 76, 85, 136, 155, 190, 200, 289
　expansion of, 100
flood control projects, 167
forced labor
　environmental impact of, 103
Ford Foundation, the, 305
Forest and Steppe (journal), 125
Forest Club, the, 310
Forest Industry (journal), 125
forestry, 5, 7, 8, 24, 25, 27, 30, 34, 35, 36, 37, 56, 71, 73, 76, 81, 100, 102, 103, 105, 114, 122, 123, 124, 129, 136, 156, 158, 159, 160, 161, 162, 168, 170, 176, 180, 188, 207, 235, 237, 238, 239, 288, 289, 291, 294, 306, 309, 310, 317
　and environmental degradation, 184, 237, 250
　and nature control, 125, 126
　and the gulag, 102, 104
　clear-cutting, 121
　department of, 31, 58
　sustainable yield, 30
　under Peter the Great, 30
　under Stalin, 120
Forestry Committee of the USSR, 162
forests, 8, 9, 10, 11, 17, 25, 26, 27, 29, 31, 35, 36, 37, 41, 65, 66, 100, 144, 156, 159, 237, 250, 288, 289
　conservation of, 25
　degradation of, 23, 35, 49, 69, 113, 114, 123, 156, 157, 159, 161
　forest belts, 120, 123, 124, 125, 126
　forest corps, 31
　forest fires, 43, 142, 319
　nationalization of, 29, 61
　nuclear contamination, 264
　regulation of, 37
　study of, 25
　sustainable, 124, 306
　under peasant management, 37
Foundations of Marxism-Leninism, 143
France, 9, 24, 54, 86, 131, 196, 239
　air pollution, 221
Free Economic Society, the, 33, 34, 47, 58
fur, 10, 27

garbage, 17, 84
　disposal of, 214
Gary, Indiana, 84, 85, 216

genetics, 67, 125, 143, 174, 247, 267
    and Lysenko, 173
Geneva, 221
geoengineering, 8, 9, 71, 119, 165, 172, 289
geography, 2, 4, 51, 54
    of natural landscapes, 14
    of plants, 107
    physical, 14, 51
Georgia, 158, 166, 212, 223, 288, 293, 306, 307, 309, 311
Germany, 9, 24, 34, 51, 54, 63, 86, 116, 117
    conservation efforts, 57
    forestry, 34
    industrialization, 216
    scientific expeditions, 52
    slave labor, 118
Germany, East, 115, 191, 221
Germany, West, 191, 221
Gidroenergoproekt, Central Asian Branch of, 166
Gidromet, 204, 205, 206
Gidroproekt, 89
Giproshakht, 131
glasnost, 254, 255, 257, 258, 267, 274, 276, 291
    impacts of, 292
Glavsevmorput, the, 102
global warming, 312
GOELRO, 60, 79
Gorbachev, Mikhail, 3, 11, 145, 198, 250, 251, 254, 255, 256, 257, 258, 267, 268, 269, 273, 275, 276, 283, 285, 286, 291, 300
    environmental issues, 219, 245, 247, 256, 298
Gorky, Maxim, 100
Goskomgidromet. *See* State Committee on Hydrometeorology
Goskompriroda. *See* State Committee for Nature Protection, the
GOSPLAN, 60, 176, 188, 213, 224, 226, 270, 272
    and environmental protection, 190
Grand Coulee Dam, the, 167
Great Britain, 51, 86, 131, 196, 220, 239
    air pollution, 221
Great Terror, the, 137
Green City, the, 92, 93

Green Party, the, 277, 304, 305
Greenpeace, 302, 305, 310
gulag labor camps, the, 8, 72, 99, 100, 101, 102, 103, 127, 129, 132, 141, 142, 166
    and environmental degradation, 102, 103
    and nature manipulation, 120
Gulf of Finland, the, 32

habitat destruction, 34, 88, 138, 311
hazardous waste, 5, 138, 197, 217
Helsinki Accord, the, 196, 197
herbicides, 38, 126, 211, 223, 274, 291
Hindus, Maurice, 96
Hiroshima, 114, 259, 267
Hungary
    emissions, 221
    Hungarian Revolution, the, 141
    steppe grasslands of, 18
hunting, 10, 17, 30, 36, 37, 41, 61, 174, 192, 194, 245, 246, 311, 312
hydrobiology, 107
hydroelectric power, 2, 5, 8, 63, 79, 80, 82, 120, 126, 128, 130, 163, 164, 165, 167, 168, 169, 171, 186, 231, 249, 256, 280
    and nature transformation, 163
    environmental impact, 168, 169, 170
    in Canada, 166
    in Siberia, 235
    opposition movements, 282
hydrology, 24, 25, 51, 81, 172, 249
Hydrometeorological Agency, the. *See* Gidromet

Ianshin, A. L., 203
Iasnaia Poliana, 29
icebreakers, 239
Igarka, 100
Ignalina Nuclear Plant, 281, 283, 284, 292, 313
Ilf, Ilya, 76
Illarionov, Andrei, 301
Imperial Academy of Sciences, the, 25, 30, 31, 55, 58, 59, 60, 65, 125, 138, 173, 175, 176, 203, 228, 296, 300
    and environmental thought, 58
Imperial Forest Department, the
    scientific expeditions, 47
Imperial Forestry Institute, the, 35

Imperial Moscow Society of Agriculture, the, 34
Imperial Russian Geographical Society, the, 58
Imperial Society of Fishing and Fish Farming, the, 52
India, 1, 19
  Bhopal chemical explosion, 258, 259
indigenous people, 11
  Komi, 16
  modernization of, 118
  Nenets, 16
  Saami, 16
  Yakut, 16
Indirgika River, the, 105
industrial pollution, 75
industrial revolution, the, 43, 56
industry and industrialization, 2, 3, 4, 5, 7, 11, 23, 24, 25, 27, 28, 36, 42, 56, 57, 61, 62, 63, 68, 71, 72, 75, 77, 79, 83, 86, 90, 92, 93, 94, 96, 101, 109, 110, 113, 117, 132, 133, 146, 148, 160, 180, 185, 186, 187, 208, 225, 230, 233, 249, 272, 289, 292
  and the gulag, 102
  environmental impacts, 32, 73, 75, 184, 250
  environmental issues, 89
  in the Arctic, 105
  in urban areas, 218
  military, 74
  nationalization of, 71
  pulp and paper, 36, 229, 238
  under Stalin, 12, 77
infant mortality, 185, 233, 251, 255
Inguri GES, 166
Institute for the Study of Occupational Diseases, 90
Institute of Economic Problems of the Kola Science Center, the, 106
Institute of Global Climate and Ecology, 301
intelligentsia, the, 4, 9, 10, 43, 53, 54, 56, 65, 106, 108, 109, 110, 144, 174, 240, 245, 247, 277
  environmentalism among, 181, 302
Intergovernmental Panel and Climate Change, the, 205
International Atomic Energy Agency (IAEA), the, 260, 266

International Institute of Applied System Analysis (IIASA), the, 242
International Union for Conservation of Nature and Natural Resources, the, 192, 305
International Whaling Commission, the, 142, 194
Ipatieff, Vladimir, 59, 60
Iran, 1, 230, 306
Irkutsk, 129, 227
Irkutsk Hydropower Station, the, 226
iron, 73, 76, 113, 116, 129, 144, 186
irrigation, 18, 19, 72, 78, 81, 96, 120, 138, 146, 164, 172, 212, 232, 233, 249, 289
  and land reclamation, 210
  and tractor industry, 95
  environmental impacts, 167, 171, 223
  irrigation projects, 25, 49, 98, 137, 163, 309
Irtysh River, the, 1, 14, 129, 158, 177
Israel
  water use, 232
Italy, 221, 283
Izmailovo Forest Park, the, 28
Izrael, Yuri, 205, 301

Japan, 1, 33, 114, 207, 239, 308
  air pollution, 214
  and whaling, 194
Jews, 55, 117, 248
Joint Committee on Environment Protection, the, 195
journalists, 10, 158, 168, 248, 277, 280
  censorship of, 249
  environmentalism among, 248

Kaliningrad
  environmental protests, 306
Kama River, the, 103, 160, 172, 177
Kamennaia Steppe, the
  and soil research, 47
Kansk-Achinsk combine, the, 186, 206, 220
Kara Kum desert, the, 164
Kara-Bogaz Gulf Bay, the, 231
Karaganda Coal Basin, the, 79
Karakalpak Autonomous Republic, the
  infant mortality, 233
Kara-Kum Canal, the, 233

Karelia, region of, 17, 104
  reforestation programs, 238
Kazakhstan, 19, 79, 81, 119, 126, 128, 130, 131, 146, 148, 149, 150, 152, 180, 221, 232, 283, 293, 299, 303, 305, 306
  air pollution, 308, 314
  climate, 19
  emissions controls, 215
  famine, 74, 79
  nuclear testing, 142, 283
  Virgin Lands Campaign, 151, 309
Kazan, 58, 79, 279, 280
  civic clubs, 246
Kemerovo
  air pollution, 219
  reforestation programs, 238
KGB, the, 141, 201, 203
Kharkiv, 43, 58, 91
Kharkiv Tractor Plant, 95
Khmelnitsky Nuclear Power Station, 284
Khorezm oasis, the, 19
Khrushchev, Nikita, 11, 129, 136, 137, 138, 139, 140, 141, 143, 146, 147, 148, 151, 152, 154, 159, 163, 164, 165, 168, 172, 173, 174, 176, 179, 180, 184, 208, 212, 226, 243, 247, 257, 279, 290, 309. *See also* Virgin Lands Campaign
  agriculture and, 146, 147, 149, 151, 153, 154, 161, 211
  and conservation, 155, 174
  and crop failure, 151
  and Lysenkoism, 174
  corn campaign, 153, 154
  environmental degradation, 176
  environmental policies, 132, 143, 144, 158, 165, 173, 174, 175, 176, 179, 184, 208
  forest management, 155, 156, 159
  hydroelectric projects, 165, 167, 235
  nature preserves, dissolution of, 144, 176, 189
  opposition to hydroelectricity, 166, 168
  peaceful coexistence, 141, 144
  removal of, 141, 152, 162, 184
  Secret Speech, 290
  Thaw period, 136, 137, 139, 141, 180, 185, 245, 249, 290
Kiev, 32, 55, 77, 260, 261, 299, 309
  air pollution, 308
  hydroelectric power, 169
  war, 114
Kiev Ecological-Cultural Center, the, 303
Kirovsk, 105
Kishinev, 202
Knipovich, Nikolai Mikhailovich, 51, 52, 53, 54, 57, 62, 63, 64
Kola Peninsula, the, 118
Kolva River, the, 103
Kolyma, 101, 103
Kolyma River, the, 105
Kolyma Tales, 101
Komarov Botanical Institute, the, 193
Komarov, Boris, 193, 200, 229
Komi Republic, the, 101, 172
Komi, the, 118
Komsomol, the, 101
  and railroad construction, 236
Kondo-Sosvinsk. *See* zapovedniks
Kosygin, Alexei, 242
Kotlas–Vorkuta Railroad, the, 101
Kozhevnikov, Grigorii, 57, 61
kraevedenie, 45, 245, 248
Krzhizhanovskii Energy Institute, the, 129
Krzhizhanovskii, Gleb, 92
Kuban River, the, 233
Kuibyshev, 164
Kuibyshev Hydroelectric Power Station (GES), the, 120, 164, 165
kulaks, 72, 73, 97, 100, 105, 176
Kura River, the, 223
  pollution, 223
Kurchatov, Igor, 142
Kurosawa, Akira (filmmaker), 57
Kursk Province, 18
  forests of, 37
Kursk tank battle, the, 116
  and environmental destruction, 116
Kuzbass Region, the, 129
Kuznetsk coal basin, the, 130
Kyoto Protocol, the, 283, 300, 301
  Russian ratification of, 300, 301, 318
  shortcomings, 301
Kyrgyzstan, 19, 20, 212, 293, 312, 314, 317
Kyshtym, 132, 142, 268

Lake Baikal, 129, 138, 144, 160, 180, 224, 225, 226, 227, 235, 241, 244, 249, 256
  and paper mills, 225
  conservation efforts, 290
  environmental decrees, 227
  industrialization of, 228, 229, 249
  pollution, 138, 160, 173, 191, 227, 228
  restoration of, 190, 229
Lake Balkash, 19
Lake Lagoda, 227
Lapland
  nature preserve, 175
Latvia, 255, 280, 282, 313
  protests, 281
Lazovsky
  nature preserve, 175
Le Corbusier, 92
Le Havre, 114
lead, 15, 44, 49, 82, 179, 263
Lena River, the, 1, 14, 101, 105, 235
Lengidroenergoproekt Engineering Institute, the, 129
Lenin, Vladimir, 3, 5, 11, 26, 61, 63, 64, 71, 72, 77, 90, 92, 94, 106, 109, 127, 162, 211, 289
  and nature conservation, 65
  collectivization, views on, 94
Leningrad, 17, 64, 77, 89, 104, 113, 114, 214, 227, 247, 280
Leonov, Leonid, 155, 249, 250
Leopold, Aldo, 226
Limits to Growth, 242
Lithuania, 255, 282, 284, 305, 313
  independence movement, 279, 283
  nuclear power, 281, 292
livestock, 146
  industrialization of, 208
  killing of, 97
  production of, 208
  slaughter of, 74
logging. *See* foresty
Lovozero, 105, 118
Lunacharskii, Anatolii, 92
Lysenko, Trofim, 124, 125, 126, 137, 173, 174, 230, 291
  and forestry, 125, 126
Lysenkoism, 124, 173, 247

Machine Tractor Stations, 95, 96, 127, 209
Magadan, 101
Magnitogorsk, 77, 84, 85, 235
Main Hydrological Construction Agency, the, 166
Main Scientific Administration (Glavnauka) of Narkompros, the, 64
Main Turkmen canal, the, 164, 165
Malenkov, Georgy, 136, 147
Mangyshlak Peninsula, the, 229
Marsh, George, 54
Marx, Karl, 5, 240
Marxism, 4, 5, 68, 75, 128, 240
  and living standards, 290
  labor theory of value, 207
Materialism and Empiriocriticism, 127
Mechnikov, L. I., 54
media, 192, 225, 249, 251, 279
  environmentalism and, 249, 307
  state control of, 13, 181
medicine, 30, 72, 136, 266
Medvedev, Dmitry, 111, 316
Mendeleev, Dmitri, 47, 54
metallurgy, 16, 27, 29, 31, 32, 68, 71, 76, 83, 86, 129, 133, 137, 138, 167, 188, 234, 256, 289, 291, 294
  and forced labor, 100
  in the arctic, 105
  pollution, 75
Metsamor Nuclear Plant, the, 282
Michurin, Ivan, 125
mining, 8, 29, 31, 32, 56, 58, 71, 73, 76, 101, 105, 115, 118, 133, 136, 138, 208, 250, 256, 289, 293
  and environmental degradation, 186
  and prison labor, 83, 100
  pollution, 75
Ministry of Atomic Power (Rosatom), the, 268. *See also* nuclear power
Ministry of Defense, the, 227
Ministry of Education, the, 59
Ministry of Electrification, the, 8, 129
Ministry of Ferrous Metallurgy, the, 199
Ministry of Fisheries, the, 199, 206
Ministry of Health, the, 91, 199, 203
Ministry of Oil and Gas Industry, the, 206
Ministry of the Environment, the, 8, 45, 162, 193, 206, 218, 274, 295, 304, 318
Ministry of Water Management (Minvodkhoz), the, 8, 199, 201, 203, 206, 270
  and environmental damage, 203

Minsk, 248
  Blitzkrieg, 113
Monchegorsk, 105
Moldova, 15, 18, 202, 293, 299, 307, 318
Molotov, Viacheslav, 136
Mongolia, 18, 130, 238
monocultures, 34, 38, 212, 232
Moon, David, 39, 41, 42
Morgun, Fedor, 273, 274
Moscow, 21, 32, 64, 77, 78, 79, 89, 113, 116, 129, 147, 164, 173, 188, 194, 195, 199, 203, 218, 232, 246, 247, 262, 283, 298, 300, 301, 308, 319
  environmentalism, 58, 282
  forestry authority, 157
  pollution, 89, 91, 204, 206, 214, 218
  response to Chernobyl, 282
Moscow Agricultural Institute, the, 45
Moscow Agricultural Society, the, 33
Moscow River, the, 33, 177, 216
Moscow Society for the Admirers of Nature, the, 58, 68, 138, 173, 174
Moscow State University, 33, 55, 57, 59, 141
  environmental movements, 245
Moscow Zoo, the, 245
Moscow-Volga canal, the, 100
Muir, John, 56
Murgab delta, the, 19
Murmansk, 51, 59, 105, 239

Nagasaki, 259, 267
Nagorno-Karabakh
  ethnic conflicts, 293
Narkompros, 62, 65, 66
National Environmental Protection Act, the, 189
national parks
  establishment of, 10
national security, 2, 10, 131
natural gas, 165, 186, 235, 316
natural resources, 1, 3, 9, 11, 24, 25, 48, 53, 59, 60, 64, 65, 67, 71, 105, 109, 110, 115, 140, 188, 207, 213, 246, 272, 289, 293
  conservation of, 157
  exploitation of, 12, 37, 74, 100, 122, 188, 209, 236, 251, 294, 295
  management of, 3, 5, 12, 24, 25, 35, 66, 99, 251, 288
  state ownership of, 3

naturalists, 10, 12, 56, 65, 144. *See also* conservation
  and environmental activisim, 245
nature, 2, 4, 7, 8, 9, 10, 12, 23, 26, 29, 47, 58, 61, 62, 68, 109, 117, 123, 128, 133, 172, 173, 180, 190, 267, 272, 274, 291
  control of, 64, 71, 143, 210, 289
  government policies toward, 8, 10, 12, 24, 30, 33, 57, 60, 65, 107, 109, 197, 205
  human interaction with, 8, 9, 11, 14, 28, 31, 55, 56, 74, 92, 93, 119, 123, 126, 232, 288
  subjugation of, 100
  transformation of, 2, 68, 121, 124, 136, 163
  value of, 67, 109
Nature and the Socialist Economy, the renaming of, 68
nature conservation, 176
  investments in, 200
  laws, 175
  urban, 219
nature monuments, 57
nature preserves, 2, 3, 5, 7, 10, 12, 28, 57, 58, 64, 67, 74, 107, 109, 110, 144, 174, 175, 189, 288, 310, 311. *See also* zapovedniks
  and economic development, 111
  illegal logging, 310
nature protection
  laws, 176
Nazis, the, 85, 104, 113, 117
Nenets Autonomous Okrug. *See* tundra
Nenets, the, 118, 131
Neporozhnyi, Petr, 206
Netherlands, the, 54, 126
Neva River, the, 32, 89, 177, 275
Nevada-Semipalatinsk, 283
New Economic Policy, the, 65, 66, 71, 72, 122, 288, 289
Newly Independent States (NIS), 255, 266, 279, 293, 294, 297, 299, 302, 305, 313, 314, 315, 316
  Earth Summit participation, 299
  environmental issues, 298, 307, 308, 309, 310, 315, 316, 317
  negative decoupling, 315
Nicholas II, 1, 25, 45, 60

Nikel, 101, 105
  sulfur emissions in, 308
Nikitin, Alexander, 299
nitrogen oxides, 217, 219
Nixon, Richard, 195
noosphere, 54, 67, 289
Norgaiskaia Steppe
  and agriculture, 126
Norilsk, 83, 101, 104, 105, 311
North America, 2, 4, 11, 12, 16, 44, 54, 258
  forestry certification, 306
North Crimean Canal, the, 164
Northern Sea Route, the, 86
Norway, 49, 51, 52, 192
Novaia Zemlia, 131
Novocherkassk
  protests, 140
Novo-Dniester dam, the, 202
Novokuznetsk
  steel production, 235
Novosibirsk, 78, 248
nuclear fallout, 188, 263, 265
nuclear power, 131, 141, 185, 239, 260, 268, 282
  pollution, 131
  regulation of, 206
nuclear radiation, 216
nuclear testing, 142
Nurek, 166
Nurek Reservoir, 195

Ob River, the, 1, 14, 105, 128, 129, 187, 222
  hydroelectricity, 129
  pollution, 177
Obninsk, 298
occupational safety, 73, 84
oil, 73, 105, 112, 129, 130, 136, 140, 165, 186, 188, 207, 208, 223, 229, 234, 235, 250, 254, 293, 294, 306, 316, 317, 318
  and environmental degradation, 69
One Day in the Life of Ivan Denisovich, 142
One-Storey America, 76
Ordzhonikidze, Sergo, 87
ore, 10, 28, 73, 74, 75, 83, 105, 113, 186, 223, 250
  and environmental degradation, 69
Orel, 58

overfishing, 63
ozone depletion, 251

Pacific Ocean, the, 1, 14, 130
Palchinsky, Peter, 109
  and employee welfare, 80
Pamir Mountains, the, 14
Pamyat', 248
  goals of', 248
Pasternak, Boris, 142
Pauling, Linus, 241
peasant communes
  dissolution of, 44
peasants, 10, 11, 13, 24, 27, 29, 32, 41, 42, 43, 44, 46, 47, 48, 61, 67, 73, 74, 77, 94
  and agriculture, 42, 46, 48, 54
  relationships to nature, 28
Pechanganikel, 308
Pechora River, the, 16, 104, 172
  degradation of, 172
Pechoro-Ilych. *See* zapovedniks
People's Commissariat of Agriculture, the, 122
People's Commissariat of Internal Affairs (NKVD), the, 100, 101, 104, 166
perestroika, 254, 255, 258, 259, 267, 268, 269, 275, 276, 281, 283, 285, 291, 292
  and environmentalism, 257, 274, 275
permafrost, 16, 17, 236
  and agriculture, 17
  and construction, 195
  and cultivation, 41
  building challenges, 104
pesticides, 211, 223, 233, 234, 274, 291
Peter the Great, 9, 23, 26, 29, 30, 31, 37, 63
Petersburg Medical Institute, 52
Petersburg Society of Naturalists, the, 52
Petersburg University, 52
Pinchot, Gifford, 38, 51
Pinega River, the, 160
Pittsburgh, Pennsylvania, 85, 216
planned economy, 5, 210, 318
poaching, 121, 155, 174, 224, 245, 310, 311
  of timber, 38, 161
Poland, 117, 196, 221, 313
polar bears, 16, 312

pollution, 1, 2, 3, 4, 6, 7, 17, 24, 69, 71, 73, 74, 75, 76, 80, 83, 85, 89, 93, 101, 115, 133, 137, 138, 145, 176, 177, 179, 180, 184, 185, 188, 189, 191, 192, 198, 204, 207, 214, 215, 222, 225, 230, 232, 234, 237, 239, 244, 246, 250, 254, 256, 272, 275, 280, 290, 292, 296, 307, 309, 311, 315
  abatement of, 9, 89, 201, 222, 251
  and chemical weapons, 114
  control of, 91, 190, 200
  in conservation areas, 173
  industrial, 88, 104, 178, 215, 217, 289
  nuclear waste, 142, 188, 261, 264, 269
  regulation of, 9, 11, 12, 91, 187, 199, 203, 205
  under Stalin, 89
  water, 63, 171, 179, 202, 215, 222, 223, 230
Poltava, 47, 54, 62, 273
population growth, 42
  decline of, 188
  pressure on agriculture, 40
  urban, 87
Portugal
  emissions, 221
poverty, 33, 34, 48, 49, 56, 184, 191, 293
preservation, 3, 11, 84, 107, 109, 110. See also conservation
  continuity in, 3
  of forests, 30, 39
  of Siberia, 56
  under the Bolshevik regime, 26
Pripiat, 262
  evacuation of, 262
Pripiat River, damming of, 80
protests, 140, 144, 275, 279, 280, 281, 282, 305, 306
  written, 280
public health, 2, 89, 90, 91, 92, 101, 139, 142, 216, 233, 257, 303
  and industrialization, 56
  and pollution, 178, 256
  problems, 32, 33, 226, 290
Putin, Vladimir, 13, 295, 296, 297, 301, 316
  environmental policies, 292, 318

railroads, 59, 73, 127
Rainbow Keepers, the, 306

Rasputin, Valentin, 249, 250
Red Army, the, 62, 116, 118
Red Book of Endangered Species, the, 312
reforestation, 12, 38, 61, 158, 201, 237, 238, 310
reindeer, 16, 118, 195
  domesticated, 195
  herding of, 118
reservoirs, 207, 222, 223. See also hydroelectric power
  impacts on fisheries, 171
Riazanov, V. A., 92
Richta, Radovan, 240, 241
river diversion, 187, 201, 203, 206, 210, 231, 232, 249, 256, 257, 270, 275
  opposition to, 248
roads
  and industrialization, 73
  construction of, 25, 235
  investment in, 216
Romania, 221
Roosevelt, Theodore, 25
Ruhr region, the, 216
Russia, 15, 17, 18, 23, 29, 31, 37, 47, 50, 61, 90, 97, 101, 102, 123, 127, 166, 188, 262, 263, 266, 283, 284, 293, 295, 305, 307, 308, 313, 314, 315
  deforestation, 56, 311
  environmental movements, 300, 303, 306
  environmental policies, 308, 318
  industrialization, 56, 71
  pollution, 215, 301, 314
Russian Academy of Sciences, the, 301, 304
Russian Far East, the, 57
  and nature preservation, 56
Russian Federation, the, 5, 11, 13, 122, 255, 268, 294, 299, 307
  and natural resource management, 17
  environmental policies, 316, 318
  Environmental Protection Agency, 13, 292
Russian Forest Service, the
  disbandment of, 318
Russian Green Party, the, 277
  beliefs, 277
Russian Physical Chemical Society, the, 58
Russian Revolution, the, 4, 9, 25, 26, 44, 46, 50, 55, 59, 60, 62, 63, 66, 77, 107, 224, 288

sable, 27, 65
    market for, 194
Saino-Shushenskaia GES, the, 168
Sakharov, Andrei, 142, 241
    environmental beliefs, 241
Salekhard, 105
salinization, 81, 190, 233, 309
salmon, 50, 80
    decline of, 224
Samarkand, 19
Sami, the, 118
Sand County Almanac, 226
Saratovsk
    GES, 169
Sarpinskaia Steppe
    and agriculture, 126
Scandinavia, 220
    air pollution, 219
    forestry certification, 306
science, 30, 47, 51
    and agricultural improvement, 67
    and drought prevention, 47
    and human welfare, 64
    attitudes towards, 143
scientists, 2, 4, 6, 10, 11, 12, 13, 24, 25,
    26, 46, 54, 56, 61, 62, 66, 67, 110,
    111, 125, 131, 138, 141, 168, 171,
    173, 174, 178, 184, 190, 239, 242,
    247, 248, 249, 251, 264, 270, 272,
    288, 290, 291
    and conservation, 57, 107, 174, 175,
        219
    and environmentalism, 56, 200, 301,
        303
    and forestry, 122
    environmentalism among, 227
    views on climate change, 40
Scott, John, 84
Sechenov, Ivan, 47
seismology, 59
Selenga Pulp Mill, the, 227
Semenov-Tian-Shanskii, A.P., 61
Semipalatinsk, 131, 283
serfdom, 34, 43, 44
Sergiev-Posad, 32
sewage, 177, 215, 223, 228
Shalamov, Varlam, 101
shamanism, 28, 57, 118
Shelgunov, N. V., 54
Sholokhov, Mikhail, 225
Shvarts, Evgenii, 203, 245

Siberia, 1, 8, 17, 18, 23, 41, 57, 73, 76, 79,
    101, 102, 124, 128, 129, 130, 138,
    143, 144, 148, 152, 166, 195, 222,
    256, 290
    agriculture, 119, 146, 235
    crop failure, 151
    development of, 27, 128, 177, 188, 235,
        291
    forests, 36, 41, 159, 237, 239
    industry, 71, 116, 129, 186, 220,
        235
    railroad construction, 127, 236
    resource development in, 4, 129,
        137
    river diversion, 187, 224
Sierra Club, the, 173
Silent Spring, 179, 191, 226, 291
    censorship of, 179
Simferopol, 58
Slavutich, 263
    birth defects, 263
Slovakia
    energy industry, 315
Slovenia
    energy production, 314
Smolensk, 113
smychka, 78
snowmobiles, 145, 195
socialism, 3, 72, 84, 90, 137, 138, 143,
    144, 148, 213, 250, 254, 278, 288,
    289, 292, 318
    achievements of, 164
    environmental issues, 13, 125, 191, 257,
        279
    role of nature in, 7
Socialist Geography (journal), 125
socialist realism, 76
Socio-Ecological Union, the, 303
soil, 47
    conservation of, 25
    erosion, 38, 155
    science, 46
soldiers, 32, 113, 115, 246, 262
    exposure to radioactivity, 142, 260
Solikamsk, sulphite plant, 103
Solov'ev, V. S., 56
Solovetsk Biological Station, the, 52
Soros, George, 305
South Ukraine Canal, the, 164
Soviet Academy of Sciences, the, 228
Soviet Peace Committee, the, 232, 248

Index   337

Soviet Union, 2, 3, 4, 5, 7, 8, 9, 10, 11, 13,
    14, 15, 17, 26, 69, 72, 73, 74, 76, 77,
    79, 80, 81, 82, 83, 84, 85, 86, 89, 90,
    91, 96, 97, 98, 99, 109, 111, 113,
    114, 119, 127, 129, 130, 131, 132,
    136, 137, 138, 141, 145, 147, 148,
    153, 156, 162, 163, 166, 171, 184,
    185, 187, 191, 196, 211, 213, 215,
    222, 279, 283, 284, 285, 289, 292,
    298, 307, 318
  agriculture, 146, 151, 152, 153, 165,
    169, 212, 232, 309
  air pollution, 221
  and UN membership, 192
  censorship, 302
  cold war, 72, 196
  collapse of, 6, 11, 13, 255, 257, 259,
    274, 288, 292, 293, 294, 309, 316
  conservation, 68, 172, 173, 198, 244
  environmental issues, 111, 122, 142,
    179, 189, 190, 191, 192, 193, 196,
    200, 201, 202, 207, 221, 270, 271,
    277, 297, 299, 300, 306
  forestry, 161, 237, 239
  hydroelectricity, 164, 168
  impact of Chernobyl, 259
  industrial output, 186
  industrialization, 93, 130, 308
  military, 196
  mining, 105, 186
  nuclear power, concerns about, 282
  nuclear testing, 283
  pollution, 199, 220, 221, 223, 230
  pollution standards, 200
  reforms, 291
  United Nations conference participation,
    192
  whaling, 192, 194
Soviet Union, European region
    pollution, 179
sovnarkhozy, 139, 159, 160, 162
space, 185
Sputnik, 141, 164
Sredazgiprovodkhlopok, 131
Sredazvodkhoz. *See* Central Asian Water
    Trust, the
St. Petersburg, 32, 33, 55, 58, 308
St. Petersburg Forestry Institute, the, 35, 37
St. Petersburg University, 47
Stalin, Joseph, 2, 4, 5, 11, 12, 61, 65, 68,
    71, 72, 74, 79, 85, 86, 87, 94, 99,
    106, 110, 118, 121, 128, 131, 132,
    133, 138, 141, 143, 144, 162, 164,
    168, 173, 174, 184, 185, 226, 239,
    243, 256, 289, 290
  agriculture, 18
  Blitzkrieg, response to, 113
  collectivization, 48, 95
  conservation, 110, 123, 124
  Constitution, 198
  death of, 101, 118, 126, 136, 147
  environmental issues, 69, 74, 127, 145,
    198
  Great Break, the, 68, 71, 72, 73, 75,
    289
  industrialization, 68, 77, 122, 132, 186
  nature manipulation, 71, 119, 279
Stalingrad, 114, 120, 164
  Battle of, 116
  GES, 165
Stalingrad Tractor Plant, the, 95
Stalinism, 5, 71, 73, 74, 75, 80, 110, 132,
    136, 289, 290
Stalinist Plan for the Transformation of
    Nature, the, 12, 72, 119, 123, 275,
    289
Stanchinskii, Vladimir, 66, 107
State Committee for Fisheries, the, 171,
    224
State Committee for Hydrometeorology
    and Environmental Regulation, the,
    205
State Committee for Nature Protection
    (Goskompriroda), the, 271, 272, 273,
    274, 294, 295
State Committee for Science and
    Technology, the, 242
State Committee on Environmental
    Protection, the, 209, 295, 297
State Committee on Geology and Mining,
    the, 295
State Committee on Hydrometeorology
    (Goskomgidromet), the, 295
  censorship, 269
State Committee on Statistics
    (Goskomstat), the, 204, 269
State Construction Committee (Gosstroi),
    the, 270
State Electrification Program, the, 92
steppe, 23, 35, 40, 126, 130, 232
  black-earth region, 47
  destruction of, 41, 120

Stockholm, 191, 192, 219
  environmental conferences, 195
Strumilin, Stanislav, 92
sturgeon, 50, 63, 80, 82, 120, 230, 311
  decline of, 171, 222, 224, 231
Sukachev, Vladimir Nikolaevich, 125
sulfur dioxide, 178, 179, 197, 219, 220, 221, 308
sustainability, 34, 123, 139, 317
  threats to, 13
Sverdlovsk, 245
  environmental dangers, 216
Swajian, Leon, 95
Sweden, 196, 227
  acid rain, 196
  air pollution, 221
  and Chernobyl, 259
Syr Darya River, the, 19, 20, 81, 98, 99, 232

Taganrog Bay, the, 234
Tagil, Nizhny, 280
taiga, 1, 8, 17, 102, 105, 130, 186, 235, 236, 237
  as a carbon sink, 317
  destruction of, 106
Taiga Rescue Network, the, 303
Taimyr
  and reindeer, 195
Taishet, 101
Tajikistan, 19, 20, 166, 212, 293, 311
  environmental movements, 303
Taklamakan desert, 19
Tale of Two Cities, 56
Tarr, Joel, 88
Tashkent, 78, 98, 199
Tatarstan, 279, 280
Tbilisi, 223, 282, 308, 309
technocracy, 108, 109, 124
Tennessee Valley Authority, the, 8, 80, 131
Thaw, The, 136
The Forest, 155
The Trace of the Black Wind (Sled Chornaha Vetru), 282
Thoreau, Henry David, 56
tigers, Siberian, 111, 195
Tigrovaia Balka nature preserve, the, 311
Timiriazev, K. A., 46
Timofeev-Resovsky, N. V., 243
Tiumen region, the, 186
Tolstoy, Leo, 218

tractors, 73, 94, 95, 96, 97, 99, 103, 146, 149, 151, 186
Transcaucasian Water Trust, the, 97
Trans-Siberian Railroad, the, 25, 235
Trotsky, Leon, 62, 77
Tsimlianskoe Reservoir, the, 164, 170
Tsvetkov, M. A., 37
tundra, 15, 16, 130, 186, 236
  and environmental degradation, 118
  and global warming, 16
Tunguska River, the
  gold mining, 101
Turkey, 59, 221
Turkmenistan, 19, 212, 231, 233, 293, 306, 307
  emissions, 307, 314
Turksib Railroad, the, 76, 77, 78, 79, 98, 235
Tver, 32

Ufa, 280
Ukraine, 15, 28, 49, 50, 76, 113, 117, 119, 126, 151, 157, 188, 212, 223, 258, 260, 262, 263, 284, 286, 288, 292, 293, 299, 305, 307, 308
  Chernobyl, 265, 266, 267
  conservation, 173, 174, 244
  econationalism, 292
  environmental issues, 282, 299, 305, 318
  famine, 74, 79
  forests, 18, 157, 158, 311
  independence movement, 279, 283, 304
  pollution, 89, 215, 308, 314
Ukrainian Academy of Sciences, the, 55, 169
Ukrainian Popular Front (Rukh), the, 283
UN, 191
  Conference on Environment and Development, 299, 300
  Conference on the Environment, 219, 240
  Economic Comission for Europe, 192
  environmental issues, 192, 196, 220
  Framework Convention on Climate Change, 299
Uncle Vanya, 39
UNESCO, 197
  biosphere project, 190
  Soviet participation, 192

Index

Union of Soviet Writers, the, 171, 205
United Kingdom, the, 9, 141
United States, 8, 9, 14, 24, 25, 33, 38, 40, 49, 72, 73, 74, 81, 119, 123, 131, 132, 151, 162, 167, 195, 196, 197, 199, 220, 287, 289, 293, 297, 301
   agricultural practices, 29, 46, 150
   conservation efforts, 57, 173
   drought, 210
   environmental issues, 144, 189, 191, 193, 207, 269, 306
   Environmental Protection Agency, 297
   Kyoto Protocol, rejection of, 302
   pollution issues, 200, 214, 220, 222, 302
   public transportation, 215
Ural Mountains, the, 1, 14, 15, 28, 73, 76, 79, 85, 89, 102, 116, 124, 126, 128, 130, 131, 132, 160, 207, 235, 259, 280
   logging, 156
   metallurgy, 315
Ural River, the, 16, 121
   pollution mitigation, 222
Urals-Kuznetsk Combine, the, 186
   coal production, 235
urbanization, 32, 42, 85, 87, 90, 92, 93, 109
   of the tundra, 17
Usa River, the, 16
Uspenskii, Gleb, 56
Ust-Kut, 235
Uzbekistan, 19, 81, 212, 293, 307
   environmental regulations, 199

Vere River, the
   pollution, 223
Vernadsky, Vladimir, 9, 54, 55, 57, 59, 62, 67, 243, 289
Villiams, V. R., 127
Vilyuy River, the
   gold mining, 101
Virgin Lands Campaign, the, 147, 148, 149, 151, 152, 169, 174, 180, 208, 209, 212, 290, 309
Vladimir, 32
Vladivostok, 239
Volga region, the, 126
   industrial production, 116

Volga River, the, 1, 14, 15, 33, 63, 71, 76, 81, 90, 119, 120, 121, 151, 160, 172, 222, 232, 275
   destruction of, 278
   fisheries, 52, 171, 230
   hydroelectric stations, 78, 163, 164, 165, 169, 171
   pollution, 177, 199, 222, 230
Volga-Don Canal, the, 76, 164, 165
Volgadonsk, 268
Vologda, 30, 35, 154, 199
Vorkuta, 101, 104, 105
Vorkutstroi, 104
Vorontsov, Nikolai, 274
Vychegda River, the, 172

Warsaw Treaty Organization, the, 255
Weiner, Douglass, 6, 7, 57, 68, 110, 112, 126
whaling, 142, 193, 194
   population declines,
White Sea, the, 8, 30, 52, 166
wilderness, 2, 6, 10, 11
   expansion of, 145
   ideas of, 7
   preservation, 56
   value of, 109
wildlife
   bears, 41
   beavers, 41
   birds, 16, 111, 115, 195, 202, 287
   bison, 60, 111
   caribou, 195
   deer, 41, 65, 118
   elk, 41
   foxes, 41
   geese, 65
   hares, 41
   mink, 41
   otters, 195
   porpoises, 194
   rabbits, 16
   seal, 192
   seals, 228
   wild boar, 41
   wildcat, 65
   wolverines, 65
   wolves, 41
Witte, Sergei, 25
World Bank, the, 294, 297

World Climate Change Conference, the, 301
World Trade Organization, 301
World War I, 1, 25, 33, 44, 50, 57, 58, 59, 60, 62, 66, 77
World War II, 50, 73, 74, 89, 100, 103, 105, 111, 114, 116, 117, 118, 128, 136, 158, 160, 162, 225, 245
  and environmental destruction, 72, 112, 114, 115, 156, 289
World Wildlife Federation, the, 111, 302, 305, 306

Yablokov, Alexei, 194, 245, 276, 296, 303
Yanitsky, Oleg, 277, 304, 307
Yeltsin, Boris, 13, 255, 276, 296, 300
Yerevan, 282
Yugoslavia, 221

Zabaikal region, 58
Zakvodhoz. *See* Transcaucasian Water Trust, the
Zalygin. S. P., 110, 248
Zapolyarnyi, 105
zapovedniks, 2, 28, 49, 56, 57, 58, 61, 62, 63, 64, 65, 66, 67, 109, 111, 112, 128, 173, 174, 175, 311
  and research, 110
  establishment of, 65
zemstvos, 44, 45, 46
Zhuk Gidroproekt Institute, the, 129, 166, 172

Other Books in the Series (*continued from page iii*)

Myrna I. Santiago *The Ecology of Oil: Environment, Labor, and the Mexican Revolution, 1900–1938*

Frank Uekoetter *The Green and the Brown: A History of Conservation in Nazi Germany*

James L. A. Webb, Jr. *Humanity's Burden: A Global History of Malaria*

Richard W. Judd *The Untilled Garden: Natural History and the Spirit of Conservation in America, 1740–1840*

Edmund Russell *Evolutionary History: Uniting History and Biology to Understand Life on Earth*

Alan Mikhail *Nature and Empire in Ottoman Egypt: An Environmental History*

Gregory T. Cushman *Guano and the Opening of the Pacific World: A Global Ecological History*